Frontier Studies in Soil Science

Series Editor

Avelino Núñez-Delgado, Lugo, Spain

This series aims to illuminate the latest research, insights, and advancements in the ever-evolving world of soil science. This series is dedicated to fostering a comprehensive understanding of soils, from their micro to macro characteristics, their role in ecosystems, to their significance in global challenges. Our commitment is to present groundbreaking studies, carried out by esteemed experts and emerging scholars, ensuring that readers are consistently provided with the most up-to-date and relevant information in the realm of soil science.

Avelino Núñez-Delgado
Editor

Frontier Studies in Soil Science

Springer

Editor
Avelino Núñez-Delgado
Department of Soil Science
and Agricultural Chemistry, Engineering
Polytechnic School
University of Santiago de Compostela
Lugo, Spain

ISSN 3004-9865 ISSN 3004-9873 (electronic)
Frontier Studies in Soil Science
ISBN 978-3-031-50502-7 ISBN 978-3-031-50503-4 (eBook)
https://doi.org/10.1007/978-3-031-50503-4

This Springer imprint is published by the registered company Springer Nature Switzerland AG
The registered company address is: Gewerbestrasse 11, 6330 Cham, Switzerland

Paper in this product is recyclable.

Preface

Our world is currently affected by a variety of environmental and social problems that need to be taken into account, and humans should make the highest efforts to help mitigate or solve these crucial issues. Within them, with a more specific focus, working to achieve substantial advances in Soil Science research would promote significant positive effects for the environment in general, and particularly for living beings that inhabit it.

This book has been conceived with this objective, specifically focusing on new research in Soil Science, but also constituting the first volume in a tentative Book Series dealing with new research proposals in sciences (mainly environmental), in which the aim will be that different problems and proposals for research will be progressively addressed to control them as far as possible.

The current book includes chapters written by research teams working on soil and environmental sciences in different countries, showing specific particularities, but all of them searching for improving our knowledge on soils and the overall environment, and on ways to increase wellness in our planet (Fig. 1).

Fig. 1 Images showing examples of forests and soils, where humans could find paths to sustainability

The scientific editor would like to thank all authors, reviewers and the staff of Springer Nature involved. We hope that this work will be interesting and useful for researchers, as well as for the society, contributing to the study and understanding of the soil and the specific environmental aspects covered in the book.

Lugo, Spain

Dr. Avelino Núñez-Delgado
Scientific Editor

Contents

Introduction Chapter for the Book "Frontier Studies in Soil Science"

Avelino Núñez-Delgado, Esperanza Álvarez-Rodríguez, María J. Fernández-Sanjurjo, Ana Barreiro-Buján, Andrés Rodríguez-Seijo, Juan José Villaverde, Montserrat Díaz-Raviña, and Jorge Mataix-Solera

Abstract In this introductory chapter the authors show their views on what they consider current and future fundamental issues to advance knowledge and research in Soil Science. Each of the authors gives responses to a question posed by the scientific editor of the book. Furthermore, additional data is included to show a picture corresponding to the current situation of the theme, as per different scientific searching tools.

Keywords New research · Environmental sciences · Environmental issues · Research proposals · Soil science

1 Justification

This introductory chapter has been designed following a peculiar structure, with the reason explained below.

The origin of this chapter is closely related to a scientific meeting held in Guimaraes (Portugal), with the participation of the so-called "Guimaraes Group", taking place on May 17th, 2023, to talk about soil science, among other tasks. There, the scientific editor of this book asked a question to those in the meeting, as well as

A. Núñez-Delgado (✉) · E. Álvarez-Rodríguez · M. J. Fernández-Sanjurjo · A. Barreiro-Buján
Department of Soil Science and Agricultural Chemistry, Engineering Polytechnic School, University of Santiago de Compostela, 27002 Lugo, Spain
e-mail: avelino.nunez@usc.es

A. Rodríguez-Seijo
Soil Science and Agricultural Chemistry, Faculty of Sciences, University of Vigo, 32004 Ourense, Spain

J. J. Villaverde · M. Díaz-Raviña
Departamento de Bioquímica de Suelos, Ecosistemas y Ecología Agroforestal, Misión Biológica de Galicia (MBG-CSIC Sede Santiago), Santiago de Compostela, Spain

J. Mataix-Solera
GETECMA-Soil Science and Environmental Technologies Group, Department of Agrochemistry and Environment, Miguel Hernández University, Elche, Spain

A. Núñez-Delgado (ed.), *Frontier Studies in Soil Science*, Frontier Studies in Soil Science, https://doi.org/10.1007/978-3-031-50503-4_1

Fig. 1 A sight of Guimaraes, where the Guimaraes Group was constituted in June 2022, and held a scientific meeting on May 2023

to a limited number of additional colleagues which were contacted by email. As a result of the talks and email exchanges, both the editor and the researchers providing responses decided that it could be interesting to include the question and the answers to it as main aspects of this Introduction chapter (Figs. 1 and 2).

2 Question to the "Guimaraes Group"

The question was as follows:

What do you see as the current (and/or future) fundamental issues to advance knowledge and research in Soil Science?

You could consider any geographic scale, as some issues may be seen as more relevant or meaningful when formulated at the local level, while others become more important with another spatial dimension.

Please, argue sufficiently, giving all the details needed, in a way that you see (and the readers could see) your response clearly depicting what you intend to convey.

Figure 3 shows some soil profiles, just illustrating classical starting points for soil research.

Fig. 2 *Brugmansia arborea* at Guimaraes (on the left), and *Magnolia grandiflora* in Lugo university campus (on the right), symbols of the Guimaraes Group

Fig. 3 Three soil profiles in the North of Galicia

3 The Answers

Ordered alphabetically by names, the responses by the authors of the chapter were as follows.

3.1 Response by Álvarez-Rodríguez, Esperanza

In the field of Soil Science, several fundamental issues currently exist that pose challenges to the advancement of knowledge and research. These issues are crucial to address in order to effectively understand and manage soils, both at local and global scales. Below, I will outline some significant challenges:

1. *Soil degradation and loss*. Soils are fundamental to life and to maintain their health is vital for agricultural development and ecological sustainability [2]. Soil degradation it is a change in soil health condition that reduces the capacity of the ecosystem to provide goods and services [23]. Land degradation has become a global issue that affects most terrestrial biomes and agro-ecological systems. The decline in soil quality and productivity due to factors such as erosion, compaction, nutrient depletion, pollution, acidification, desertification, and salinization, is a major challenge for future generations [59]. This issue is particularly critical due to unsustainable agricultural practices, urbanization, industrial activities and forest fires that contribute to soil degradation. Erosion, both water- and wind-induced, leads to loss of fertile topsoil, reducing soil productivity and negatively impacting food security [8]. It is essential to study the underlying causes and implement effective conservation practices to mitigate soil erosion. Soil pollution is another important problem that leads to soil degradation and have severe implications for food security and ecosystem health. Anthropogenic activities, such as the use of agrochemicals, industrial discharges, and improper waste disposal, lead to soil contamination [52]. The pollutants may persist in the soil for extended periods, adversely affecting soil quality, ecosystem health, and potentially entering the food chain [44]. In this sense, in recent years there has been a growing concern about emerging pollutants, chemical compounds that are not traditionally monitored or regulated but have been recognized as potentially harmful to ecosystems and human health. These contaminants include pharmaceuticals, personal care products, pesticides, and industrial chemicals, and they may enter the soils through various pathways such as wastewater irrigation or application of contaminated biosolids as fertilizers [13, 60]. Research is needed to assess the fate, transport, and persistence of emergent contaminants in soils, as well as their potential impacts on soil ecosystems and the wider environment. Developing analytical techniques and monitoring programs to identify and quantify these contaminants in soils, and develop legislation with limit values for these contaminants, is crucial for effective risk assessment and management. It is also essential to study the use of low-cost materials, such as residues or by-products from different activities, to prevent the entry of these contaminants into the food chain. It is well known that the generation of a huge amount of waste is one of the major problems facing today's society, including agricultural, industrial, and municipal waste, or plastics from different activities, that may have detrimental effects on soils [52]. Improper disposal of wastes can lead to soil contamination, nutrient imbalances, and the release of harmful substances. Some waste may contain heavy metals and other chemical pollutants, which can persist in soils and pose long-term risks to ecosystems and human health [44]. However, it is necessary to analyze these

residues because many of them could be valued in the agricultural field, providing organic matter and nutrients, or as adsorbents for pollutants such as heavy metals or emerging contaminants. Understanding the impacts of different waste management practices on soil quality, developing effective remediation strategies, and promoting sustainable waste management systems are crucial research areas to mitigate the negative consequences of waste generation on soils. Some of them could be a valuable resource for soil when managed correctly, since they can improve fertility, soil structure, water retention, aeration, and nutrient availability for plants, reducing the need for chemical fertilizers. Research is also needed to use them as pollutant bio-adsorbents. Addressing waste generation requires a comprehensive approach involving waste reduction, recycling, proper waste management, and the development of sustainable waste-to-energy technologies. Emphasizing the importance of recycling and composting can reduce the volume of waste entering landfills and positively impact soil health through organic matter recycling.

So, advancing knowledge and research on the mechanisms, impacts, and prevention of soil degradation are crucial for developing sustainable land management practices and promoting soil conservation. Implementing erosion control measures, reducing and recycling waste and restoring degraded soils through reclamation and rehabilitation efforts are important strategies to combat soil degradation.

2. *Climate change and soil carbon sequestration.* Soil plays a significant role in carbon sequestration, regulating atmospheric CO_2 levels. However, climate change, with its increased temperatures, altered precipitation patterns, and extreme weather events, can impact soil organic carbon content and decomposition rates [33]. Understanding how climate change affects soil carbon dynamics and finding ways to enhance soil carbon sequestration are critical for climate change mitigation. Developing a comprehensive understanding of the factors controlling soil organic matter decomposition, carbon stabilization mechanisms, and the impacts of land management practices on soil carbon storage is crucial. This research can inform strategies for enhancing soil carbon sequestration, promoting sustainable agriculture, and mitigating climate change.

The warming of the Earth's atmosphere also enhances evaporation rates, which, in turn, depletes moisture from the soil and exacerbates drought conditions. Droughts are intrinsically linked to water scarcity, as diminished precipitation reduces surface water levels, depletes groundwater reserves, and dries up rivers and lakes. Water stress reduces the availability of water resources for agriculture, further compromising food production and exacerbating famine risks. Addressing climate change requires global cooperation and action to mitigate greenhouse gas emissions, promote sustainable practices, and enhance resilience to its impacts. By taking proactive measures, societies may mitigate the severity of drought, famine, pandemics, and other environmental problems linked to climate change.

3. *Soil–Plant interactions and nutrient cycling.* Fertile soil is a non-renewable resource in terms of human time since and its production and regeneration can take hundreds or thousands of years [45]. Soils provide essential nutrients for plant growth, and their availability and cycling processes are fundamental for sustainable agricultural production. However, there are still significant knowledge gaps regarding the

complex interactions between soils and plants. Understanding the mechanisms of nutrient uptake, nutrient cycling dynamics, and the impact of soil properties on plant health and productivity is crucial. This knowledge can lead to the development of improved fertilizer management practices, precision agriculture techniques, and the promotion of sustainable nutrient cycling in agroecosystems. Additionally, studying the influence of soil microbiota on nutrient availability and plant health is an emerging field that holds promise for advancing sustainable soil management practices.

4. *Forest Fires*. The increase in frequency and severity of forest fires in recent years have become a significant environmental concern [15]. Some impacts on soils are those indicated by Roshan and Biswas [49]: (a) Soil erosion: forest fires can lead to the destruction of vegetation and organic matter of soil. Without the vegetation cover, the soil becomes vulnerable to erosion by wind and water. On the other hand, when soil organic matter burns, it is transformed in CO_2 and released into the atmosphere, contributing to global warming and climate change. This loss of organic matter also impacts soil structure and nutrient cycling processes. Soil erosion also increase surface runoff after a forest fire that may lead to sediment and ash entering nearby water bodies, negatively affecting water quality, clog waterways, and harm aquatic life. (b) Alteration of nutrient cycling: forest fires may alter the nutrient cycling processes in the soil. Some nutrients, like nitrogen and sulfur, are volatilized and lost during combustion, while others become more available for plant uptake after the fire. (c) Loss of soil microbial diversity: Forest fires may have severe impacts on soil microorganisms, which are essential for nutrient cycling and ecosystem functioning. High temperatures and combustion by-products may lead to the death or reduction of soil microbial populations, disrupting the delicate balance of the soil microbiome. To address the challenges posed by forest fires, both prevention and post-fire management strategies are essential. Preventive measures and post-fire restoration efforts can focus on erosion control, reseeding native vegetation, and promoting the recovery of soil microbial communities. Additionally, promoting sustainable forest management practices and public awareness and community engagement in fire safety and prevention are also crucial to minimize the human impact on forest fires. Researching holistic practices to forest fire management can protect soil health, preserve biodiversity, and ensure the long-term resilience of forest ecosystems.

Addressing these fundamental issues requires interdisciplinary research collaborations, increased funding for soil science research. Moreover, knowledge exchange and capacity building initiatives can help bridge the gap between research findings and practical applications, ensuring that scientific advancements positively impact soil management practices at various geographic scales.

3.2 *Response by Barreiro-Buján, Ana*

The detailed analysis of the soil microbial community is, currently and will be in the future, fundamental for the advance in the knowledge and research in Soil Science. Previous research has extensively addressed the determination and characterization of the different microorganisms that inhabit soil ecosystems meanwhile the study of the soil microbial community functionality remains a challenge [39]. We know different families, species and genes that are part of the soil microbial community, but there is a lack of knowledge about what are their roles and functions in the complex soil ecosystem. Soil microbes are a key factor for the soil quality and health since they are largely responsible for soil fertility, and structure, biochemical cycling and soil C sequestration, even though for the latest they can act as both sink and source [46]. The decoded of the soil microbial functions in different ecosystems could contribute for example to improve crop yield and quality in agricultural soils [35]. For the study of the soil microbial community both large and local scale are important. Local scale studies are indispensable since climatic and physic-chemical properties can vary largely, even within small distances, and are determinant for the soil microbial community structure [6]. On the other hand, the results from the local studies are required to feed global models to obtain reliable results and predictions in subjects of major importance such as climate change and the role of the soil microbes.

3.3 *Response by Díaz-Raviña, Montserrat*

In the last two decades, specific concepts and ideas have been stressed to highlight the importance and role of soils in a healthy environment. These concepts refer to soil quality, soil health, soil sustainability and soil ecosystems as well as sustainable development goals. Soil health can be seen as the capacity of soil to function as a vital living system with an ecosystem and land-use boundaries, to sustain plant and animal productivity, maintain or enhance water and air quality and promote plant and animal health. Besides conceptually, soil health can be defined operationally by providing a key list of soil health indicators that includes physical, chemical, and biological properties. Studies concerning soil quality assessment and monitoring tools have been growing exponentially since the 1990s. Although a huge amount of papers (including original research, review papers, opinion papers, etc.) addressing many aspects concerning this topic have been performed worldwide (such as selection of best individual indicators used both individually, as simple indices, or in combination using complex equations derived from mathematical combinations or the application of statistical programs), the results are contradictory, questioning the validity of the tested indicators [7, 31].

The complexity and site-specificity of soil as well as the many linkages between soil functions and soil ecosystem services require highly demanding research focused on health assessment. Thus, the first step to address this known gap is the systematic compilation of the results of the studies performed over the last two decades in the same area and situations in which health assessment has been described. This will allow us to have a view of the art-state concerning how soil health can be effectively assessed, monitored, restored (if necessary) and—most importantly—used to guide land use and soil management decisions. Currently, there is an increasing interest of farming communities, researchers, industry, and policy makers towards a farmer-feasible soil health assessment. Since soils vary greatly, approaches must be defined at local areas and cannot be extrapolated to other environment contexts. For this purpose, research should be focused on more practical methodologies and protocols, including the existent feasible, effective, not expensive, and easily interpretable indicators, as well as in situ novel methods combining the measures of different soil properties. Soil functions, ecosystems quality and quantity, and the sustainable development goals are intrinsically related. At the end of last century, ecosystems services approaches emerged and have been implemented in some soils [43]. Studies of soil ecosystem services assessment, which is dependent on market values changing between countries or regions as well as over time, should be carried out in the future to encourage better dialogue between science and decision-making and hence for giving appropriate stakeholder recommendations and facilitate soil governance.

In the discipline of soil microbial ecology, numerous studies concerning the high-resolution description of soil microbial communities, detection of taxa with minor abundance, screening of gene expression or the detailed characterization of metabolomes have been performed. However, the fault of enough effective description tools and limited possibilities to assign traits to community members make it difficult to link microbes to functions, and the analysis of processes related to enzyme activity measurements is still imperfect [4, 24]. Further studies should fill these knowing gaps as well as should focus on the observation of microorganisms at different levels of organization, ranging from a small scale, in order to look at individual microorganisms (properties, in situ interactions, metabolic rates and activity), to a large scale, in order to look at the microorganisms at the community, ecosystem or biome levels. The recent development of novel tools including metabolomics, identification of genomes in metagenomics sequencing datasets or collection of trait data has the potential to bring soil ecology further. However, considering the extremely high soil complexity (scales, spatial heterogeneity, and temporal dynamics) and the small percentage of its biodiversity which is known (only around 5%), it will remain a highly demanding scientific discipline over the time.

There is much literature exploring the response of soil microbial communities (mass, activity, structure, and function) to different perturbations of both natural and anthropogenic origin (erosion, salinity, drought, climate change, agricultural expansion, conversion of natural habitats in rural areas and urban sprawl, soil management linked to agroforestry practices, presence of inorganic and organic contaminants, wildfires, etc.) and it is clear that these communities often do not recover their original state. But our understanding of how soil microbial communities and their collective metabolic activities resist and recover following these disturbances, and the consequences of abrupt transitions in soil microbial communities for ecosystem function, is scarce. Given the important role of soil microbial communities for ecosystem restoration and sustainability, this represents an important knowledge gap. Recently, promising frameworks for quantifying resistance and resilience, and for diagnosing abrupt ecological change, have been developed [5]. Further research on this topic is needed to improve our knowledge of the soil functioning and agroforestry ecosystems restoration.

In agricultural and forest ecosystems, biodiversity is a complex and multi-faceted concept that includes both above-ground (e.g., plant species, structure, evolutionary and functional trait diversity and composition) and below-ground diversity (e.g., soil macroscopic animals and microorganisms). The studies carried out to date have mainly focused on biodiversity within a single trophic group neglecting the fact that ecosystems functioning depends strongly on complex interactions between trophic levels and between above-ground and below-ground biodiversity [61]. Further research is needed to explore the relative effects of above-ground and below-ground biodiversity on ecosystems multifunctionality, while considering the direct and indirect effects of environmental conditions. In this regard, it could be clearly useful to optimize the functioning of the soil-microorganism-plant system and hence promoting the development and validation of environmentally friendly strategies based on nature for the protection, maintenance, and improvement of agricultural and forest soil health.

3.4 *Response by Fernández-Sanjurjo, María J.*

In recent decades, the overall society has become aware of the fundamental functions of soils in the ecosystem, in addition to that related to providing food and water, storing carbon and, in general, in the development of life. However, despite a rather slow awareness of the evident importance of the soil, some activities of direct or indirect anthropic origin are becoming more numerous and intense, causing a decrease, or even promoting the total destruction of edaphic functionalities. Many of these activities, such as pollution, overexploitation, erosion, or compaction, have been studied since the past century. However, the current strong pressure affecting ecosystems, and specifically the soil, including the effects of climate change, highlights the need to continue investigating the evolution of soil properties, and specially methods or procedures to mitigate these impacts.

From just two centuries back to now, coinciding with the beginning of the so-called Anthropocene, the impact of human activities on the environment has experienced a clear and progressive rise, which is causing that the soil receives a cocktail of chemical substances foreign to it, or that are found naturally in lower concentrations. These chemicals include heavy metals, plastics, rare earths, and drugs such as antibiotics and anti-inflammatories, both for veterinary and human use. For example, global production of synthetic chemicals has increased exponentially, rising from 1 million tons in 1903 to 400 million tons in 2000 [55]. Many of these pollutants are relatively recent but their speed of accumulation is very worrying; they are the so-called "emerging pollutants". Given the magnitude of this problem, it is necessary to elucidate in detail their interactions with the soil, especially regarding microplastics and rare earths. In relation to microplastics, they come from the partial degradation of the huge number of plastics that are produced annually and that are dispersed throughout the environment, among other reasons due to their low recycling rate worldwide ($<40\%$ [37]).

In fact, the term "plastisphere" has been coined as a new environmental compartment [62]. As they are complex and very small particles (<5 mm), their study entails clear difficulties. Most research focuses on its behavior in the marine environment, while little is known about contamination in fresh water and soils. As for rare earths, they are the group of lanthanides of the periodic table of elements, which are being called "the vitamins of modern industry" [30] given their wide use in sectors such as technology (which generates millions of tons of waste annually), agriculture (in fertilizers and sewage sludge) and medical-pharmaceutical fields (contrasts, antibiotics), which also causes a remarkable dispersion in the ecosystem, with serious effects on microorganisms, plants, and human health. It is essential to investigate the behavior of these new contaminants in the environment, specifically their biogeochemistry, their bioavailability, degradability, and their effects on the edaphic living phase, especially in relation to microorganisms, in addition to seeking alternatives to decrease their dispersion in the ecosystem.

Another serious impact affecting soils is the agricultural overexploitation of some areas, with associated progressive loss of the arable layer, increased pollution due to the use of certain fertilizers and phytosanitary products, etc. Due to that, it is essential to intensify research on improving soil management, especially in aspects such as the search for alternatives to chemical fertilizers, biological control, improvement in crop rotation and tillage techniques. Furthermore, it is urgent to develop rigorous territorial planning, keeping in mind the conservation of the most fertile soils and biodiversity, avoiding actions that increase erosion and soil destruction.

In addition, climate change, whose scientific evidence is compelling, is causing a magnification of impacts on the soil environment. In this sense, it is essential to determine how certain edaphic processes may be evolving due to the changes in the temperature, precipitation and sunshine regimes, to subsequently try mitigating these impacts. Specific studies should focus on how this phenomenon affects the availability of water, nutrient cycling, soil C storage capacity, and physical properties, as well as erosive processes. Furthermore, the influence of climate change on the alterability of rocks, clay-humic interactions, mineral formation and, in general, on the rate of soil formation, would be another key and interesting field of knowledge, since edaphogenetic trends could be modified in the medium term, in a detectable way, if the climatic variations currently observable continue to take place. On the other hand, climate change and forest fires are two closely linked processes feeding each other. Extreme fires are expected to increase by 50% by 2100 [53]. The increase in its frequency and intensity will aggravate the drastic impacts on the ecosystem due to fires. Although the effects on soils are well studied, it is necessary to delve deeper into the behavior of fires, how to introduce new and effective restoration techniques and preventive measures and, in general, how to improve land management, which again is related to the actual application of large-scale territorial planning.

Numerous investigations have been carried out regarding soil–plant relations. However, one aspect that remains not enough explored is the rhizosphere zone, which is still not well known due to the difficulty of investigating it and the lack of accepted standard methodologies [28]. It should be relevant to better understand the type of compounds that are excreted and synthesized in the root environment, as well as the type of microorganisms that proliferate in this area, and also their relations with processes such as mineral alteration and nutrient uptake. Researching on rhizosphere processes in scenarios with greater and more varied pollution and with the presence of extreme values of temperature and precipitation will be a remarkable future challenge, mostly taking into account that these impacts will hinder food production, facilitating general soil degradation. These studies will be key for the advancement in the development of new microbiological methods are needed to improve seed germination, to, totally or partially, replace chemical fertilization and the use of pesticides, as well as for their use both in the field of phytoremediation and regarding plant colonization of marginal areas.

To achieve these objectives and provide relevant advances in knowledge, collaboration will be necessary, not only referred to experts in the different disciplines of soil science, but also to those in the fields of health, communication, new technologies, etc. This broad scientific synergy would allow the achievement of the proposed goals in the field of soil science and, in general in the environment, that help minimize the impact of the human activities mentioned above.

3.5 Response by Mataix-Solera, Jorge

In recent decades there were significant advances in the knowledge of soil science filling gaps, developing new techniques for a faster analysis of soil parameters, and collaborating from different disciplines understanding that soil is interdisciplinary by nature [11], but still we have much to do, especially in a world exposed to the effects of climate change modifying very fast the environments and the living conditions, which include soil. The study about how the soils can be adapted to these changes will be crucial. Soil is a living system where a 50% of earth biodiversity is inside it, much of them still unknown. Soil microbial communities are very sensible to perturbations. Understanding how they response to modifications climate will allow to take decisions to combat ecosystems degradation and to mitigate the impact of this global changes in agriculture, for example by improving the effectiveness of plant nutrition, the use of water, soil carbon sequestration, etc. As an example, at present, one of the most important threats to the health of our forest soils are wildfires that jeopardize the provision of soil ecosystem services. Certainly, at present, the escalating problem of significant forest fires is widely acknowledged as one of the most critical environmental challenges exacerbated by global warming. This issue is particularly pronounced in areas characterized by adverse climatic conditions, including extended periods of drought. Today wildfires are becoming more intense and bigger (mega-fires) and have unprecedented environmental and socioeconomic consequences (e.g. economical losses, air pollution, greenhouse gas emission and land degradation).

Over the past few decades, extensive research has been dedicated to examining the impacts of forest fires on soil systems [17]. These studies have revealed that, for instance, Mediterranean regions often display a degree of resilience in their soils following such events [3]. However, the perturbations caused by wildfires are now encroaching upon areas that historically faced fewer fire risks and affecting soil much more vulnerable to degradation in ecosystems less adapted to fires. Furthermore, there is growing concern that this issue may become endemic on a global scale in the near future due to the influence of climate change. Exploring the impact of wildfires in diverse environments, particularly in semi-arid and arid regions, holds particular significance. Soil microorganisms in these areas have developed adaptations to withstand stress conditions, and understanding their response to fire can provide valuable insights for decision-making in other fire-affected ecosystems.

In addition, it is imperative to emphasize the necessity for further research into the effects of various post-fire management approaches on soil. Recent studies have highlighted the significant role of soil biocrust in the initial stages of post-fire recovery and its vulnerability to certain post-fire practices like salvage logging or mulching [26, 38].

The soil science community still has work to do in terms of effectively disseminating knowledge to the general public. Increased efforts in this regard, coupled with the publication of more research papers in high-impact journals, can help elevate the importance of soil on the agenda of environmental policymakers.

3.6 Response by Rodríguez-Seijo, Andrés

Soil sciences are facing a changing and uncertain future due to the influence of different factors such as population growth, unequal resource distribution and the impact of global warming, a fact that is increasingly present in our daily lives through extreme weather phenomena (e.g., floods, droughts, etc.) and accentuation of adverse effects such as reduced precipitation rate and increased temperature. The combination of these factors, especially climate change, will play a major role in soil health, such as increased wildfires (both in intensity and frequency), impacts on soil microbial ecology, and, among other aspect, agricultural production in the face of the challenge of feeding an estimated 9.8 billion people in 2050, and under a scenario with less water availability and reduced soil fertility due to soil degradation [34, 47, 51].

We cannot forget environmental issues caused by inorganic and organic contamination due to heavy metals, pesticides, PAHs, etc. that still represent a major environmental concern. However, under these new scenarios, it will also be necessary to pay attention to soil contamination by emerging contaminants such as antibiotics, viruses, nano- and microplastics or per- and polyfluoroalkyl substances (PFAS) due to their wide distribution in all kinds of environments and especially due to their understudied impacts on soil properties, biota and therefore, their potential transfer to the food chain [12, 13, 16, 60]. In this regard, plastic pollution has gained attention in recent years due to its impacts on soil's physical, chemical, and biological properties. Due to their hydrophobic and porous surface, large surface area, small size or lightweight, they can be easily dispersed to all kinds of environments, could act as carriers of inorganic and organic contaminants and pathogens, and also can be hotspots for antimicrobial resistance genes and modify disease transmission process and drug resistance pathogens [36].

In response to this, cooperation is needed with other disciplines in the search for common solutions, such as nanotechnology with chemistry and physics sciences to the nanomaterials' development that allow the farmers to achieve controlled release of agrochemicals through bio-nanocomposites or encapsulated nanoparticles for fertilization or fungicide applications due to their ability to smart, controlled and targeted release of active elements, but also for their use for soil decontamination in a dual action (e.g., [48, 51]). The development of green technologies is also needed for an integrated and holistic restoration of ecosystem functions in degraded areas. A similar situation could be related to the development of agriculture 4.0, with drone technology for smart and more applied agriculture or with the combination of sensors (optical, electrochemical, mechanical sensors, etc.) and artificial intelligence for improved field cultivation, greenhouse automation or early detection of agricultural diseases and environmental impacts by soil contaminants [32, 42].

3.7 *Response by Villaverde, Juan José*

Among all issues to advance knowledge and research in soil science, two of them are key in current times:

(a) To develop legal frameworks based on the most recent advances, in which the challenges to overcome are included.
(b) To perform disruptive studies able to take advantage of the synergies generated by the combination of experimental and computational approaches.

The scope of the next argue to support the above issues could be unapproachable, because the challenges that soil science must face are many and very complex, as can be seen from the above responses. Therefore, henceforth the attention is focused on the challenge to control agricultural soil contamination with emerging pollutants.

A legal framework, that delimits the placing on the market and use of products that contain emerging pollutants or of products from which emerging pollutants are generated, is required to limit soil exposure to these substances of concern. The objective of this framework should be to guarantee the protection of human and animal health and the environment (especially the quality of soil and drinking water) and always supported by existing scientific and technical knowledge. This objective is important to promote progress in knowledge and research on topics related to soil science. In addition, it guarantees dynamic Regulations, which can be adapted to new needs that arise. In the European Union (EU), for example, the marketing of plant protection products (from now named as pesticides to simplify) is currently governed by Regulation (EC) No. 1107/2009, while their sustainable use is promoted by Directive 2009/128/EC. Both documents are the basis for the use of pesticides in the EU through an integrated pest management strategy, within the framework of sustainable agriculture, aimed at keeping production while respecting the environment, being economically viable and socially acceptable. In this sense, the European Commission performs the risk management for establishing the use of a pesticide. However, the European Food Safety Authority (EFSA) has previously established how risky is the use of this pesticide, based on scientific data obtained by guidance documents and analyzed by means of a complex process of risk assessment. In this way, studies on the impact of pesticides on soils are beginning to appear in the scientific literature [54], driven by the existing legal framework and under criteria approved by groups of experts and committees designated by EFSA for this purpose [20, 21]. These two facts guarantee the harmonization of studies and avoid wrong conclusions due to methodological reasons.

On the other hand, currently, experimental studies are beginning to have difficulties in explaining the effects caused by some types of chemical compounds (e.g. nanomaterials [19]), for example, on the functioning of edaphic ecosystems [1], or evaluating their effects on soil quality in accordance with the new regulatory requirements [41]. Furthermore, most experimental studies typically focus on the parent compound, paying less attention to the Transformation Products (TPs) generated. However, the importance of TPs should not be underestimated, since they often present different physicochemical properties compared to the parent (e.g., solubilities are usually increased with the consequent risk of water contamination [10]). TPs also usually show different environmental behavior (e.g., greater persistence, different toxicity, ecotoxicity and biological activity [9]) and their concentration depends on numerous factors and processes [21]. Despite all this, TPs have historically been less studied due to (i) the impossibility of performing a complete study of such a broad spectrum of TPs and (ii) the impossibility of isolating TPs in sufficient quantities to study them in detail. Current computational-based tools, such as those derived from Quantum Mechanics (QM) or Molecular Mechanics (MM), or based on Quantitative Structure–Activity/Structure–Property Relationships (QSAR/QSPR), have great capacity to obtain accurate information at microscopic (the first two) and macroscopic (the later) scale [57]. In this way and thanks to these innovative computational-based tools, the new challenges can be addressed and resolved in a comprehensive, practical and pragmatic way, for example, by supporting experimental data that may raise doubts or providing data reliable enough to reduce the number of experimental studies needed [14]. In this sense, ab initio calculations have been used as cutting-edge methodology to identify TPs that are difficult to observe at analytical level during the risk assessment of the emerging pollutants [56, 63]. On the other hand, proteins compose an important fraction of soil organic matter and there is an incipient interest to study the proteins interactions within this surrounding at a structural level. This information can contribute to improve their action regards a potential application, such as on health of crops. The use of Molecular Dynamics (MD) is beginning to pay to this objective [27]. In this line, QM/MM simulations, in which a part of the system (usually small) is modeled using quantum mechanical calculations and the other one by MD simulations, have also great potential to study these interactions in the soil [50]. Finally, QSAR/QSPR models have also helped to fill the knowledge gap relative to the properties of the TPs when it is difficult to determine them experimentally [18, 58].

In summary: to develop studies at the frontier of knowledge is increasingly necessary, combining different but complementary work areas (e.g., legislation and computation), as well as interdisciplinary tools and studies, which not only allow the generation of knowledge, but also promote a radical change in the way to face the starting hypotheses. In this way, it is possible to overcome the existing limitations in each branch of knowledge and advance in each of the challenges.

4 Additional Data

Some of the most relevant scientific searching tools provided the following data as regards "frontier research in soil science".

Google scholar (GS) provided 759 results for the searching string "frontier-research soil-science", while the number was 430,000 for "frontier research soil science" (with 2,990 results just for the first six months of the year 2023). Ordered by relevance, the most recent references deal with a variety of themes, going from nanotechnology to soil erosion, farming, and others that have been continuously considered as key in soil research.

Web of Science provided 2,909 results for the string "frontier research soil science", dealing with soil health, soil microbiology, and various other themes.

Scopus found 128 results for "frontier research soil science", dealing with soil health, soil restoration and a variety of other themes.

Searching for "new research soil science" on GS, the first reference (sorted by relevance) is a paper by Hartemink and McBratney [29], where the authors start indicating that "*The renaissance was an intellectually-rich period following a period of stasis in the medieval period. Something analogous appears to be currently taking place in soil science where novel approaches to thought are combined with a revival of ideas from the past.*"

When restricted to the period 2022–23, one of the most relevant papers reported is the one by Evans et al. [22], who indicate "*The importance of soils to society has gained increasing recognition over the past decade, with the potential to contribute to most of the United Nations' Sustainable Development Goals (SDGs). With unprecedented and growing demands for food, water and energy, there is an urgent need for a global effort to address the challenges of climate change and land degradation, whilst protecting soil as a natural resource.*" Another interesting paper is that by Friedrichsen et al. [25], indicating that "*Innovation for soil research can advance multidimensional outcomes for production and environmental quality through consideration of human well-being components. Integrating components from social and soil sciences builds the capacity to foster research and innovations that promote human well-being in conjunction with soil health.*"

Fig. 4 Two sights of coastal zones in Galicia (NW Spain) with vegetation growing on different soils

5 Final Remarks

Considering that soil is of main and fundamental importance in our planet, the authors of this introductory chapter would like to support current and future research on new aspects of Soil Science that could help to shed light on and solve crucial environmental issues affecting the Earth, as previously done for specific geographic areas (see for instance [40]). This book has as objective to contribute to this task (Fig. 4).

Acknowledgements Funding: This work was supported by the Spanish "*Agencia Estatal de Investigación*" (State Investigation Agency) [grant number PID2021-122920OB-C21].

The sponsors had not involvement in study design; in the collection, analysis and interpretation of data; in the writing of the report, and in the decision to submit the article for publication.

References

1. Ahmed A, He P, He P, Wu Y, He Y, Munir S (2023) Environmental effect of agriculture-related manufactured nano-objects on soil microbial communities. Environ Int 173:107819. https://doi.org/10.1016/j.envint.2023.107819
2. Amundson R, Berhe A, Hopmans J, Olson C, Sztein AE, Sparks D (2015) Soil and human security in the 21st century. Science 348:6235. https://doi.org/10.1126/science.1261071
3. Añó Vidal C, Sánchez Díaz J, Carbó Valverde E (2022) Efectos de los incendios en los suelos forestales de la Comunidad Valenciana. Revisión bibliográfica. Cuaternario y Geomorfología 36(1–2):53–75. https://doi.org/10.17735/cyg.v36i1-2.92407
4. Baldrian P (2019) The known and the unknown in soil microbial ecology: minireview. FEMS Microbial Ecol 95:fiz005. http://orcid.org/0000-0002-8983-2721
5. Barget RD, Caruso T (2020) Soil microbial communities responses to climate change: resistance, resilience and transitions to alternative states. Philos Trans R Soc B 375:20190112. https://doi.org/10.1098/rstb.2019.0112

6. Barreiro A, Fox A, Jongen M, Melo J, Musyoki M, Vieira A, Zimmermann J, Carlsson G, Cruz C, Lüscher A, Rasche F, Silva L, Widmer F, Dimitrova Martensson LM (2022) Soil bacteria respond to regional edapho-climatic conditions while soil fungi respond to management intensity in grasslands along a European transect. Appl Soil Ecol 170:104264. https://doi.org/10.1016/j.apsoil.2021.104264
7. Bhaduri D, Sihi D, Bhowmik A, Verma BC, Munda S, Dari B (2022) A review on effective soil health bio-indicators for ecosystem restoration and sustainability. Front Microbiol. https://doi.org/10.3389/fmicb.2022.938481
8. Borrelli P, Alewell C, Alvarez P, Anache JAA, Baartman J, Ballabio C et al (2021) Soil erosion modelling: a global review and statistical analysis. Sci Total Environ 780:146494. https://doi.org/10.1016/j.scitotenv.2021.146494
9. Boxall A (2009) The handbook of environmental chemistry, vol XIV, Part 2P, Transformation products of synthetic chemicals in the environment. Springer, Berlin, p 249. https://doi.org/10.1007/978-3-540-88273-2
10. Boxall ABA, Sinclair CJ, Fenner K, Kolpin D, Maund SJ (2004) Peer reviewed: when synthetic chemicals degrade in the environment. Environ Sci Technol 38(19) 368A–375A. https://doi.org/10.1021/es040624v
11. Brevik EC, Cerdà A, Mataix-Solera J, Pereg L, Quinton JN, Six J, Van Oost J (2015) The interdisciplinary nature of soil. Soil 1:117–129. https://doi.org/10.5194/soil-1-117-2015
12. Brusseau ML, Anderson RH, Guo B (2020) PFAS concentrations in soils: background levels versus contaminated sites. Sci Total Environ 740:140017. https://doi.org/10.1016/j.scitotenv.2020.140017
13. Büks F, Kaupenjohann M (2020) Global concentrations of microplastics in soils—a review. Soil 6:649–662. https://doi.org/10.5194/soil-6-649-2020
14. Clark RD (2018) Predicting mammalian metabolism and toxicity of pesticides in silico. Pest Manag Sci 74(9):1992–2003. https://doi.org/10.1002/ps.4935
15. Collins L, Clarke H, Clarke MF, McColl Gausden SC, Nolan RH, Penman T, Bradstock R, Varner M (2022) Warmer and drier conditions have increased the potential for large and severe fire seasons across South-Eastern Australia. Glob Ecol Biogeogr 31:1933. https://doi.org/10.1111/geb.13514
16. Cycoń M, Mrozik A, Piotrowska-Seget Z (2019) Antibiotics in the soil environment-degradation and their impact on microbial activity and diversity. Front Microbiol 10:338. https://doi.org/10.3389/fmicb.2019.00338
17. Doerr SH, Santín C, Mataix-Solera J (2022) Fire effects on soil. Ref Modul Earth Syst Environ Sci. https://doi.org/10.1016/B978-0-12-822974-3.00106-3
18. Duchowicz PR, Aranda JF, Bacelo DE, Fioressi SE (2020) QSPR study of the Henry's law constant for heterogeneous compounds. Chem Eng Res Des 154:115–121. https://doi.org/10.1016/j.cherd.2019.12.009
19. EFSA (2009) The potential risks arising from nanoscience and nanotechnologies on food and feed safety. EFSA J 7(3):958. https://doi.org/10.2903/j.efsa.2009.958
20. EFSA (2014) EFSA guidance document for evaluating laboratory and field dissipation studies to obtain DegT50 values of active substances of plant protection products and transformation products of these active substances in soil. EFSA J 12(5):3662. https://doi.org/10.2903/j.efsa.2014.3662
21. EFSA (2017) EFSA guidance document for predicting environmental concentrations of active substances of plant protection products and transformation products of these active substances in soil. EFSA J 15(10):e04982. https://doi.org/10.2903/j.efsa.2017.4982
22. Evans DL, Janes-Bassett V, Borrelli P, Chenu C, Ferreira CSS, Griffiths RI, Kalantari Z, Keesstra S, Lal R, Panagos P, Robinson DA, Seifollahi-Aghmiuni S, Smith P, Steenhuis TS, Thomas A, Visser SM (2022) Sustainable futures over the next decade are rooted in soil science. Eur J Soil Sci 73(1):e13145. https://doi.org/10.1111/ejss.13145
23. FAO (2021) FAO soil portal. https://www.fao.org/soils-portal/soildegradation-restoration
24. Fierer N, Wood SA, Bueno de Mesquita CP (2021) How microbes can, and cannot, be used to assess soil health. Soil Biol Biochem 153:108111. https://doi.org/10.1016/j.soilbio.2020.108111

25. Friedrichsen CN, Mizuta K, Wulfhorst JD (2022) Advancing the intersection of soil and well-being systems science. Soil Secur 6:100036. https://doi.org/10.1016/j.soisec.2022.100036
26. García-Carmona M, Arcenegui V, García-Orenes F, Mataix-Solera J (2023) The recovery of soils after post-fire management: the role of biocrusts and soil microbial communities. Span J Soil Sci 13:11388. https://doi.org/10.3389/sjss.2023.11388
27. Gotsmy M, Escalona Y, Oostenbrink C, Petrov D (2021) Exploring the structure and dynamics of proteins in soil organic matter. Proteins: Struct Funct Bioinform 89(8):925–936. https://doi.org/10.1002/prot.26070
28. Gregory PJ, George TS, Paterson E (2022) New methods for new questions about rhizosphere/plant root interactions. Plant Soil 476(1–2):699–712. https://doi.org/10.1007/s11104-022-05437-x
29. Hartemink AE, McBratney A (2008) A soil science renaissance. Geoderma 148(2):123–129. https://doi.org/10.1016/j.geoderma.2008.10.006
30. Humphries M (2010) Rare earth elements: the global supply chain. CRS-Report for Congress-4137, Diane Publishing
31. Karlen DL, Veum KS, Sudduth KA, Obrycki JF, Nunes MR (2019) Soil health assessment: past accomplishments, current activities, and future opportunities. Soil Tillage Res 195:104365. https://doi.org/10.1016/j.still.2019.104365
32. Kumar P, Singh A, Rajput V, Singh Yadav AK, Kumar P, Singh AK, Minkina T (2022) Chapter 36—Role of artificial intelligence, sensor technology, big data in agriculture: next-generation farming. In: Sharma P, Yadav D, Kumar Gaur R (eds) Bioinformatics in agriculture, pp 625–639. https://doi.org/10.1016/B978-0-323-89778-5.00035-0
33. Kumara TM, Pal S, Chand P, Kandpal A (2023) Carbon sequestration potential of sustainable agricultural practices to mitigate climate change in Indian agriculture: a meta-analysis. Sustain Prod Consum 35:697. https://doi.org/10.1016/j.spc.2022.12.015
34. Lal R, Bouma J, Brevik EC, Dawson LA, Field DJ, Glaser B, Hatano R, Hartemink AE, Kosaki T, Lascelles B, Monger CH, Muggler CC, Ndzana GM, Norra S, Pan X, Paradelo R, Reyes-Sánchez LB, Sandén T, Singh BR, Spiegel H, Yanai J, Zhang J (2021) Soils and sustainable development goals of the United Nations: an International Union of Soil Sciences perspective. Geoderma Reg 25:e00398. https://doi.org/10.1016/j.geodrs.2021.e00398
35. Liu Q, Cheng L, Nian H, Jin J, Lian T (2023) Linking plant functional genes to rhizosphere microbes: a review. Plant Biotechnol J 21:902–917. https://doi.org/10.1111/pbi.13950
36. Loiseau C, Sorci G (2022) Can microplastics facilitate the emergence of infectious diseases? Sci Total Environ 823:153694. https://doi.org/10.1016/j.scitotenv.2022.153694
37. Maqsood T, Dai J, Zhang Y, Guang M, Li B (2021) Pyrolysis of plastic species: a review of resources and products. J Anal Appl Pyrol 159:105295. https://doi.org/10.1016/j.jaap.2021.105295
38. Mataix-Solera J, Arellano EC, Jaña JE, Olivares L, Guardiola J, Arcenegui V, García-Carmona M, García-Franco N, Valenzuela P (2021) Soil vulnerability indicators to degradation by wildfires in Torres del Paine National Park (Patagonia, Chile). Span J Soil Sci 11:10008. https://doi.org/10.3389/sjss.2021.10008
39. Muhammad S, Hu J, Jousset A (2019) More than the sum of its parts: microbiome biodiversity as a driver of plant growth and soil health. Annu Rev Ecol Evol Syst 50:145–168. https://doi.org/10.1146/annurev-ecolsys-110617-062605
40. Núñez-Delgado A, Otero-Pérez XL, Álvarez-Rodríguez E (2023) Editorial: current research on soil science and related aspects of environmental sciences in Galicia. Span J Soil Sci 13.https://doi.org/10.3389/sjss.2023.11485
41. Ockleford C, Adriaanse P, Berny P, Brock T, Duquesne S, Grilli S, Hernandez-Jerez AF, Bennekou SH, Klein M, Kuhl T, Laskowski R, Machera K, Pelkonen O, Pieper S, Stemmer M, Sundh I, Teodorovic I, Tiktak A, Topping CJ, Wolterink G, Craig P, de Jong F, Manachini B, Sousa P, Swarowsky K, Auteri D, Arena M, Rob S (2017) Scientific opinion addressing the state of the science on risk assessment of plant protection products for in-soil organisms. EFSA J 15(2):e04690. https://doi.org/10.2903/j.efsa.2017.4690

42. Paul K, Chatterjee SS, Pai P, Varshney AK, Juikar SJ, Prasad V, Bhadra B, Dasgupta S (2022) Viable smart sensors and their application in data driven agriculture. Comput Electron Agric 198:107096. https://doi.org/10.1016/j.compag.2022.107096
43. Pereira P, Bogunovic I, Muñoz-Rojas M, Brevik EC (2018) Soil ecosystem services, sustainability, valuation and management. Curr Opin Environ Sci Health 5:7–13. https://doi.org/10.1016/j.coesh.2017.12.003
44. Rai PK, Lee SS, Zhang M, Tsang YF, Kim KH (2019) Heavy metals in food crops: health risks, fate, mechanisms, and management. Environ Int 125:365. https://doi.org/10.1016/j.envint.2019.01.067
45. Ramola GC, Sahu L, Rathod D, Kukreti A (2023) Soil degradation and deterioration: causes, effects and case studies. In: Uniyal A, Sharma I, Tiwari I (eds) Soil restoration: assessment and reclamation, pp 107–123. ISBN: 978-93-94380-21-9
46. Raza T, Qadir MF, Khan KS, Eash NS, Yousuf M, Chatterjee S, Manzoor R, Rehman SU, Oetting JN (2023) Unrevealing the potential of microbes in decomposition of organic matter and release of carbon in the ecosystem. J Environ Manag 344:118529. https://doi.org/10.1016/j.jenvman.2023.118529
47. Rodrigo-Comino J, López-Vicente M, Kumar V, Rodríguez-Seijo A, Valkó O, Rojas C, Reza Pourghasemi H, Salvati L, Bakr N, Vaudour E, Brevik EC, Radziemska M, Pulido M, Di Prima S, Dondini M, de Vries W, Santos ES, Mendonça-Santos ML, Yu Y, Panagos P (2020) Soil science challenges in a new era: a transdisciplinary overview of relevant topics. Air Soil Water Res 13:1–17. https://doi.org/10.1177/1178622120977491
48. Rodríguez-Seijo A, Soares C, Ribeiro S, Amil BF, Patinha C, Cachada A, Fidalgo F, Pereira R (2022) Nano-Fe_2O_3 as a tool to restore plant growth in contaminated soils—assessment of potentially toxic elements (bio)availability and redox homeostasis in *Hordeum vulgare* L. J Hazard Mater 425:127999. https://doi.org/10.1016/j.jhazmat.2021.127999
49. Roshan A, Biswas A (2023) Fire-induced geochemical changes in soil: implication for the element cycling. Sci Total Environ 868:161714. https://doi.org/10.1016/j.scitotenv.2023.161714
50. Ryde U (2016) QM/MM calculations on proteins. In: Voth GA (ed) Methods in enzymology. Academic Press, pp 119–158. https://doi.org/10.1016/bs.mie.2016.05.014
51. Santás-Miguel V, Arias-Estévez M, Rodríguez-Seijo A, Arenas-Lago D (2023) Use of metal nanoparticles in agriculture. A review on the effects on plant germination. Environ Pollut 334:122222. https://doi.org/10.1016/j.envpol.2023.122222
52. Shi T, Guo L, Chen Y, Wang W, Shi Z, Li Q, Guofen W (2018) Proximal and remote sensing techniques for mapping of soil contamination with heavy metals. Appl Spectrosc Rev 53:783. https://doi.org/10.1080/05704928.2018.1442346
53. UNEP (2022) https://www.unep.org/news-and-stories/press-release/number-wildfires-rise-50-2100-and-governments-are-not-prepared
54. Urionabarrenetxea E, Casás C, Garcia-Velasco N, Santos MJG, Tarazona JV, Soto M (2023) Environmental risk assessment of PPP application in European soils and potential ecosystem service losses considering impacts on non-target organisms. Ecotoxicol Environ Saf 266:115577. https://doi.org/10.1016/j.ecoenv.2023.115577
55. Vargas-Berrones K, Bernal-Jácome L, de León-Martínez LD, Flores-Ramírez R (2020) Emerging pollutants (EPs) in Latin América: a critical review of under-studied EPs, case of study-nonylphenol. Sci Total Environ 726:138493. https://doi.org/10.1016/j.scitotenv.2020.138493
56. Villaverde JJ, Sandín-España P, Alonso-Prados JL, Lamsabhi AM, Alcamí M (2018) Pesticide byproducts formation: theoretical study of the protonation of alloxydim degradation products. Comput Theor Chem 1143:9–19. https://doi.org/10.1016/j.comptc.2018.08.006
57. Villaverde JJ, Sevilla-Morán B, López-Goti C, Alonso-Prados JL, Sandín-España P (2020) Contributions of computer-based chemical modeling technologies on the risk assessment and the environmental fate study of (nano)pesticides. In: Shukla V, Kumar N (eds) Environmental concerns and sustainable development: Volume 1: Air, water and energy resources. Springer, Singapore, pp 1–27. https://doi.org/10.1007/978-981-13-5889-0_1

58. Villaverde JJ, Sevilla-Morán B, Alonso-Prados JL, Sandín-España P (2022) A study using QSAR/QSPR models focused on the possible occurrence and risk of alloxydim residues from chlorinated drinking water, according to the EU Regulation. Sci Total Environ 839:156000. https://doi.org/10.1016/j.scitotenv.2022.156000
59. Wang J, Zhen J, Weifang H, Chen S, Lizaga I, Zeraatpisheh M, Yang X (2023) Remote sensing of soil degradation: progress and perspective. Int Soil Water Conserv Res 11:429. https://doi.org/10.1016/j.iswcr.2023.03.002
60. Wang Y, Munir U, Huang Q (2023) Occurrence of per- and polyfluoroalkyl substances (PFAS) in soil: sources, fate, and remediation. Soil Environ Health 1(1):100004. https://doi.org/10.1016/j.seh.2023.100004
61. Yuan Z, Ali A, Ruíz-Benito P, Jucker T, Mori AS, Wnag S, Zhang X, Hao Z, Wang X, Loreau M (2020) Above- and below-ground biodiversity jointly regulate temperate forest multifunctionality along a local-scale environment gradient. J Ecol 108:2012–2024. https://doi.org/10.1111/1365-2745.13378
62. Zettler ER, Mincer TJ, Amaral-Zettler LA (2013) Life in the "plastisphere": microbial communities on plastic marine debris. Environ Sci Technol 47(13):7137–7146. https://doi.org/10.1021/es401288x
63. Zheng X, Chen S, Gao L, Liu Y, Shen F, Liu H (2020) Experimental and theoretical study of kinetic and mechanism of hydroxyl radical-mediated degradation of sulfamethazine. Environ Sci Pollut Res 27(32):40504–40511. https://doi.org/10.1007/s11356-020-10072-z

Antibiotics as Emerging Pollutants of Soil Ecosystems

Paola Grenni, Andrea Visca, and Anna Barra Caracciolo

Abstract Soil is a non-renewable resource and maintaining its quality (including absence of contaminants) and fertility is fundamental for ensuring the safety of food derived from soil-based agriculture for both animals and human ("One Heath concept"). In this context, native soil microbial communities have a key role in providing several ecosystem regulating and provisioning services. Antibiotics are emerging contaminants, which can reach soils through different ways. They cause concern due to their possible direct biocide effects at very low concentrations. Moreover, environmental antibiotic contamination can significantly contribute to the selection of antibiotic resistance bacteria (ARB) carrying antibiotic resistance genes. Owing to the tight relationships between bacteria and plants, both antibiotics and ARB can enter plants and reach animal and human microbiomes through their ingestion. In this chapter, the main ways antibiotics enter soils, together with the effects on soil microbial communities and plants, are reported. A special focus is dedicated to sulphonamides and fluoroquinolones, which are largely used for both human and animal health.

Keywords One Health concept · Antibiotic resistance genes · Sulphonamides · Fluoroquinolones

1 Introduction

In accordance with the "One Health concept" [16], environmental, animal and human health are directly connected. For example, terrestrial ecosystems are fundamental for providing several ecosystem regulating (e.g., disease and pest regulation, decontamination), supporting (e.g., soil formation, plant growth) and provisioning services.

P. Grenni (✉) · A. Visca · A. B. Caracciolo
National Research Council, Water Research Institute, Rome, Italy
e-mail: paola.grenni@irsa.cnr.it

P. Grenni
NBFC, National Biodiversity Future Center, 90133 Palermo, Italy

A. Núñez-Delgado (ed.), *Frontier Studies in Soil Science*, Frontier Studies in Soil Science, https://doi.org/10.1007/978-3-031-50503-4_2

Provisioning services include products obtained from ecosystems, such as food, raw materials, fuel, genetic resources and various chemicals [68, 74]. Consequently, it is fundamental to maintain soil fertility and ensure that food derived from soil-based agriculture is of good quality, because it can influence directly and indirectly human and animal health [61].

Pharmaceuticals are substances used in the diagnosis, treatment, or prevention of disease and for restoring, correcting, or modifying organic functions in human and animals. They include many drug classes (analgesics, anti-inflammatory, cardiovascular, antidiabetics, estrogens, antiepileptic, psychiatric drugs, and antibiotics) with various biochemical activities and chemical–physical properties. Since pharmaceuticals are only partially metabolized (10–90%) by treated organisms [91], a large amount of them is excreted unaltered or as active metabolites via urine and faces, reaching wastewater treatment plants in the case of human [79] and soil in the case of veterinary drugs [34]. For example, from 50 to 90% of antibiotic doses such as tetracyclines, erythromycin and lincomycin are excreted in urine and faces [18]. Several substances not degraded in wastewater treatment processes can accumulate in significant amounts in sewage sludge. For instance, a total concentration of about 11,800 ng/g dry matter and 45 pharmaceuticals, illicit drugs and their metabolites were recently found in a sewage sludge from a municipal wastewater treatment plant [40].

Pharmaceuticals and their metabolites can directly reach agricultural soil. In fact, several antibiotics (e.g. doxycycline, 728.4 ng/g; oxytetracycline, 139 ng/g; chlortetracycline, 1,079 ng/g) and antimicrobials (e.g. triclocarban, 51.8 μg/kg) have been found in in soil as emerging contaminants [34, 50, 63]. This could be due to the use of organic amendments such as animal manure applied as a fertilizer, reclaimed wastewater used for irrigation purposes [7, 34, 65, 66] and biosolids application (treated sewage sludge which meets pollutant and pathogen requirements for land application). The potential inclusion of pharmaceuticals, and in particular antibiotics, with the use of digestate as a soil amendment is currently under investigation. Recent studies have shown significant decreases in antibiotic concentrations (sulfamethoxazole, enrofloxacin and ciprofloxacin) in agro-zootechnical waste after its anaerobic digestion in biogas plants [84]. Moreover, digestate was found to stimulate a higher soil microbial biodiversity, introducing and promoting more bacteria related to antibiotic degradation [6].

The degradability of these pollutants depends on their chemical structure and abiotic and biotic site-specific conditions (e.g., temperature, soil type etc.). Some compounds have an intrinsically low degradability. A recent review of antibiotic persistence reports fluoroquinolones (in particular ciprofloxacin), macrolides (e.g., azithromycin) and tetracyclines (doxycyclines and tetracycline) as the most persistent antibiotics in soils [18]. Due to their persistence, these chemicals can have a serious potential impact and adversely affect terrestrial ecosystems [11]. As they are bioactive substances with high effects at low concentrations, they can influence the structure and functioning of microbial communities in soil with consequences for the ecosystem services provided by them [32]. Pharmaceuticals and their metabolites occurring in agricultural soil from animal manure, reclaimed wastewater or biosolids, can be adsorbed by plants. For example, in a recent study [6] with *Lactuca sativa* grown in a soil with manure or digestate and spiked with a mixture of two sulfonamide (sulfamethoxazole, SMX) and two fluoroquinolones (sulfamethoxazole, SMX and enrofloxacin, ENR) showed that plants partially up-taken CIP and ENR from soil (bioaccumulation factors > 1). Manure stimulates the bioaccumulation more than digestate. Interestingly, in lettuce leaves, the fluoroquinolone-resistance gene (*aac-(6′)-Ib-cr*) was the most abundant.

The octanol–water partition coefficient (Kow) is considered a tool for predicting plant uptake of non-ionizable pharmaceuticals. On the other hand, polar and ionizable pharmaceuticals (the majority of them are in this category) are found in soil pore water and are, consequently, the major fractions bioavailable for plant uptake and those most likely bioconcentrated in plants [51]. In a recent review, three ionizable antibiotics were identified as those with the highest concentrations in plant tissues (trimethoprim: 270 ng/g; chlortetracycline 532 ng/g; oxytetracycline 330 ng/g) [63]. Other studies found pharmaceuticals in plant tissue, with concentrations of up to 487 μg/kg for the antibiotic Sulfadiazine in roots of winter wheat grown on soil treated with liquid pig manure.

Once in plants, these pollutants may adversely affect plant growth and development [13, 15]. Phytotoxic effects were found in laboratory conditions in a varied range of plants exposed to concentrations ranging from μg/L to g/L of individual compounds. However, in some cases plant metabolism can achieve their detoxification, inactivation, or excretion.

The combined effects of simultaneously occurring compounds have been rarely studied, and it was found that chemical mixtures can exert synergistic or additive effects resulting in enhanced phytotoxicity [15]. In the case of antibiotics, negative effects on plants can be due to direct damage or to interference with the symbiotic interactions with soil bacteria [31]. For example, amoxicillin was found to have ecotoxicological effects (growth and germination) on Alfalfa, carrot and lettuce in a range of 0.001–10 mg/L. Sulfadiazine was found to have negative effects on Zea mays, depending on its concentration (10 mg/kg: root growth decline; 200 mg/kg: death of plants).

Although a low risk for humans of exposure to individual substances through the consumption of vegetables was calculated by some authors [12, 72], a long-term risk exposure to pharmaceutical mixtures or to their active metabolites cannot be excluded.

Due to the lack of comprehensive data on toxicological and ecotoxicological effects and environmental concentrations, most pharmaceuticals are not regulated in either water or soil and for this reason are considered emerging pollutants.

2 Antibiotics

Antibiotics currently in use are natural, synthetic, and semi-synthetic molecules. Natural antibiotics are produced by bacteria and fungi (e.g., benzylpenicillin and gentamicin) to inhibit or kill other competitor microorganisms (with bacteriostatic or bactericidal effects). Semi-synthetic compounds are natural antibiotics chemically altered by inserting an additive within the drug formulation, which improves its effectiveness (more stable and less biodegradable). Antibiotics are complex molecules that can have different functional groups within their chemical structure and can be classified in different classes based on their action mechanism: inhibition of cell wall synthesis, alteration of cell membranes, protein or nucleic acids synthesis inhibition, inhibition and metabolic or anti-competitive antagonism [32].

Antibiotics can persist in soil, being transported to surface and groundwater (in the case of polar, acidic and low Kow compounds) or absorbed by plants after being introduced into the environment through manure or biosolid application and/or wastewater irrigation. Wu et al. [89] investigated the accumulation of antibiotics and other pharmaceuticals in four major vegetables grown hydroponically for 21 days; they discovered that antibiotic transport was positively connected with the octanol/water partition coefficient (log Kow) of the chemicals and the amount and plant part in which pharmaceuticals can be adsorbed; it also depended on the vegetable and pharmaceutical types. Hu et al. [37] studied the incidence and migration of antibiotics in organic farming employing biosolids as fertilizers and found that 11 antibiotics were distributed and accumulated in plants. Sabourin et al. [73] investigated the absorption of antibiotics from municipal biosolid-treated agricultural soils in tomatoes, carrots, potatoes, and sweet corn, and found that sulfonamide, quinolone and trimethoprim concentrations in the edible part of the crops ranged from 0.02 to 14 ng/g (dw). Moreover, antibiotics and their related resistance genes (ARGs) have been found in surface water, groundwater (including drinking and reclaimed water), soil and sediments [86].

Due to their intrinsic biocide action, antibiotics can kill or inhibit native soil prokaryotic communities, affecting key ecosystem processes such as contaminant degradation and nutrient cycling; an alteration in the latter can influence soil fertility and primary production [32]. It is generally found that antibiotics in soil negatively affect soil microbial composition more than that of plants [80]. For example, Cyanobacteria, a diverse bacterial group of photoautotrophs, ubiquitous in terrestrial ecosystems and playing important ecological functions (e.g., primary production and nitrogen fixation), can be sensitive to antibiotic effects. A lab experiment with eight Cyanobacteria species exposed to antibiotics showed quite variable growth inhibition (with NOEC calculated from 1.5 to 157 μg/L), depending on the compound mode of action and the species tested. These results make it difficult to establish predictable non-effective concentrations [47].

Environmental antibiotic contamination due to extensive and increasing use causes particular concern for human and animal health due to the risk of the development and spreading of multi-resistant bacteria [32]. The selection and spread of ARGs endow drug-resistant bacteria (ARB) with the ability to overcome the effects of antibiotics. This is a well-known phenomenon in hospitals, where drug-resistant pathogens can persist and infect patients and make infection treatment ineffective [87]. The scientific community has only recently become aware that the proliferation of ARBs and ARGs is a complex phenomenon involving not only the nosocomial environment, but also natural ecosystems, where they are present as micro-pollutants.

The phenomenon of antibiotic resistance is a bacterial natural defense mechanism, involved in competitive interactions (e.g. for space or nutrients) with other bacteria and fungi [4, 9, 53], defense against predators (e.g. Protists) [17, 43, 78], and protection versus other environmental stressors [21, 81]. Not all bacterial strains can produce antibiotics, but all bacteria (both antibiotic and non-antibiotic producers) can develop resistance [60].

This capacity is a microorganism peculiarity. The development of resistance can occur through a vertical gene transfer (VGT) inside the same bacterial species and/or a horizontal gene transfer (HGT) between different species. The HGT between microorganisms and/or pathogens is possible thanks to mobile genetic elements (MGEs), such as plasmids, integrons and transposons [90]. HGT is a key process rendering possible a large distribution of ARGs in the environment [9] and it is promoted by a widespread environmental antibiotic presence [60].

Bacteria use regulatory mechanisms that selectively stimulate the expression of the appropriate antibiotic resistance genes only during exposure to antibiotics, avoiding the fitness burden and facilitating the long-term maintenance of such genes in the genome. Moreover, some resistance genes can have negative effects on bacterial cells. For example, resistance genes that provide methylation of specific antibiotic residues in 23S ribosomal RNA, protect bacteria from macrolide antibiotics; however, they can also cause cell-wide disruption in protein synthesis, leading to major fitness defects [19, 33]. In the case of the tetracycline efflux pump gene (*tet*A), the transcriptional repressor TetR constitutively binds the *tet*A promoter and inhibits the expression of the *tet*A resistance gene in the absence of antibiotics [36]. When tetracycline is present in the cell, direct binding of tetracycline to the TetR repressor leads to its dissociation from the DNA and drives *tet*A expression, leading to antibiotic resistance. In this case, it could be useful to analyze not only ARGs but also their promoter/repressors.

Qian et al. recognized that multidrug resistance and efflux pumps are the dominant class genes and mechanisms in native soils and the resistome profile is mainly driven by bacterial community composition. No significant differences were observed for ARG diversity and abundance between native soil and a long-term cultivated agriculture soil.

In animals treated with antibiotics, intestinal bacteria can act as a reservoir for ARGs, transferring genetic material to soil microorganisms, and from the latter to humans directly or indirectly, for example by eating fresh vegetables [5].

A recent paper reports that fertilization of *Lactuca sativa* with swine manure or digested sewage sludge results in a total hazard quotient calculated for antibiotics of <1, suggesting that *Lactuca* consumption amended with organic fertilizers does not pose a risk to human health due to the presence of antibiotics [52]. The same study found concentrations ranging from 0.67 ng/g (for lincomycin) to 14.2 ng/g (for ciprofloxacin) in *Lactuca* leaves, but the potentially harmful combined effects were not assessed. Another recent study reports that pig manure application to soil increased the bioaccumulation of ARGs and mobile genetic elements (MGEs) in carrot roots [55] and the peeling of the tubers is an effective strategy for reducing the risk of human intake of ARGs.

Välitalo et al. [83] reviewed the ecotoxicological effects of several antibiotics on aquatic microorganisms (Green algae, cyanobacteria and some *Proteobacteria* such as *Vibrio fischeri* and *Pseudomonas putida*); high effect concentrations (EC_{50}) were reported for various antibiotics and *V. fischeri* with the standard acute toxicity test, and this may underestimate the risk of antibiotics to environmental bacteria but can have long-term negative effects. For example, *Daphnia magna* was found to have a reduction of fertility with 5,000 μg/L of ciprofloxacin and ofloxacin in a 42-days test [60].

There is currently limited knowledge about the ecotoxicology and sub-lethal effects on aquatic and soil organisms. In any case, the ecotoxicology of antibiotics is different from that of conventional pollutants because they are specifically designed for a biological activity at low concentrations. Moreover, they act on specific targets that can be widely preserved in multiple life forms.

3 Examples of Commonly Used Antibiotics: Ciprofloxacin, Sulfamethoxazole and Enrofloxacin

Among antibiotics, sulfonamides and fluoroquinolones are the most widely used in human and veterinary medicine and frequently detected as environmental micropollutants [6, 77, 84].

Sulfonamides are one of the first antibiotics used in medicine. Gerhard Domagk, a Nobel Laureate in 1939, was the first to discover their antibacterial characteristics, observing that prontosil, a sulfonamide dye, restrained pathogenic bacteria cells selectively. Prontosil is a pro-drug that the human body converts to sulfanilamide, an antibacterial active ingredient. Subsequently, other antibacterial compounds derived from this chemical group, with the same primary core, but with different bioactivities, have were produced [67].

Sulfonamides have a structure extremely similar to para-amino benzoic acid; for this reason, they can operate as competitive inhibitors by interfering in its role in the production of folic acid, purine and DNA [3]. The resistance is due to genes that encode alternative versions of dihydropteroate synthase enzymes (DHPS), which are drug-resistant and exclude prokaryotic cells. In fact, the sulfonamide target in bacteria is the dihydropteroate synthase enzymes (DHPS) which compete with the para-aminobenzoic acid (PABA) precursor. The synthesis of folic acids by the reaction of 6-hydroxymethyl-dihydropterin 1′-diphosphate with 4-aminobenzoate catalyzed by DHPS, produces 7,8-dihydropteroate and inorganic pyrophosphate, which are involved in DNA synthesis. This process is hampered by sulfonamides [27] (Fig. 1). Bacteria can spread resistance to sulfonamides to other bacteria through plasmids.

Sulphamethoxazole (SMX), a sulfonamide with an aniline and an isoxazole group, is one of the antibiotics most prescribed and used in both human and veterinary medicine; it can be used to treat and prevent respiratory infections and mastitis in cattle combined with the antibiotic trimethoprim. SMX has a bacteriostatic effect,

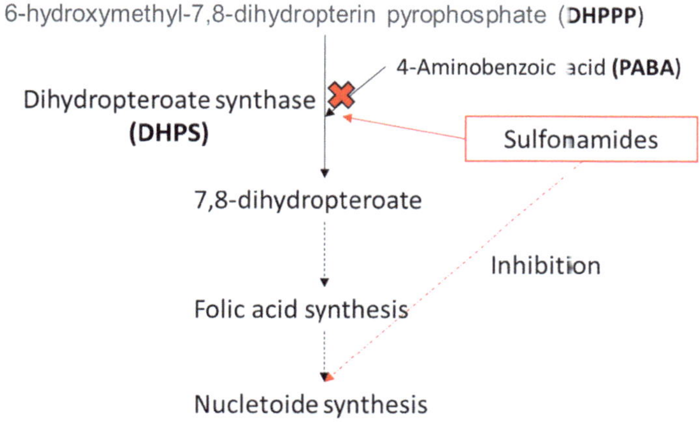

Fig. 1 The Sulfonamide target is the folate pathway. Sulfonamide antibiotics can operate as competitive inhibitors by interfering in its role in the production of folic acid, purine and DNA. DHPS: dihydropteroate synthase enzymes; PABA: para-aminobenzoic acid (modified from [5])

stopping the bacterial synthesis of folic acid starting from para-aminobenzoic acid, required for nucleotide synthesis.

In treated organisms, about 43% of SMX is metabolized to N_4-acetyl-sulfamethoxazole and 9–15% to sulfamethoxazole N1-glucuronide, and 15–25% is excreted unchanged. Once in the environment, SMX degradation can occur with variable degradation rates. Some authors report a degradation in biologically active soils of between 50 and 80% of its initial concentration in 20 days, depending on anaerobic/aerobic conditions. Other studies report a halving of its initial concentration (20 mg/kg) in 7 days, thanks to the presence of antibiotic resistance bacteria able to degrade it. Several studies have shown the phenomenon of ARB and ARG spreading in soil with the use of manure and biosolids as organic fertilizers. Consequently, potential SMX, ARB and ARG accumulation in fresh edible plants needs to be investigated [64, 89]. A recent study showed possible SMX accumulation in lettuce, and this was positively correlated with antibiotic concentration in soil. Moreover, SMX contamination strongly reduced bacterial diversity and changed the composition of bacterial and fungal communities in the soil. SMX accumulation in edible plants can subsequently involve ARB and ARG acquisition in humans and farm animals and this phenomenon needs to be explored.

The resistance-plasmid enzymes for SMX are encoded by two genes, *sul1* and *sul2* [75]. The sul1 gene is frequently found in *Tn21* type integrons with other resistance genes, whereas *sul2* is usually found on tiny plasmids such as the highly mobilizable IncQ family (RSF1010) and the pBP1 family [45]. Both *sul1* and *sul2* DHPS products have para-aminobenzoic acid (PABA) Km (Michaelis constant) values of about 0.6 M, resulting in resistance to high sulfonamide concentrations [5]. In fact, Km is the concentration of a substrate, which permits an enzyme to achieve half Vmax (Vmax: maximum rate of reaction). An enzyme with a high Km has a low affinity for its substrate and requires a higher concentration of substrate to achieve Vmax.

For most enzymes, Km lies between 10^{-1} and 10^{-7} M. The DHPS of *sul2* seems to be extremely selective in discriminating between its regular PABA substrate and sulfonamide.

Perreten and Boerlin [69] also identified a third resistance gene, *sul3*, which codes for a 263-amino-acid protein that is identical to a dihydropteroate synthase expressed by the 54-kb conjugative plasmid pVP440 from *E. coli*. Sánchez et al. found that SMX resistance can be additionally due to genes encoding efflux pumps, such as *sme*DEF.

Quinolones are another group of the most commonly prescribed antimicrobials in the world, and they are used to treat a variety of bacterial illnesses in humans and animals [70]. Quinolones can convert gyrase and topoisomerase IV, which are primary targets, into hazardous enzymes that fragment bacterial chromosomes [25]. The first broad-spectrum quinolone, norfloxacin, was only approved for treating urinary tract infections and sexually transmitted illnesses. Since the 1990s, the number of quinolone-resistant bacterial strains has progressively increased due to the widespread use of these antibiotics [23], threatening their clinical utility.

There are several resistance mechanisms in bacterial cells against quinolones [30], including those involving chromosomal mutations in genes encoding topoisomerase IV or gyrase IV (involved in DNA replication), or drug accumulation reduction. The latter can occur thanks to overexpression of bacterial efflux pumps (which push the drug outside a cell) or down-regulation of porin proteins (membrane protein), preventing ciprofloxacin passive diffusion inside cells. Quinolone resistance mediated by plasmids has also been identified. The *qnr* gene, which encodes a pentapeptide capable of binding chromosomal DNA and shielding it from drug action, was the first plasmid-mediated resistance mechanism identified [41]. The cr variation of the *aac(6′)-lb* gene, which encodes for an aminoglycoside acetyltransferase that acetylates ciprofloxacin, is another plasmid-mediated resistance mechanism [28].

Fluoroquinolones (FQs) are wholly synthetic and broad-spectrum antibiotics, commonly used in animal husbandry. They derive from quinolones by adding fluorine to the molecule. The mode of action of FQs is described in Fig. 2. Briefly, fluoroquinolones specifically inhibit DNA gyrase and topoisomerase IV. DNA gyrase catalyzes the separation of daughter chromosomes by introducing negative superhelical twists in the bacterial DNA double-helix (upstream from the replication fork). This action is required for DNA replication to begin and allows initiation proteins to bind. The gyrA and gyrB genes, respectively, encode two GyrA and two GyrB monomeric subunits that make up DNA gyrase. Topoisomerase IV oversees decatenation, or the removal of interlinking between daughter chromosomes, permitting segregation into two daughter cells at the end of a replication round. Topoisomerase IV is composed of four homologous monomeric subunits, two ParC subunits and two ParE subunits encoded by the parC and parE genes, respectively. Fluoroquinolones cause conformational changes in the enzyme-bound DNA complex (i.e., DNA gyrase with bacterial DNA or topoisomerase IV with bacterial DNA) that limit normal enzyme function. As a result, the combined drug–enzyme–DNA complex stops the replication fork from moving forward, preventing normal bacterial DNA synthesis and eventually leading to bacterial cell death.

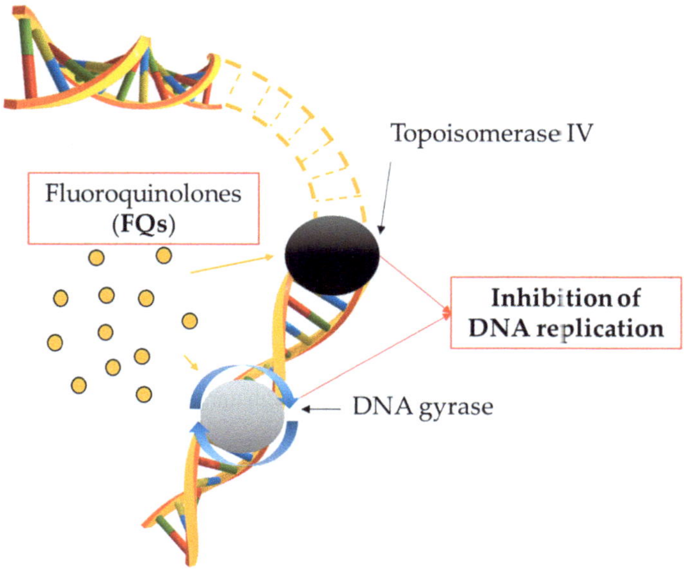

Fig. 2 Mode of action of Fluoroquinolones. They inhibit DNA gyrase and topoisomerase IV, causing conformational changes in the enzyme-bound DNA complex, limiting normal enzyme function (modified from [5])

Enrofloxacin (ENR) is a broad-spectrum antibiotic belonging to the fluoroquinolone family [82]. It has been widely used in many countries to treat a number of poultry infections, primarily those caused by *Escherichia coli* and *Pasteurella multocida*, as well as avian mycoplasmosis. Although the use of enrofloxacin in chickens was outlawed in 2005 in the United States, FQ-resistant bacteria are still being detected [39].

Ciprofloxacin (CIP) is the first fluoroquinolone with antibiotic activity in body areas other than the urinary tract [76]. Ciprofloxacin is used to treat a variety of Gram-negative bacteria and, to a lesser extent, Gram-positive infections. The therapeutic success of CIP generated a slew of newer-generation quinolones with an even larger spectrum of activity, particularly against Gram-positive bacteria. CIP was included in the European Watch List in 2018 [24] as a result of its widespread use and environmental detection.

Ciprofloxacin is also the main metabolite of ENR [29]. Because both ENR and CIP are adsorbed to soil and have slow biodegradation rates, considerable levels of FQ can be found in agricultural soils [48, 92]. FQ stability in soil is influenced by abiotic (light, soil organic matter) and biotic (bacterial populations that can degrade them) factors [22].

Strong adsorption, on the other hand, results in decreased absorption of FQs residues by plants [12]. The fate of this family of antibiotics in the ecosystem food chain (including the human food web) is almost unknown in soils [71].

A recent project ("Evaluation of the presence of Antibiotics in Zootechnical waste and in the digestate of biogas plants: study of strategies for their RemOval—AZeRO antibiotics"), supported by the Lazio Region (Lazio Innova, Italy) evaluated the presence and concentration of some antibiotics (sulfamethoxazole, ciprofloxacin and enrofloxacin) in zootechnical waste and in the digestate of full-scale biogas digesters (Fig. 3). The aim of the project was to assess how an antibiotic presence can influence the anaerobic digestion process and which conditions can favor antibiotic degradation during anaerobic digestion treatment of cattle manure and slurry. Antibiotic degradation is a fundamental process for preventing or reducing the environmental spreading of residual antibiotics and ARGs through agricultural practices such as the use of digestate as fertilizer [5].

Lab experiments were performed using digestate spiked with 5 mg/L of sulfamethoxazole. Interestingly, the microbial community was not negatively influenced by the antibiotic in terms of biogas production; on the contrary, the acidogenic and acetogenic phases were more efficient in terms of CH_4 and H_2 production. Moreover, SMX was significantly degraded and, at the end of the 70-day experiment, only 20% of its initial concentration was detected. At the same time, the relative abundance (ARG/16S) of the SMX resistance genes (*sul1*, *sul2*, and the proxy *intI*1) found initially decreased during the digestion process in both spiked and control batches [54]. Additionally, annual analyses of input and output samples from a full-scale anaerobic plant showed that sulfamethoxazole, ciprofloxacin and enrofloxacin degraded during the anaerobic digestion process, although with different percentages (84–100%), depending on the specific season and on antibiotic type. In line with the decrease in antibiotics, all ARGs detected (*sul1*, *sul2*, *qnr*S, *qep*A, *aac-(6′)-Ib-cr*) decreased significantly (up to 80%) in the digestate samples. This proved that the anaerobic digestion process can be an effective tool for lowering antibiotic residues and ARGs in organic amendments and consequently in soil if they are used as organic fertilizers [84]. Digestate is known to stimulate soil microbial biodiversity, promoting bacterial genera involved in soil fertility, but was also found that its presence can decrease antibiotic bioaccumulation in plants [6].

4 Biodegradation of Antibiotics

The native microbial community of soil is fundamental for ecosystem functioning. For example, it has an important role in regulating the environmental fate of pollutants, by controlling ecosystem quality [32]. In fact, microorganisms are key degraders of organic matter, offering nutrients to other organisms [57] and removing contaminants (including pharmaceuticals). Microorganisms contribute to ecosystem self-purification processes by degrading contaminants via metabolic and/or co-metabolic pathways. In this sense, they provide an ecological regulating service.

Fig. 3 Pictures of an open-air tank where manure is stored before entering the digester (**a**), exit pipe of digestate after the digestion process (**b**) and storage of solid digestate after solid/liquid separation process (**c**)

This function could be hampered by changes in microbial functional groups, due to the presence of pollutants if their toxicity inhibits microbial activity [1, 2]. As a result, contamination can be critical in overall natural ecosystem quality.

Degradation of pharmaceutical residues in soil, and in particular of antibiotics could consequently prevent potential risks to human health and soil ecosystems. Antibiotic degradation depends on several factors: among others, adsorption, transport and the presence of an active, abundant and diverse microbial community. Adsorption depends on antibiotic properties (e.g., molecular structure, hydrophobicity, polarity, polarizability, and spatial configuration), soil characteristics (e.g., texture, mineralogy, pH) soil organic matter (influenced by the presence of biosolids or other organic amendments); moreover, potential chemical mixture interactions can also influence it [58]. Adsorption and transport influence antibiotic bioavailability in soil. Antibiotic concentrations and any repeated treatments and, most of all, the presence of bacteria able to resist them and to develop strategies to degrade these substances are of fundamental importance.

Among pharmaceuticals, antibiotics have very variable half-lives, ranging from 0.43 to 3,466 days, although, in most cases, they do not exceed 100 days [18]. Microorganisms can respond to an antibiotic presence with resistance mechanisms (i.e. resistance genes) and remove it, thus also showing resilience capabilities. Biodegradation of antibiotics can occur through metabolism or co-metabolism [85]. Microorganisms can metabolically degrade antibiotics, using them as carbon, nitrogen and energy sources [88]. On the other hand, microorganisms can unintentionally transform (co-metabolism) a pollutant using enzymes or cofactors produced for microbial metabolism of other molecules.

Antibiotic degradation, considered indirect resistance, can lessen the selective pressure caused by high concentrations of these pharmaceuticals, minimizing the need for bacterial populations to evolve resistance genes. Biodegradation can merely lead to an antibiotic modification (biotransformation), or to complete reduction to inorganic molecules (mineralization).

Sulfonamides are antibiotics found in soils up to mg/kg and show halving times (DT_{50}) varying from 7 days to 59 days [18]. They have a weak soil adsorption and a high mobility in line with their low log Koc values [18]. Crane et al. report that antibiotics with a Koc >4,000 L/kg are non-mobile and very persistent in soils (DT_{50} > 60 days) and those with Koc values <15 L/kg are highly mobile and easily degraded (DT_{50} < 5 days). The biodegradation of sulfonamide antibiotics is due mainly to bacteria and can be favored by nitrogen availability. For example, some authors report that in the presence of additional nitrogen sources, sulfamethoxazole was fully converted into 3-amino-5-methylisoxazole, with no antibiotic activity. Several soil bacterial strains, such as *Bacillus licheniformis* ATCC 14580 (*Firmicutes*), *Aquamicrobium defluvium* (*Alphaproteobacteria*), *Pseudomonas putida* (*Gammaproteobacteria*), *Alcaligenes* sp. (*Betaproteobacteria*), *Microbacterium* sp. C448, *Microbacterium lacus* SDZm4 and *Terrabacter* sp. 2APm3 (*Actinobacteria*), have been found to degrade sulfonamides. In any case, sulfonamides are generally reported to be resistant to microbial mineralization in soils. Hirth et al. reported an increase in sulfamethazine mineralization in soil thanks to bioaugmentation with an indigenous soil microbial community, including the *Microbacterium* sp. C448 strain.

Quinolones and fluoroquinolones have long half-lives in soils (e.g., ciprofloxacin was found to have a half-life of more than 3,400 days) [13]. Ciprofloxacin has frequently been detected in soil due to its high sorption to organic matter, and this phenomenon promotes its accumulation with concentrations that reach 7.2 mg/kg dry soil. Fluoroquinolones can be transformed by photolysis, but this abiotic degradation happens little in soil. Some authors reported that they are recalcitrant to degradation in soil with non-acclimatized microbial communities. However, microbial transformation can occur, and N-acetylation is considered the most common transformation pathway for fluoroquinolones. This reaction can be catalyzed by numerous bacterial and fungal strains.

Interestingly, a variant of the enzyme initially linked to aminoglycoside resistance (aminoglycoside acetyltransferase (AAC(6′)-lb) was recently shown to be able to N-acetylate ciprofloxacin. Different fungal strains can transform ciprofloxacin, but only two fungi (*Curvularialunata, Ascomycetes*; *Gloeophyllum striatum, Basidiomycetes*) were able to mineralize danofloxacin, enrofloxacin or ciprofloxacin.

It has been recently found that ciprofloxacin degradation in soil can be promoted when it is incorporated into soil together with manure. Finally, two cultivars of *Brassica parachinensis* L. were found to promote ciprofloxacin degradation in soil and one bacterial genus *Spirochaeta* (*Spirochaetes*) and one fugal genus (*Trichosporon, Basidiomycota*) were associated with this antibiotic degradation in the plant rhizosphere.

5 Environmental Concentrations of Antibiotics Improving ARG Spreading

A large percentage of antibiotics used in clinical practices are either direct products or modifications of chemicals generated by soil bacteria. As a result, identifying the ecological role of antibiotics in soil ecosystems is a key issue for understanding antibiotic resistance. Soil bacteria can produce antibiotics not only for competitive and defensive interactions, but also at sub-inhibitory concentrations for inter- and intra-domain communication. It is difficult to define common native levels of resistance genes in pristine environments and to determine safe release levels of antibiotic concentrations, which do not cause AMR propagation.

Experimental findings reveal that AMR genes are selected in complex communities at antibiotic doses (the minimum selective concentration; MSC) that are substantially lower than those that prevent sensitive bacteria from growing (minimum inhibitory concentration; MIC). Most studies on MSC are performed on single species. Studies of single species imply that resistant cells can protect susceptible cells by degrading antibiotics both intracellularly and extracellularly, enhancing the relative fitness of susceptible strains and hence the MSC. Excreted metabolites, on the other hand, might increase or reduce antibiotic efficacy, resulting in a decrease or increase in MSCs.

The impact on mixed bacterial communities has been little studied so far. The real effect of other species most likely varies depending on the circumstances, the existence of the community normally raises the MSC. A recent study showed that SMX at some environmental concentrations tested (20 mg/Kg) did not have effects on microbial community growth and did not select for antibiotic-resistant gene (ARG) maintenance or propagation. SMX concentrations which inhibited microbial growth were higher (>4.75 mg/L) than those generally found in natural settings. This was presumably due to a previous adaptation to the antibiotic by the microbial community analyzed. In fact, the microbial populations were not only unaffected by the antibiotic but were also capable of rapidly degrading SMX in liquid cultures (within 24 h).

The scientific community and industry stakeholders have been making efforts to estimate antibiotic concentrations that, based on current empirical knowledge, should provide safety limits for protecting human health from risks of AMR selection [10, 46, 59]. Two international frameworks have established PNECs that might protect from AMR selection: (i) Bengtsson-Palme and Larsson [10] proposed the establishment of compound-specific safe antibiotic emission limits (PNECs) derived from MICs for clinically relevant bacteria; and (ii) Le Page et al. [46] proposed a single-value production discharge limit of 100 ng/L, based on MICs for clinically relevant bacteria.

Bengtsson-Palme and Larsson [10] compared antibiotic concentrations measured in municipal sewage treatment plant effluents to PNECs to evaluate if their proposed PNECs were ecologically relevant. They discovered that effluent concentrations surpassed PNECs in 28% of cases. These findings suggested that more advanced treatment could be required to reduce AMR selection in treatment plants. Nonetheless, post-discharge environmental fate and transport mechanisms, such as dilution and degradation, could influence effluent antibiotic concentration of receiving water bodies at the watershed scale. However, neither PECs nor measured ambient concentrations (MECs) of antibiotics in surface waters have been investigated for the proposed PNECs. Furthermore, there has been no simultaneous co-consideration of current international frameworks for guiding antibiotic PNECs in order to reduce AMR proliferation and human health risk.

Antibiotic residues consumed by drinking water or contaminated food may interact with the human microbiome, which is made up of a large number of different bacteria that live in the human body. Antibiotic residues from the environment can infiltrate the human gastrointestinal tract, which has around 800–1000 distinct bacterial species and over 7000 different strains. About 95% of the microorganisms are beneficial bacteria, while the rest are dangerous and opportunistic infections. *Bacteroidetes* and *Firmicutes* dominate the micro-ecological balance between these bacteria and the human body over time, with minorities of *Actinobacteria*, *Proteobacteria*, *Verrucomicrobia*, and other bacteria phyla that have not yet been discovered or characterized. Antibiotic exposure is closely related to changes in gut microbiome composition, as measured by observational, clinical and epidemiological investigations, since antibiotics have a broad-spectrum effect on the host-associated microbial community rather than a single target species. Antibiotic therapy may alter the composition of the intestinal microbiota, resulting in an increase in *Firmicutes* and a decrease in *Bacteroidetes*, as well as the formation of antibiotic-resistant bacteria that can live for years in the human intestine.

6 Phytotoxicity of Antibiotics

The phytotoxicity of antibiotics varies between antibiotic compounds and plant species. Research on the phytotoxic effects of different kinds of antibiotics is still lacking, and most of them are measured in vitro rather than under soil conditions. The phytotoxicity endpoints most studied are germination, growth and development. Antibiotics have a two-phase effect on plant growth, characterized by the excitatory effects caused by low-dose stimulation and by high-dose inhibition [49]. Tetracyclines reduce the production of pinto beans (*Phaseolus vulgaris*), but they promote the nutrient absorption of wheat (wheat) and corn (*Zea mays*) [8]. Tetracycline significantly increased the activity of the stress proteins glutathione S-transferase and peroxidase in corn plants but not in pinto beans [26]. Pan and Chu [62] report that the phytotoxic effects of antibiotics on seed germination are lower than those on root/stem elongation. Antibiotics may have difficulty penetrating the seed coat, so they cannot be absorbed by the seed and affect the growth of the radicle [56]. Root elongation is a more sensitive endpoint than bud elongation and seed germination [14]. Jin et al. [42] found that the inhibitory effect of sulfonamides on shoot elongation was significantly higher than that on root elongation. Some plants are more susceptible than others; for example, Hillis et al. [35] found that carrots are more sensitive to tetracycline than lettuce and alfalfa. Hydrophobicity is the most important factor in controlling phytotoxicity [62]. However, most effective antibiotic concentrations measured in in vitro phytotoxicity tests are unlikely to occur in environmental soils. In addition, the potential environmental impact of antibiotic metabolites has not been extensively studied. It is necessary to study the chronic phytotoxicity of antibiotics in soil using wastewater, reclaimed water and organic amendments (e.g., manure, biosolids), which contain real antibiotic residues in greenhouse experiments [20, 44, 73, 89]. Some of these works found accumulation of 13 antibiotics in plants. Hydroponic experiments found lettuce, spinach, cucumbers and peppers accumulated sulfonamides in their roots [89]; however, although this experimental approach can be more intuitive for understanding absorption mechanisms, it does not represent the complexity of real agricultural environments (e.g., the physical and chemical properties of soils are not considered. In any case, data collected on antibiotic phytotoxicity are still limited. Mesocosm and field experiments are desirable for assessing the ability of plants to actually absorb and accumulate potential antibiotics. Realistic data can be used for risk assessment evaluations [38, 64].

7 Conclusions and Future Perspectives

Antibiotics together with their metabolites and transformation products have been detected in natural environments, including soil, at residual concentrations. Since they can have effects at low concentrations, they can pose a risk for ecosystems and human health. Antibiotics can persist in soil, depending on their intrinsic characteristics and initial amount, abiotic (e.g., adsorption to soil, water availability, temperature) and above all biotic factors (the presence of a microbial community able to resist their biocide effects and/or to remove or detoxify them).

An antibiotic presence in soil can have direct effects: death or inhibition of bacterial populations involved in key ecosystem services; indirect effects: promoting bacterial resistance mechanisms (e.g., ARGs), which can be transferred from soil bacteria to plants, and selecting bacterial populations able to degrade them. The link between antibiotic degradation and resistance still has to be better investigated, however the more an antibiotic is promptly degraded, the less the risk of resistance spreading. The concentrations able to maintain and select antibiotic resistance or degradation among mixed native microbial populations have not currently been determined.

Moreover, few studies are focused on metabolites, transformation products or mixture of pharmaceuticals and other pollutants that can co-occur in soil (e.g., heavy metals, antibiotic degradation products). Finally, their long-term effects are still to be investigated and are a key focus for future research on this complex issue.

References

1. Abatenh E, Gizaw B, Tsegaye Z, Wassie M (2017) The role of microorganisms in bioremediation—a review. Open J Environ Biol 2(1):038–046. https://doi.org/10.17352/ojeb.000007
2. Abdelsalam NA, Ramadan AT, ElRakaiby MT, Aziz RK (2020) Toxicomicrobiomics: the human microbiome vs. pharmaceutical, dietary, and environmental xenobiotics. Front Pharmacol 11. https://doi.org/10.3389/fphar.2020.00390
3. Arivazhahan A (2021) Sulfonamides, quinolones, and agents for urinary tract infections. In: Paul A, Anandabaskar N, Mathaiyan J, Raj GM (eds) Introduction to basics of pharmacology and toxicology. Springer, Singapore, pp 807–819. https://doi.org/10.1007/978-981-33-6009-9_53
4. Bahram M, Hildebrand F, Forslund SK, Anderson JL, Soudzilovskaia NA, Bodegom PM, Bengtsson-Palme J, Anslan S, Coelho LP, Harend H, Huerta-Cepas J, Medema MH, Maltz MR, Mundra S, Olsson PA, Pent M, Põlme S, Sunagawa S, Ryberg M, Tedersoo L, Bork P (2018) Structure and function of the global topsoil microbiome. Nature 560(7717):233–237. https://doi.org/10.1038/s41586-018-0386-6
5. Barra Caracciolo A, Visca A, Massini G, Patrolecco L, Mazzurco Miritana V, Grenni P (2020) Environmental fate of antibiotics and resistance genes in livestock waste and digestate from biogas plants. Environ Sci Pollut Res Manag 2020(01). https://doi.org/10.37722/ESPRAM.20201
6. Barra Caracciolo A, Visca A, Rauseo J, Spataro F, Garbini GL, Grenni P, Mariani L, Mazzurco Miritana V, Massini G, Patrolecco L (2022) Bioaccumulation of antibiotics and resistance genes in lettuce following cattle manure and digestate fertilization and their effects on soil

and phyllosphere microbial communities. Environ Pollut 315:120413. https://doi.org/10.1016/j.envpol.2022.120413
7. Bártíková H, Podlipná R, Skálová L (2016) Veterinary drugs in the environment and their toxicity to plants. Chemosphere 144:2290–2301. https://doi.org/10.1016/j.chemosphere.2015.10.137
8. Batchelder AR (1982) Chlortetracycline and oxytetracycline effects on plant growth and development in soil systems. J Environ Qual 11(4):675–678. https://doi.org/10.2134/jeq1982.00472425001100040023x
9. Bengtsson-Palme J, Kristiansson E, Larsson DGJ (2018) Environmental factors influencing the development and spread of antibiotic resistance. FEMS Microbiol Rev 42(1):fux053. https://doi.org/10.1093/femsre/fux053
10. Bengtsson-Palme J, Larsson DGJ (2016) Concentrations of antibiotics predicted to select for resistant bacteria: proposed limits for environmental regulation. Environ Int 86:140–149. https://doi.org/10.1016/j.envint.2015.10.015
11. Berendonk TU, Manaia CM, Merlin C, Fatta-Kassinos D, Cytryn E, Walsh F, Bürgmann H, Sørum H, Norström M, Pons M-N, Kreuzinger N, Huovinen P, Stefani S, Schwartz T, Kisand V, Baquero F, Martinez JL (2015) Tackling antibiotic resistance: the environmental framework. Nat Rev Microbiol 13(5):310–317. https://doi.org/10.1038/nrmicro3439
12. Boxall ABA, Johnson P, Smith EJ, Sinclair CJ, Stutt E, Levy LS (2006) Uptake of veterinary medicines from soils into plants. J Agric Food Chem 54(6):2288–2297. https://doi.org/10.1021/jf053041t
13. Carballo M, Rodríguez A, De la Torre A (2021) Phytotoxic effects of antibiotics on terrestrial crop plants and wild plants: a systematic review. Arch Environ Contamin Toxicol. https://doi.org/10.21203/rs.3.rs-433396
14. Cheng W-H, Endo A, Zhou L, Penney J, Chen H-C, Arroyo A, Leon P, Nambara E, Asami T, Seo M, Koshiba T, Sheen J (2002) A unique short-chain dehydrogenase/reductase in Arabidopsis glucose signaling and abscisic acid biosynthesis and functions. Plant Cell 14(11):2723–2743. https://doi.org/10.1105/tpc.006494
15. Christou A, Michael C, Fatta-Kassinos D, Fotopoulos V (2018) Can the pharmaceutically active compounds released in agroecosystems be considered as emerging plant stressors? Environ Int 114:360–364. https://doi.org/10.1016/j.envint.2018.03.003
16. Conrad PA, Meek LA, Dumit J (2013) Operationalizing a One Health approach to global health challenges. Comp Immunol Microbiol Infect Dis 36(3):211–216. https://doi.org/10.1016/j.cimid.2013.03.006
17. Cruz-Loya M, Kang TM, Lozano NA, Watanabe R, Tekin E, Damoiseaux R, Savage VM, Yeh PJ (2019) Stressor interaction networks suggest antibiotic resistance co-opted from stress responses to temperature. ISME J 13(1):12–23. https://doi.org/10.1038/s41396-018-0241-7
18. Cycoń M, Mrozik A, Piotrowska-Seget Z (2019) Antibiotics in the soil environment-degradation and their impact on microbial activity and diversity. Front Microbiol 10:338. https://doi.org/10.3389/fmicb.2019.00338
19. Dar D, Sorek R (2017) Regulation of antibiotic-resistance by non-coding RNAs in bacteria. Curr Opin Microbiol 36:111–117. https://doi.org/10.1016/j.mib.2017.02.005
20. Dolliver H, Gupta S, Noll S (2008) Antibiotic degradation during manure composting. J Environ Qual 37(3):1245–1253. https://doi.org/10.2134/jeq2007.0399
21. Dragosits M, Mozhayskiy V, Quinones-Soto S, Park J, Tagkopoulos I (2013) Evolutionary potential, cross-stress behavior and the genetic basis of acquired stress resistance in *Escherichia coli*. Mol Syst Biol 9(1):643. https://doi.org/10.1038/msb.2012.76
22. Drillia P, Stamatelatou K, Lyberatos G (2005) Fate and mobility of pharmaceuticals in solid matrices. Chemosphere 60(8):1034–1044. https://doi.org/10.1016/j.chemosphere.2005.01.032
23. Emmerson AM (2003) The quinolones: decades of development and use. J Antimicrob Chemother 51(90001):13–20. https://doi.org/10.1093/jac/dkg208
24. European Commission (2018) Updated surface water Watch List adopted by the Commission. https://ec.europa.eu/jrc/en/science-update/updated-surface-water-watch-list-adopted-commission

25. Fàbrega A, Madurga S, Giralt E, Vila J (2009) Mechanism of action of and resistance to quinolones. Microb Biotechnol 2(1):40–61. https://doi.org/10.1111/j.1751-7915.2008.00063.x
26. Farkas A, Erratico C, Viganò L (2007) Assessment of the environmental significance of heavy metal pollution in surficial sediments of the River Po. Chemosphere 68(4):761–768. https://doi.org/10.1016/j.chemosphere.2006.12.099
27. Fischer M, Thöny B, Leimkühler S (2010) The biosynthesis of folate and pterins and their enzymology. In: Mander L, Lui HW (eds) Comprehensive natural products II chemistry and biology. Elsevier, Oxford, pp 599–648. https://doi.org/10.1016/B978-008045382-8.00150-7
28. Frasson I, Cavallaro A, Bergo C, Richter SN, Palù G (2011) Prevalence of *aac(6')-Ib-cr* plasmid-mediated and chromosome-encoded fluoroquinolone resistance in *Enterobacteriaceae* in Italy. Gut Pathogens 3(1):12. https://doi.org/10.1186/1757-4749-3-12
29. Garcia MA, Solans C, Calvo A, Hernandez E, Rey R, Bregante MA, Puig M (2005) Determination of enrofloxacin and its primary metabolite, ciprofloxacin, in pig tissues. Application to residue studies. Biomed Chromatogr 19(1):27–31. https://doi.org/10.1002/bmc.411
30. Georgina Solano-Gálvez S, Fernanda Valencia-Segrove M, José Ostos Prado M, Berenice López Boucieguez A, Abelardo Álvarez-Hernández D, Vázquez-López R (2021) Mechanisms of resistance to quinolones. In: Antimicrobial resistance—a one health perspective. IntechOpen. https://doi.org/10.5772/intechopen.92577
31. Grassi M, Rizzo L, Farina A (2013) Endocrine disruptors compounds, pharmaceuticals and personal care products in urban wastewater: implications for agricultural reuse and their removal by adsorption process. Environ Sci Pollut Res 20(6):3616–3628. https://doi.org/10.1007/s11356-013-1636-7
32. Grenni P, Ancona V, Barra Caracciolo A (2018) Ecological effects of antibiotics on natural ecosystems: a review. Microchem J 136:25–39. https://doi.org/10.1016/j.microc.2017.02.006
33. Gupta P, Sothiselvam S, Vázquez-Laslop N, Mankin AS (2013) Deregulation of translation due to post-transcriptional modification of rRNA explains why erm genes are inducible. Nat Commun 4(1):1984. https://doi.org/10.1038/ncomms2984
34. Gworek B, Kijeńska M, Wrzosek J, Graniewska M (2021) Pharmaceuticals in the soil and plant environment: a review. Water Air Soil Pollut 232(4):145. https://doi.org/10.1007/s11270-020-04954-8
35. Hillis DG, Fletcher J, Solomon KR, Sibley PK (2011) Effects of ten antibiotics on seed germination and root elongation in three plant species. Arch Environ Contam Toxicol 60(2):220–232. https://doi.org/10.1007/s00244-010-9624-0
36. Hinrichs W, Kisker C, Duvel M, Muller A, Tovar K, Hillen W, Saenger W (1994) Structure of the Tet repressor-tetracycline complex and regulation of antibiotic resistance. Science 264(5157):418–420. https://doi.org/10.1126/science.8153629
37. Hu Y, Shamaei-Tousi A, Liu Y, Coates A (2010) A new approach for the discovery of antibiotics by targeting non-multiplying bacteria: a novel topical antibiotic for Staphylococcal infections. PLoS One 5(7):e11818. https://doi.org/10.1371/journal.pone.0011818
38. Hurtado C, Parastar H, Matamoros V, Piña B, Tauler R, Bayona JM (2017) Linking the morphological and metabolomic response of *Lactuca sativa* L exposed to emerging contaminants using GC × GC-MS and chemometric tools. Sci Rep 7(1):6546. https://doi.org/10.1038/s41598-017-06773-0
39. Ingram PR, Rogers BA, Sidjabat HE, Gibson JS, Inglis TJJ (2013) Co-selection may explain high rates of ciprofloxacin non-susceptible Escherichia coli from retail poultry reared without prior fluoroquinolone exposure. J Med Microbiol 62(11):1743–1746. https://doi.org/10.1099/jmm.0.062729-0
40. Ivanová L, Mackuľak T, Grabic R, Golovko O, Koba O, Staňová AV, Szabová P, Grenčíková A, Bodík I (2018) Pharmaceuticals and illicit drugs—a new threat to the application of sewage sludge in agriculture. Sci Total Environ 634:606–615. https://doi.org/10.1016/j.scitotenv.2018.04.001
41. Jacoby GA, Strahilevitz J, Hooper DC (2014) Plasmid-mediated quinolone resistance. Microbiol Spectrum 2(5). https://doi.org/10.1128/microbiolspec.PLAS-0006-2013

42. Jin C, Chen Q, Sun R, Zhou Q, Liu J (2009) Eco-toxic effects of sulfadiazine sodium, sulfamonomethoxine sodium and enrofloxacin on wheat, Chinese cabbage and tomato. Ecotoxicology 18(7):878–885. https://doi.org/10.1007/s10646-009-0349-7
43. Jousset A, Bonkowski M (2010) The model predator *Acanthamoeba castellanii* induces the production of 2,4, DAPG by the biocontrol strain *Pseudomonas fluorescens* Q2–87. Soil Biol Biochem 42(9):1647–1649. https://doi.org/10.1016/j.soilbio.2010.05.018
44. Kang DH, Gupta S, Rosen C, Fritz V, Singh A, Chander Y, Murray H, Rohwer C (2013) Antibiotic uptake by vegetable crops from manure-applied soils. J Agric Food Chem 61(42):9992–10001. https://doi.org/10.1021/jf404045m
45. Karczmarczyk M, Abbott Y, Walsh C, Leonard N, Fanning S (2011) Characterization of multidrug-resistant *Escherichia coli* isolates from animals presenting at a university veterinary hospital. Appl Environ Microbiol 77(20):7104–7112. https://doi.org/10.1128/AEM.00599-11
46. Le Page G, Gunnarsson L, Snape J, Tyler CR (2017) Integrating human and environmental health in antibiotic risk assessment: a critical analysis of protection goals, species sensitivity and antimicrobial resistance. Environ Int 109(September):155–169 https://doi.org/10.1016/j.envint.2017.09.013
47. Le Page G, Gunnarsson L, Trznadel M, Wedgwood KCA, Baudrot V, Snape J, Tyler CR (2019) Variability in cyanobacteria sensitivity to antibiotics and implications for environmental risk assessment. Sci Total Environ 695:133804. https://doi.org/10.1016/j.scitotenv.2019.133804
48. Leal RMP, Figueira RF, Tornisielo VL, Regitano JB (2012) Occurrence and sorption of fluoroquinolones in poultry litters and soils from São Paulo State, Brazil. Sci Total Environ 432:344–349. https://doi.org/10.1016/j.scitotenv.2012.06.002
49. Li A, Mao D, Yoshimura A, Rosen PC, Martin WL, Gallant É, Wühr M, Seyedsayamdost MR (2020) Multi-omic analyses provide links between low-dose antibiotic treatment and induction of secondary metabolism in *Burkholderia thailandensis*. MBio 11(1). https://doi.org/10.1128/mBio.03210-19
50. Li WC (2014) Occurrence, sources, and fate of pharmaceuticals in aquatic environment and soil. Environ Pollut 187:193–201. https://doi.org/10.1016/j.envpol.2014.01.015
51. Li Y, Sallach JB, Zhang W, Boyd SA, Li H (2019) Insight into the distribution of pharmaceuticals in soil-water-plant systems. Water Res 152:38–46. https://doi.org/10.1016/j.watres.2018.12.039
52. Margenat A, You R, Cañameras N, Carazo N, Díez S, Bayona JM, Matamoros V (2020) Occurrence and human health risk assessment of antibiotics and trace elements in *Lactuca sativa* amended with different organic fertilizers. Environ Res 190:109946. https://doi.org/10.1016/j.envres.2020.109946
53. Martinez JL (2008) Antibiotics and antibiotic resistance genes in natural environments. Science 321(5887):365–367. https://doi.org/10.1126/science.1159483
54. Mazzurco Miritana V, Massini G, Visca A, Grenni P, Patrolecco L, Spataro F, Rauseo J, Garbini GL, Signorini A, Rosa S, Barra Caracciolo A (2020) Effects of sulfamethoxazole on the microbial community dynamics during the anaerobic digestion process. Front Microbiol 11:537783. https://doi.org/10.3389/fmicb.2020.537783
55. Mei Z, Xiang L, Wang F, Xu M, Fu Y, Wang Z, Hashsham SA, Jiang X, Tiedje JM (2021) Bioaccumulation of manure-borne antibiotic resistance genes in carrot and its exposure assessment. Environ Int 157:106830. https://doi.org/10.1016/j.envint.2021.106830
56. Minden V, Deloy A, Volkert AM, Leonhardt SD, Pufal G (2017) Antibiotics impact plant traits, even at small concentrations. AoB Plants 9(2). https://doi.org/10.1093/aobpla/plx010
57. Mishra S, Lin Z, Pang S, Zhang W, Bhatt P, Chen S (2021) Recent advanced technologies for the characterization of xenobiotic-degrading microorganisms and microbial communities. Front Bioeng Biotechnol 9:632059. https://doi.org/10.3389/fbioe.2021.632059
58. Monteiro SC, Boxall ABA (2009) Factors affecting the degradation of pharmaceuticals in agricultural soils. Environ Toxicol Chem 28(12):2546. https://doi.org/10.1897/08-657.1
59. Mortimer M, Winchell A, Holden PA (2020) Evaluation of frameworks proposed as protective of antimicrobial resistance propagation in the environment. Environ Int 144(August):106053. https://doi.org/10.1016/j.envint.2020.106053

60. Nguyen B-AT, Chen Q-L, He J-Z, Hu H-W (2020) Microbial regulation of natural antibiotic resistance: understanding the protist-bacteria interactions for evolution of soil resistome. Sci Total Environ 705:135882. https://doi.org/10.1016/j.scitotenv.2019.135882
61. Oliver MA, Gregory PJ (2015) Soil, food security and human health: a review. Eur J Soil Sci 66(2):257–276. https://doi.org/10.1111/ejss.12216
62. Pan M, Chu LM (2016) Adsorption and degradation of five selected antibiotics in agricultural soil. Sci Total Environ 545–546:48–56. https://doi.org/10.1016/j.scitotenv.2015.12.040
63. Pan M, Chu LM (2017) Fate of antibiotics in soil and their uptake by edible crops. Sci Total Environ 599–600:500–512. https://doi.org/10.1016/j.scitotenv.2017.04.214
64. Pan M, Wong CKC, Chu LM (2014) Distribution of antibiotics in wastewater-irrigated soils and their accumulation in vegetable crops in the Pearl River Delta, Southern China. J Agric Food Chem 62(46):11062–11069. https://doi.org/10.1021/jf503850v
65. Papageorgiou M, Lambropoulou D, Morrison C, Kłodzińska E, Namieśnik J, Płotka-Wasylka J (2018) Literature update of analytical methods for biogenic amines determination in food and beverages. TrAC Trends Anal Chem 98:128–142. https://doi.org/10.1016/j.trac.2017.11.001
66. Pascual Aguilar JA, Andreu V, Campo J, Picó Y, Masiá A (2017) Pesticide occurrence in the waters of Júcar River, Spain from different farming landscapes. Sci Total Environ 607–608:752–760. https://doi.org/10.1016/j.scitotenv.2017.06.176
67. Patra AK (2012) An overview of antimicrobial properties of different classes of phytochemicals. In: Patra A (eds) Dietary phytochemicals and microbes. Springer, Netherlands, pp 1–32. https://doi.org/10.1007/978-94-007-3926-0_1
68. Pereira P, Bogunovic I, Muñoz-Rojas M, Brevik EC (2018) Soil ecosystem services, sustainability, valuation and management. Curr Opin Environ Sci Health 5:7–13. https://doi.org/10.1016/j.coesh.2017.12.003
69. Perreten V, Boerlin P (2003) A new sulfonamide resistance gene (*sul3*) in *Escherichia coli* is widespread in the pig population of Switzerland. Antimicrob Agents Chemother 47(3):1169–1172. https://doi.org/10.1128/AAC.47.3.1169-1172.2003
70. Pham TDM, Ziora ZM, Blaskovich MAT (2019) Quinolone antibiotics. MedChemComm 10(10):1719–1739. https://doi.org/10.1039/C9MD00120D
71. Picó Y, Andreu V (2007) Fluoroquinolones in soil-risks and challenges. Anal Bioanal Chem 387(4):1287–1299. https://doi.org/10.1007/s00216-006-0843-1
72. Prosser RS, Sibley PK (2015) Human health risk assessment of pharmaceuticals and personal care products in plant tissue due to biosolids and manure amendments, and wastewater irrigation. Environ Int 75:223–233. https://doi.org/10.1016/j.envint.2014.11.020
73. Sabourin L, Duenk P, Bonte-Gelok S, Payne M, Lapen DR, Topp E (2012) Uptake of pharmaceuticals, hormones and parabens into vegetables grown in soil fertilized with municipal biosolids. Sci Total Environ 431:233–236. https://doi.org/10.1016/j.scitotenv.2012.05.017
74. Saccá ML, Barra Caracciolo A, Di Lenola M, Grenni P (2017) Ecosystem services provided by soil microorganisms. In: Lukac M, Grenni P, Gamboni M (eds) Soil biological communities and ecosystem resilience, pp 9–24
75. Sánchez-Osuna M, Cortés P, Barbé J, Erill I (2019) Origin of the mobile di-hydro-pteroate synthase gene determining sulfonamide resistance in clinical isolates. Front Microbiol 9:3332. https://doi.org/10.3389/fmicb.2018.03332
76. Sharma PC, Jain A, Jain S, Pahwa R, Yar MS (2010) Ciprofloxacin: review on developments in synthetic, analytical, and medicinal aspects. J Enzyme Inhib Med Chem 25(4):577–589. https://doi.org/10.3109/14756360903373350
77. Snow DD, Cassada DA, Larsen ML, Mware NA, Li X, D'Alessio M, Zhang Y, Sallach JB (2017) Detection, occurrence and fate of emerging contaminants in agricultural environments. Water Environ Res 89(10):897–920. https://doi.org/10.2175/106143017X15023776270160
78. Song C, Mazzola M, Cheng X, Oetjen J, Alexandrov T, Dorrestein P, Watrous J, van der Voort M, Raaijmakers JM (2015) Molecular and chemical dialogues in bacteria-protozoa interactions. Sci Rep 5(1):12837. https://doi.org/10.1038/srep12837
79. Starling MCVM, Amorim CC, Leão MMD (2019) Occurrence, control and fate of contaminants of emerging concern in environmental compartments in Brazil. J Hazard Mater 372:17–36. https://doi.org/10.1016/j.jhazmat.2018.04.043

80. Sun Y, Guo Y, Shi M, Qiu T, Gao M, Tian S, Wang X (2021) Effect of antibiotic type and vegetable species on antibiotic accumulation in soil-vegetable system, soil microbiota, and resistance genes. Chemosphere 263:128099. https://doi.org/10.1016/j.chemosphere.2020.128099
81. Święciło A (2016) Cross-stress resistance in Saccharomyces cerevisiae yeast—new insight into an old phenomenon. Cell Stress Chaperones 21(2):187–200. https://doi.org/10.1007/s12192-016-0667-7
82. Trouchon T, Lefebvre S (2016) A review of enrofloxacin for veterinary use. Open J Vet Med 06(02):40–58. https://doi.org/10.4236/ojvm.2016.62006
83. Välitalo P, Kruglova A, Mikola A, Vahala R (2017) Toxicological impacts of antibiotics on aquatic micro-organisms: a mini-review. Int J Hyg Environ Health 220(3):558–569. https://doi.org/10.1016/j.ijheh.2017.02.003
84. Visca A, Barra Caracciolo A, Grenni P, Patrolecco L, Rauseo J, Massini G, Mazzurco Miritana V, Spataro F (2021) Anaerobic digestion and removal of sulfamethoxazole, enrofloxacin, ciprofloxacin and their antibiotic resistance genes in a full-scale biogas plant. Antibiotics 10(5):502. https://doi.org/10.3390/antibiotics10050502
85. von Wintersdorff CJH, Penders J, van Niekerk JM, Mills ND, Majumder S, van Alphen LB, Savelkoul PHM, Wolffs PFG (2016) Dissemination of antimicrobial resistance in microbial ecosystems through horizontal gene transfer. Front Microbiol 7:173. https://doi.org/10.3389/fmicb.2016.00173
86. Watkinson AJ, Murby EJ, Kolpin DW, Costanzo SD (2009) The occurrence of antibiotics in an urban watershed: from wastewater to drinking water. Sci Total Environ 407(8):2711–2723. https://doi.org/10.1016/j.scitotenv.2008.11.059
87. WHO (2017) Global priority list of antibiotic-resistant bacteria to guide research, discovery, and development of new antibiotics. http://apps.who.int/medicinedocs/documents/s23171en/s23171en.pdf
88. Woappi Y, Gabani P, Singh A, Singh OV (2016) Antibiotrophs: the complexity of antibiotic-subsisting and antibiotic-resistant microorganisms. Crit Rev Microbiol 42(1):17–30. https://doi.org/10.3109/1040841X.2013.875982
89. Wu X, Ernst F, Conkle JL, Gan J (2013) Comparative uptake and translocation of pharmaceutical and personal care products (PPCPs) by common vegetables. Environ Int 60:15–22. https://doi.org/10.1016/j.envint.2013.07.015
90. Xie W-Y, Shen Q, Zhao FJ (2018) Antibiotics and antibiotic resistance from animal manures to soil: a review. Eur J Soil Sci 69(1):181–195. https://doi.org/10.1111/ejss.12494
91. Zhang Y-L, Lin S-S, Dai C-M, Shi L, Zhou X-F (2014) Sorption–desorption and transport of trimethoprim and sulfonamide antibiotics in agricultural soil: effect of soil type, dissolved organic matter, and pH. Environ Sci Pollut Res 21(9):5827–5835. https://doi.org/10.1007/s11356-014-2493-8
92. Zhao L, Dong YH, Wang H (2010) Residues of veterinary antibiotics in manures from feedlot livestock in eight provinces of China. Sci Total Environ 408(5):1069–1075. https://doi.org/10.1016/j.scitotenv.2009.11.014

Rare Earths in Soils

Raimundo Jiménez-Ballesta, Pablo L. Higueras, and Francisco J. García Navarro

Abstract This chapter aims to review and summarize the scattered existing literature on the presence of the so-called rare earth elements (REEs) in soils by drawing on their basic characteristics and applications, contents, origin and possible sources, basic lanthanide chemistry, the chronology of their discovery, importance in soil genesis and the soil–plant system, and environmental consequences. The ultimate reason for this selection is that REEs accumulation in soil has increased in recent times not so much because of natural processes, but of anthropogenic activities. Thus, after the conceptualization and brief review of its historical identification, the origin and occurrences of REEs in soils are revealed. It is stated that they depend on: their presence in parent material; weathering of minerals; discharges from several sources (mining, industrial waste, P-fertilizers, sewage sludges, among others). Finally, attention is also paid to the potential risks that they entail. As occasional and uncontrolled REEs applications to agricultural soils are known to pose several environmental risks, it is necessary to undertake new studies on their bioavailability and subsequent possible adverse effects on microbial biomass and, therefore, on soil fertility. In this way, future research needs to: (i) include an understanding of REEs contents, their incorporation into soils by various external sources, but also their interactions with other soil constituents; (ii) promote new studies related to their interactions with plant growth; (iii) investigate impacts for the soil environment and human and animal health because of increasing REEs utilization. Therefore, future research lines should emphasize soil content control to identify the negative implications in environmental and agricultural sectors. Finally, the development of recycling processes for REEs is becoming an increasingly relevant, but challenging, issue.

Keywords Lanthanides characteristics and occurrence · Pedogeochemical behavior · Environmental fate · REEs enrichment · Soil–plant transfer · Research trends

R. Jiménez-Ballesta (✉)
Autónoma University of Madrid, Madrid, Spain
e-mail: raimundo.jimenez@uam.es

P. L. Higueras · F. J. García Navarro
University of Castilla-La Mancha, Ciudad Real, Spain

A. Núñez-Delgado (ed.), *Frontier Studies in Soil Science*, Frontier Studies in Soil Science, https://doi.org/10.1007/978-3-031-50503-4_3

1 Introduction

The International Union of Pure and Applied Chemistry, an organization that works on maintaining international consistency for the chemical nomenclature, has identified the 15 transition metals from the Periodic Table of the elements with atomic numbers from 57 (lanthanum) through to 71 (lutetium) as lanthanides or lanthanoids [99]. These 15 elements share common physio-chemical properties and are listed as follows: lanthanum (57La), cerium (58Ce); praseodymium (59Pr); neodymium (60Nd); promethium (61Pm); samarium (62Sm); europium (63Eu); gadolinium (64Gd); terbium (65Tb); dysprosium (66Dy); holmium (67Ho); erbium (68Er); thulium (69Tm); ytterbium (70Yb); lutetium (71Lu).

Rare earth elements (hereafter referred to as REEs) represent the cited 15 periodic elements of the lanthanide family, which lie at the bottom of the Periodic Table in the top horizontal row of the f-block elements (Fig. 1). Scandium (Sc) and yttrium (Y), also located in column 3 of the Periodic Table, are considered REEs because they exhibit similar properties to the lanthanide family and also because they usually appear all mixed in the same deposits.

Although they are called "rare", REEs are quite common litholfile elements in the Earth's crust, with diverse, but not particularly, low average concentrations in relation to other elements [87, 211]. The question that justifies their 'rarity' is that they lack large ore deposits, unlike other elements such as base metals or similar ones. They are also known as "earths" because of the way that some oxides were called in the nineteenth century, which is why they are also called "metals" or "oxides" [231].

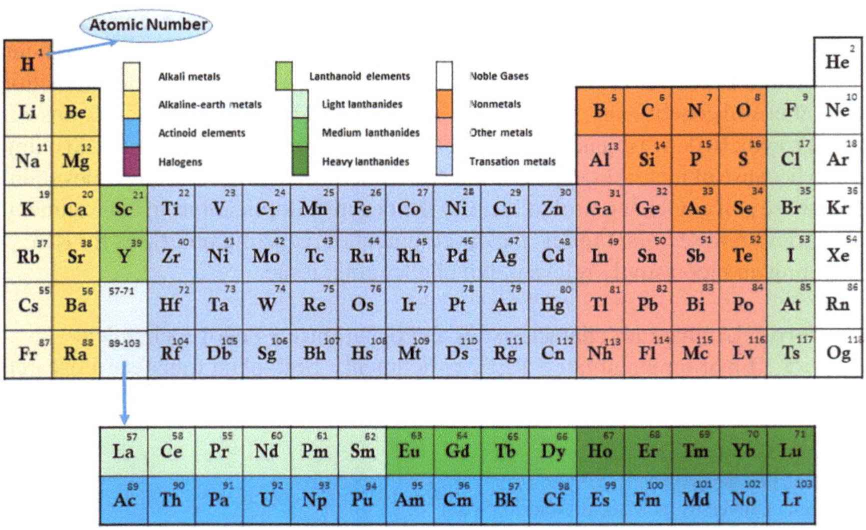

Fig. 1 REEs periodic table: rare earth elements are the 15 lanthanide series elements, plus yttrium and scandium

The set of elements making up the REEs group can be divided into three subgroups: LREEs (light REEs), MREEs (middle REEs) and HREEs (heavy REEs) [220, 251]. However, the set is usually limited to LREEs and HREEs [100, 193, 196, 256].

Some elements like Cu, Hg or Pb, among others, have been traditionally considered strategic elements because of their use in important industrial applications. The development of modern technologies has led to other elements (particularly REEs) also becoming part of this set. Indeed, in the last few decades REEs elements have drawn enormous international attention. Metallurgical processing, alloying and electronics applications (e.g., cell phones, computer components, electric motors, lenses, and a long etcetera of others) represent the most significant uses of REEs today. Many of these applications form part of the so-called green technologies. Besides, given the dependence on several of these elements for military applications, REEs are considered a national strategic resource by many countries. Hence these elements constitute an essential group of the so-called Technology Critical Elements, or TCEs [45], because of not only their fundamental role in technological applications, but also their increasingly important role in the geopolitical picture. The importance and transcendence of these elements lie in the fact that they are basic raw materials in the world of the energy transition imposed by progressive climate change, but also due to transformations toward digitalization, required by the so-called new technologies, and also by the military industry and aerospace. So, it is not surprising that REEs are known as "industrial vitamins". Nonetheless, as these elements are increasingly demanded, excess REEs, such as Gd, are considered micropollutants in the environment [20, 114], particularly those that come from wastewater treatment plants [121, 225, 226] and hospitals [115, 123], were this element is used (since 1988), as Gd-based contrast agents (GBCAs), frequently administered to patients for magnetic resonance imaging with the idea of improving tissue contrasts. Logically, after examination, they are excreted causing the accumulation of anthropogenic Gd into the environment (river and lake waters), since these Gd compounds are not removed in wastewater treatment plants (WWTP) due to their high stabilities. In addition, anthropogenic Gd chelates can also be used as tracers for emerging micro-contaminants such as steroids, pharmaceuticals and personal care products [114]. However, the potential long-term effects of exposure to Gd low doses have not yet been studied.

REEs elements are of importance in soil science and other environmental sciences. This is particularly the case of the geochemical behavior of REEs in many soil types in relation to weathering processes and the origin of the rocks containing these elements, as well as the influence of human activities on their mobilization, which have been (and still are) widely studied. Several aspects in this sense are particularly important today:

- Studies about the influence of REEs in soil formation [5, 32, 39, 120, 206, 224]
- Characterization of the variability of REEs contents in soils [37, 52, 78, 93, 97, 122, 184, 249, 255, 257]

- Fractionation, mobility and availability of elements in different edaphic contexts [1, 31, 35, 38, 93, 103, 104, 137, 148, 151, 174, 233]
- Assessment of soil–plant transferences and relations with agriculture [143, 172, 186, 197, 217, 245]
- Environmental impact of REEs, including relations with soil microbiota [126, 175]
- Health concerns about the presence of these elements in soils [68, 83, 128, 167, 168, 187, 188]

By taking into account the growing global economic importance of REEs for their use in several soil-related Earth Science fields and that most of the products/by-products that are potential sources of TRs have the soil as their first destination, this review was carried out to provide an overall environmental perspective of REEs with respect to their presence in soils driven both by natural and anthropogenic processes. This chapter attempts to offer an examination, and provides comprehensive coverage, of the geochemistry of lanthanoid elements in soils by presenting a current snapshot of their presence in soils. It also analyzes information about REEs by especially focusing on the natural and anthropogenic sources of REEs, contents and specific features in their behavior in soil. Finally, their manifestation as potential contaminants is succinctly discussed. The economic importance of these elements, and the quantification of their reserves in the world or their current supplies and demand worldwide, are not the subject of this review.

2 A Brief Chronology in the Discovery of Rare Earths

For just over three-four decades, in an international context, society has been immersed in a new revolutionary industrial process based on technological production. These advances have resulted in some elements and minerals being considered "Critical Raw Materials" by the European Commission [110] to provide crucial elements for new technologies, including REEs, which appear as the most important group of strategic elements. Although they were mostly discovered at the end of the eighteenth century, it was not until after World War II that progress was made in their possible commercial applications. Indeed, REEs were discovered long before, in 1787, by Karl Axel Arrhenius, a Lieutenant of the Swedish Army, when the black mineral ytterbite (later renamed gadolinite) from a feldspar and quartz mine near the village of Ytterby, Sweden was collected and studied in detail. In fact, the majority of REEs were originally discovered in Europe, fundamentally from the Bastnäs and Ytterby mines in Sweden [70, 94, 242]. Table 1 sumarizes the history of the discovery of the different REEs, many of which were obtained from the same 'black mineral' present in Ytterby.

It was Gadolin (chemist and mineralogist) who, in 1794, began the discovery of the different elements in this group, starting with the yttrium element. In 1907, the discovery of element lutetium is considered by some as if it were the last element

Table 1 Brief chronology of the development of rare earths appearing

Element	Name	Original name	Year	Discoverer
Y	Yttrium	Ytterby (village)	1794	Gadolin
Ce	Cerium	Ceres (asteroid)	1814	Berzelius (and Hisinger)
La	Lanthanum	To lie hidden (Gr.)	1839	Mosander
Er	Erbium	Yettery	1843	Mosander
Tb	Terbium	Yettery	1843	Mosander
Yb	Ytterbium	Ytterby	1878	Marignac
Sm	Samarium	Samarskite (mineral)	1879	Lecoq de Boisbaudran
Sc	Scandium	Scandinavia	1879	Lars Nilson
Ho	Holmium	Latin (holmia)	1879	Cleve
Tm	Thulium	Thule	1878	Cleve
Gd	Gadolinium	Johan Gadolin	1880	Marignac
Pr	Praseodymium	Gr. (green twin)	1885	von Welsbach
Nd	Neodymium	Gr. (new twin)	1885	von Welsbach
Dy	Dysprosium	Gr. from dysprositos	1886	P.E.L. de Boisbaudran
Eu	Europium	Europe	1886	Eugène Demarcay
Lu	Lutetium	Lutetia (Lt. of Paris)	1907	Urbain
Pm	Promethium	Prometheus (Gr. God)	1945	Glendenin/Marinsky

of the group to be discovered. However, the promethium element was discovered by Marinsky in 1943 due to a nuclear reaction, which truly closed the history of the discovery of REEs.

3 Some Features of the Soil Chemistry of Lanthanides and Analytical

From the chemical point of view, REEs elements share common physico-chemical properties for having a similar electronic configuration. As elements, they normally appear together in the same mineral assemblages. They also share many common characteristics and properties, which make it difficult to separate them, or to even distinguish them from one another. They are usually silver, silver-white or gray-colored metals which, in the presence of air, lose their silver-white surface and become chestnut and dark-brown due to the formation of oxides [112, 186]. REEs are chemically active metals with low ductility, although mechanical properties very much depend on purity. Their chemical bonding is predominantly ionic, and they do not tend to form covalent bonds [235].

Some chemical species, such as hydroxyl anion (OH^-), fluorine (F^-), chlorine (Cl^-), ammonia (NH_3), acetic acid ($CH_3\ COO^-$), carbonic acid (CO_3^{2-}), nitrate

(NO_3^-), sulfate (SO_4^{2-}), phosphate (PO_4^{3-}), oxide (O^{2-}), alcohols (R–OH), amines ($R\text{-}NH_2$), among others, which contain highly electronegative donor atoms like O_2 and F, are capable of forming ligands with lanthanides. REEs are not presumably as toxic as other heavy metals and metalloids, such as Hg, Cd or As, but may be chronically toxic to humans and cause long-term adverse effects [170].

The chemistry of REEs in soils has been documented by different authors like Aide [4]. In this context, most REEs exist in trivalent cations, and rarely in divalent or tetravalent cations. Although a few REEs occur in other ionic forms, such as samarium (Sa^{2+}), thulium (Tm^{2+}), and ytterbium (Y^{2+}), only some forms of cerium (Ce^{4+}) and europium (Eu^{2+}) are commonly found in natural systems [182].

3.1 Analytical Methods

Given the physico-chemical similarities that characterize REEs, their is complicated, especially if there is a mixture of them. In practice, there are several methods that allow REEs contents to be determined in soils. Today numerous techniques, such as Instrumental Neutron Activation Analysis (INAA) and Inductively Coupled Plasma Mass Spectrometry (ICP-MS), including High-Resolution Inductively Coupled Plasma Mass Spectrometry (HR-ICP-MS), are often used (probably as the most attractive [61, 74, 84]), to determine REEs in soils (Fig. 2). Their advantage lies in them having multi-element capabilities, high sensitivity, a wide linear dynamic range, fewer interferences, and easy operation and accuracy [15, 17, 18]. Other possible sensitive techniques are X-Ray Fluorescence Spectrometry (XRF) [51, 155], Inductively Coupled Plasma Optical Emission Spectrometry (ICP-OES) [156], Glow Discharge Mass Spectrometry (GD-MS), Laser-Induced Breakdown Spectroscopy (LIBS), and recently Microwave Plasma Atomic Emission Spectroscopy, (MP-AES), have been introduced. Perhaps one of the most widely used techniques is XRF, a reliable method for REEs element analyses, although this technique is not relatively sensitive, but a conventional analytical technique for REEs analyses with distinct advantages over other methods in accuracy, speed and cost terms: Wu et al. [243] presented a review on XRF analysis applications in the Chinese REEs industry. Many other authors ([17, 93, 103, 104], among others) have used XRF in vineyard soils and ore deposits.

4 Applications of Rare Earths

REEs elements form an important group of elements in applicability terms, as determined by their chemical and physical properties, which translates into vital importance in the development of the world's economy. Indeed as a consequence of REEs having unique properties, but varied in detail, this set of elements is used in a wide range of modern technology applications (Table 2), and in such a way that they

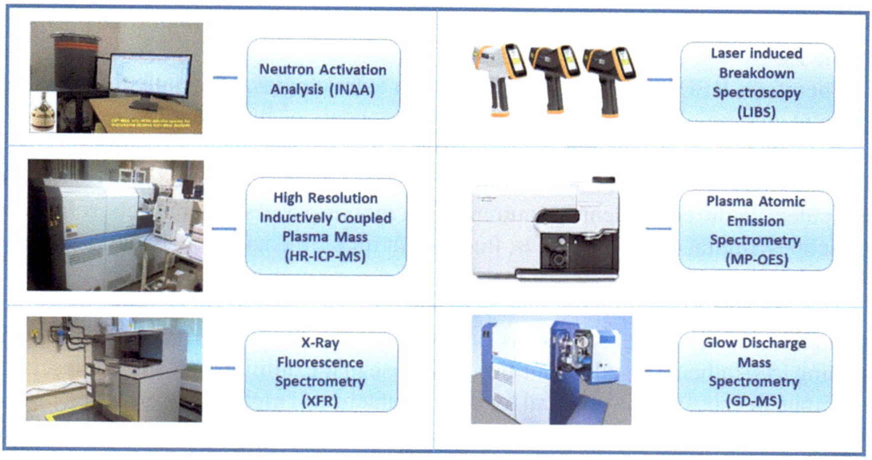

Fig. 2 Analytical methods used in the determination of REEs contents

are incorporated into various materials in medicine, imaging methods, mining and processing, lighter flints, iron and steel additives, carbon arc lighting, cell phones, catalytic converters for automobiles or rechargeable batteries [16, 36, 82, 138], and in fertilizers [40, 103, 104, 173, 200, 252, 254].

The application of REEs in various industrial fields includes non-nuclear energy production and energy-utilizing products like light bulbs, batteries and catalytic converters [14, 59, 125, 162], anti-corrosive technique development [127, 161] and magnet production. Nd is extensively applied in supermagnets for disk drives and some electric motors. Ce is a critical ingredient in autocatalyts. Almost all REEs are

Table 2 Main application of REEs (adapted from [36, 138])

Application	Most used REEs	Examples of use
Magnets	Nd, Pr, Tb, Dy	Motors, disc drive, magnetic resonance imaging, power generation, microphones and speakers, and magnetic refrigeration
Catalyst	La, Ce, Pr, Nd	Petroleum refining, catalytic converter, diesel additives, chemical processing and industrial pollution scrubbers
Alloys	La, C Pr, Nd	Batteries, fuel cells steel, lighter flints, superalloys, aluminum and magnesium
Glass and polishing	Ce, La, Pr, Nd, Gd, Er, Ho	Polishing compounds, decolorizers and colorizers, UV resistant glass and X-ray imaging
Phosphors	Eu, Tb, Nd, Er, Gd, Ce, Pr	Display phosphors CRT, LPD, LCD, fluorescent lighting, medical imaging, lasers and fiber optics
Other uses	All REEs	Nuclear, defense, water treatments, pigments (Ce), fertilizer and scientific research

used in making flat-panel TVs. These applications mean a very high REEs demand (Table 2).

Therefore, REEs are very useful in modern technologies, especially the so-called green technology [50, 57, 72, 88, 221]. Components of high-strength magnets are used in electronic equipment (very powerful permanent magnets), wind turbines, the manufacture of hybrid and electric vehicles, for precision guided weapons, in computers, audio equipment and automobiles, mobile phones, solar panels, refractory telescopes, digital cameras, DVDs, fluorescent tubes, rechargeable batteries, among others [22, 196, 201].

Their use as fluid-cracking catalysts is significant during oil refining, or in the production of optical glass and as components in phosphors for energy-efficient lighting [88], albeit in small quantities. Their use in the military industry is especially important because they are critical for the manufacture of night-vision goggles, lasers, communication devices, GPS, precision weapons, batteries, nuclear reactors, among other electronic devices. They are associated with highly radioactive and polluting elements (including Th, U and Pu), and can be used to manufacture nuclear reactors and bombs. Despite the widespread use of rare earths in the clean energy sector, the electronics sector is still the largest user of rare earths.

REEs have been used in electronics and advanced machinery for several decades, coinciding with the introduction of the first color television sets. Since then, demand for rare earths has steadily grown in relation throughout electronic products (including screens, glass, batteries, and magnets). One other potentially promising source of reduced reliance on new production of REEs is in recycling. But currently, recycling of REEs is in early stages of development.

5 Abundance of REEs in the Earth's Crust and Soils

Rare earth elements are trace elements. As shown in Table 3, the average abundances of Ce (60 mg kg^{-1}), La (30 mg kg^{-1}) or Y (24 mg kg^{-1}) in the Earth's crust are comparable to those of common metals like Cu (68 mg kg^{-1}) or Zn (79 mg kg^{-1}) according to Wedepohl [237]. The estimated average concentration of REEs in the Earth's crust ranges from around 130–240 mg kg^{-1}, which is significantly higher than other commonly exploited elements [251]. Obviously for other elements, these concentrations show very wide variations throughout the crust depending on the rock type and compositional variations [5, 81, 91, 145, 209, 227, 237].

Table 4 shows the REEs contents for the Earth's crust and soils as reported in the worldwide literature. For Bohn et al. [23], the concentrations of these elements in soils vary from 30 to 700 mg kg^{-1} (referred to the average concentration of all measurable REEs). And, for Liang et al. [130], their content in the upper soil layer (where they can interact with biota) considerably varies, reaching total average concentrations between 100 and 200 mg kg^{-1}. The most abundant REEs are Ce and La.

Table 3 Abundance of elements in the Earth's crust [237]

Element	Abundance ($mg\ kg^{-1}$)	Element	Abundance ($mg\ kg^{-1}$)
Nickel (28Ni)	90	Cesium (55Cs)	3.4
Zinc (30Zn)	79	Gadolinium (64Gd)	4.0
Copper (29Cu)	68	Dysprosium (66Dy)	3.8
Cerium (58Ce)	60	Erbium (68Er)	2.1
Lanthanum (57La)	30	Ytterbium (70Yb)	2.0
Cobalt (27Co) 30	30	Europium (63Eu)	1.3
Neodymium (60Nd)	27	Holmium (67Ho)	0.8
Yttrium (39Y)	24	Terbium (65 Tb)	0.7
Scandium (21Sc)	16	Lutetium (71Lu)	0.4
Lead (82Pb)	10	Thulium (69Tm)	0.3
Praseodymium (59Pr)	6.7	Silver (47Ag)	0.08
Thorium (90Th)	6	Gold (79Au)	0.0031
Samarium (62Sm)	5.3	Promethium (61Pm)	10^{-18}

As with the other chemical elements, the concentrations of the REEs with an even atomic number are higher than those of the elements with an odd atomic number (Fig. 3 and Table 4). Only Sc, Y La, Ce and Nd reach concentrations of >10 $mg\ kg^{-1}$.

6 Rare Earths Origin in Soils

Natural soils contain variable concentrations of REEs, depending on the particular considered element and type of soil, its geological substrate, and the interactions between REEs-bearing minerals and local soil physico-chemical conditions. These elements in soils form part of the crystalline lattice of different minerals, some of which are primary (inherited from the rock that forms the local substrate) and some are secondary, the product of the weathering of REEs-containing primary minerals and soil reactivity. The minerals that can contain REEs include carbonates, phosphates, silicates and fluorides as the most important ones.

Primary REEs are concentrated in some ore deposit types, which correspond to four geological environments, namely carbonatites, alkaline igneous systems, ion-absorption clay deposits and monazite-bearing placer deposits, and to carbonatite and alkaline igneous rocks, which constitute the majority of the world's REEs resources [55, 240].

Table 4 Mean content of REEs elements in the earth's crust and soil according to published data (expressed as mg kg^{-1})

Element	Concentrations in the Earth's crust						Soil concentrations	
	Wedepohl [237]	Taylor and McLennan [210]	Lide [132, 133]	Rudnick and Fountain [190]	Greenwood and Earnshaw [76]	Various authors[a]	Various authors[a]	Various authors[b]
Scandium (21Sc)	16	30.0	22	30	25	–	–	8.21
Yttrium (39Y)	24	20.0	33	20	31	–	–	22.0
Lanthanum (57La)	30	16.0	39	18	35	30–35	29.5–40	31.2
Cerium (58Ce)	60	33	66.5	42	66	64–66	29.5–50	66.3
Praseodymium (59Pr)	6.7	3.9	9.2	5	9.1	7.1–9.1	3–7.7	6.7
Neodymium (60Nd)	27	16	41.5	20	40	26–40	27.9–35	25.0
Promethium (61Pm)[c]	?			?				
Samarium (62Sm)	5.3	3.5	7	3.9	7	4.5–7	4.5–6.1	4.9
Europium (63Eu)	1.3	1.1	2		2.1	0.8–2.1	1–1.9	1.0
Gadolinium (64Gd)	4.0	3.3	6.2	3.6	6.1	3.8–6.1	3–4.7	4.6
Terbium (65 Tb)	0.7	0.6	1.2	0.76	1.2	0.6–1.2	0.6–0.7	0.7
Dysprosium (66Dy)	3.8	3.7	5.2	3.5	4.5	3..5–4.5	3.8–5	3.8
Holmium (67Ho)	0.8	0.8	1.3	1.2	1.3	0.8–1.3	0.4–1.1	0.8
Erbium (68Er)	2.1	2.2	3.5	2.20	3.5	2.3–3.5	2–2.8	2.3
Thulium (69Tm)	0.3	0.3	0.52	0.33	0.5	0.3–0.5	0.2–0.6	0.4

(continued)

Table 4 (continued)

Element	Concentrations in the Earth's crust						Soil concentrations	
	Wedepohl [237]	Taylor and McLennan [210]	Lide [132, 133]	Rudnick and Fountain [190]	Greenwood and Earnshaw [76]	Various authors[a]	Various authors[a]	Various authors[b]
Ytterbium (70Yb)	2.0	2.2	3.2	2	4.1	2.2–3.1	2.3–3.1	2.2
Lutetium (71Lu)	0.4	0.3	0.8	0.56	0.8	0.3–0.8	0.3–0.4	0.4
Total	**136.9**	**184.3**	**242.17**					

[a] Data according to Kotelnikova et al. [111] obtained, in turn, from the means of Taylor and McLennan [210], Greenwood and Earnshaw [76] and Kabata-Pendias and Pendias [109]

[b] Data according to Hu et al. [97] obtained, in turn, from several authors

[c] There appears to be no promethium in the Earth's crust

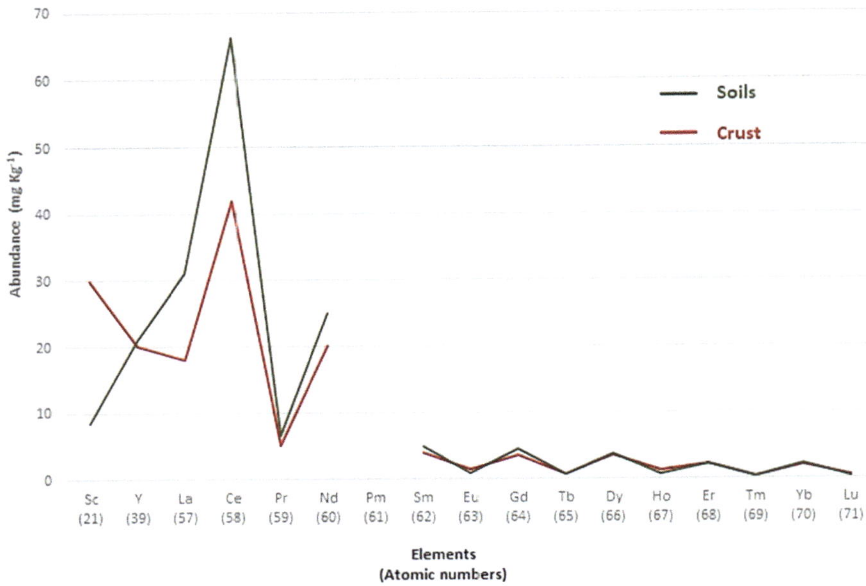

Fig. 3 Graph showing mean content of REEs elements in the earth's crust and soil. Promethium does not occur naturally on earth

6.1 Primary Mineralogy—REEs Ore Deposits

The first complete list of primary minerals containing REEs was issued by Clark [44] with the following:

- Fluorite (CaF_2), where Y and Ce replace Ca
- Allanite [$Ca(Ce,La,Y)(Al_2Fe^{2+})(O/OH/SiO_4/Si_2O_7)$], a mineral from the epidote group
- Sphene ($CaTiSiO_5$), where Y and REEs replace Ca
- Zircon ($ZrSiO_4$), where Y and HREE replace Zr
- Apatite [$Ca_5(PO_4)_3(F,Cl,OH)$, where Be, Ce, Pb, Y and REEs replace Ca
- Monazite [$(Ce,La)\ PO_4$], where most other REEs replace Ce and La
- Xenotime (YPO_4), where most other REEs replace Y
- Rhabdophane [$(Ce,La)\ PO_4\cdot(H_2O)$], where most other REEs replace La and Ce
- Bastnaesite [$(Ce,La,Y)\ CO_3F$], where most other REEs replace Y, La and Ce

The existence of more than 200 known REE-bearing minerals is presently accepted, although some authors like Walters and Lusty [231] and Dostal [55] consider that there are over 250 minerals whose chemical formula and crystal structure contain REEs as relevant constituents. Table 5 lists the most important REEs-bearing minerals and their generic formulas. Gupta and Krishnamurthy [80] state that only three of these can be considered to be REEs' ores, with the most feasible ones for the extraction of REEs bastnasite, monazite and xenotime, despite the fact that only bastnasite and monazite are usually interesting in economic terms. A dozen other minerals have also been, or could be used as, REEs resources. Specifically, bastnasite is the ore that contains the highest proportion of REEs (e.g., Ce, La, and Y). This mineral concentrates in vein deposits, contact metamorphic zones and pegmatites. Following Higueras et al. [93], monazite can be found in two profitable deposit types: (i) a diversity of hydrothermal deposits characterized by the occurrence of the so-called yellow monazite, (ii) concentrations of detrital nodules of millimetric size ("gray monazite," as defined by Donnot et al. [54]). These latter deposits are associated primarily with Ordovician shales, corresponding to secondary concentrations produced through supergene processes in recent sediments that form on the aforementioned shales. Yellow monazite appears as the main REEs ore in the huge Bayan Obo (China) deposit, formed due to the metasomatism of carbonatites in a subduction-related environment [117, 134]. Xenotime can occur along with yellow monazite, but generally occurs as a minor constituent of the corresponding type of deposits of this variety. Monazite preferentially incorporates larger LREEs (here La-Gd), whereas xenotime tends to incorporate smaller HREEs (here Tb-Lu, + Y) [159].

6.2 *REEs in Soils—Natural Origin*

The presence and abundance of REEs in soils are mainly the consequence of natural processes, but some anthropogenic activities can also alter their abundance. On these bases, anomalies with respect to background concentrations can be, therefore, indications of human-related pollution, linked to the usage of P-based fertilizants with high REEs concetrations; on the other hand, the possibility of being natural anomalies, linked to the presence of high concentrations of these elements in the substratum cannot be neglected, particularly in areas in with human activities, and in particular the use of fertilizants, is not conspicuous. High concentrations of REEs would indicate an anthropogenic rather than natural origin. The presence of REEs in soils and the natural environment depends, therefore, on not only parent material content, but also on weathering through geochemical and biological processes [64, 67, 79, 92, 149, 199, 204, 208, 234, 244]. Figure 4 illustrates a typical well-developed and representative red Mediterranean soil (profile a) identified in Torrenueva (Rhodoxeralf) in which gray monazite (b) is concentrated, coming from the weathering of Ordovician gray shales originally containing this mineral [93]. Other mineral phases than monazite can be present in common soils, explaining those variable concentrations in REEs elements.

Other natural processes seem responsible for some anomalous contents of these elements in soils [148], particularly those related to groundwaters flows [79, 199, 204, 208, 212, 222, 234], and even more particularly for the water flows related to volcanic activity [92, 244]. As a consequence, REEs are widely distributed in soils.

Table 5 Names and formulations of some important REEs-bearing minerals

Mineral	Formula	Mineral	Formula
Aeschynite[d]	$REE(Ti,Nb)_2(O,OH)_6$	Gerenite (Y)[c]	$(Ca,Na)_2(Y,REE)_3Si_6O_{18}\cdot 2H_2O$
Allanite[a,b,d]	$(Y,Ln,Ca)_2(Al,Fe^{3+})_3(SiO_4)_3(OH)$	Gittinsite[a]	$CaZrSi_2O_7$
Allanite (Ce)[c]	$(Ce,Ca,Y)_2(Al,Fe^{2+},Fe^{3+})_3(SiO_4)_3(OH)$	Gorceixite[b,c]	$(Ba,REE)Al_3(PO4)_2(OH_5\cdot H_2O)$
Allanite (Y)[c]	$(Y,Ce,Ca)_2(Al,Fe^{3+})_3(SiO_4)_3(OH)$	Goyazite[b,c]	$(Sr,REE)Al_3(PO4)_2(OH_5\cdot H_2O)$
Anatase[c]	$(Ti,REE)O_2$	Hingganite (Y)[c]	$(Y,Yb,Er)_2Be_2Si_2O_8(OH)_2$
Ancylite[b,c,d]	$Sr(REE)(CO_3)_2(OH)\cdot H_2O$	Iimoriite[a,b,c]	$Y_2(SiO_4)(CO_3)$
Apatite[a,d]	$(Ca,Ln)_5(PO_4)_3(F,Cl,OH)$	Kainosite[a,b,c]	$Ca_2(Y,Ln)_2Si_4O_{12}(CO_3)\cdot H_2O$
Bastnaesite[a,b,d]	$(Ln,Y)(CO_3)F$	**Loparite**[a,b,c]	$(Ln,Na,Ca)(Ti,Nb)O_3$
Brannerite[a,c]	$(U,Ca,Y,REE)(Ti,Fe)_2O_6$	**Monazite**[a,b,c,d]	$(Ln,Th)PO_4$
Britholite[a,c,d]	$(REE,Ca,Th)_5(SiO_4,PO_4)_3(OH,F)$	Mosandrite[a,b]	$(Na,Ca)_3Ca_3Ln\ (Ti,Nb,Zr)(Si_2O_7)_2(O,OH,F)_4$
Brockite[c,d]	$(Ca,Th,Ce)(PO_4)\cdot H_2O$	Parisite[a,b,c,d]	$Ca\ (Ln)_2(CO_3)_3F_2$
Burbankite[c,d]	$(Na,Ca)_3(Sr,Ba,Ce)_3(CO_3)_5$	Perovskite[c]	$(Ca,REE)TiO_3$
Calcio-ancylite (Ce)[c]	$(Ca,Sr)Ce_3(CO_3)_4(OH)_3\cdot H_2O$	Perrierite[d]	$REE_4(Mg, Fe,Ti)_4O_8(Si_2O7)_2$
Cerianite-(Ce)[b,c,d]	$(Ce^{4+},Th)O_2$	Pyrochlore[a,b,c,d]	$(Ca,Na,Ln)_2Nb_2O_6(OH,F)$
Cerite (Ce)[c]	$Ce_9Fe^{3+}(SiO_2)_6[(SiO_3)(OH)](OH)_3$	Rhapdophane[b,c]	$(REE)PO_4\cdot H_2O$
Cheralite[c]	$(Ca,Ce,Th)(P,Si)O_4$	Rinkite (rinkolite)[a,b,c]	$(Ca,Ln)_4Na(Na,Ca)_2Ti(Si_2O_7)_2(O,F)_2$
Chevkinite[c]	$(Ca,Ce,Th)_4(Fe^{2+},Mg)_2(Ti,Fe^{3+})_3Si_4O_{22}$	Samarskite[c]	$(REE,Fe^{2+},Fe^{3+},U,Th,Ca)(Nb,Ta,Ti)O_4$
Churchite[c]	$(Y)YPO_4\cdot H_2O$	Sphene (titanite)[c]	$(Ca,REE)TiSiO_5$
Crandallite[c]	$CaAl_3(PO_4)_2(OH)_5\cdot H_2O$	Steenstrupine (Ce)[c]	$Na_{14}Ce_6Mn_2Fe_2(Zr,Th)(Si_6O_{18})_2(PO_4)_7\cdot 3H_2O$
Dollaseite[d]	$CaLREE(Mg_2Al)[Si_2O_7][SiO_4]F(OH)$	Steenstrupine[a,d]	$Na_{14}Ln_6Mn_2Fe_2(Zr,Th)(Si_6O_{18})_2(PO_4)_7\cdot 3H_2O$
Doverite[c]	$YCaF(CO_3)_2$	Strontianite[d]	$Sr(CO)_3$
Eudialyte[a,b,c,d]	$Na_4(Ca,Ln)_2(Fe^{2+},Mn^{2+},Y)ZrSi_8O_{22}(OH,Cl)_2$	Synchysite[a,b,c,d]	$Ca(Ln)(CO_3)_2F$

(continued)

Table 5 (continued)

Mineral	Formula	Mineral	Formula
Euxenite (Y)[c,d]	$(Y,Ca,Ce,U,Th)(Nb,Ta,Ti)_2O_6$	Thalenite-(Y)[b,c]	$Y_3Si_3O_{10}OH$
Fergusonite[a,b,d]	$(Ln,Y)NbO_4$	Thorite[c]	$(Th,U)SiO_4$
Fergusonite (Ce)[c]	$(Ce,La,Y)NbO_4$	Törnebohmite[d]	$REE_2Al(SiO_4)_2(OH)$
Fergusonite (Y)[c]	$YNbO_4$	Uraninite[c]	$(U,Th,Ce)O_2$
Florencite[b,c,d]	$(REE)Al_3(PO_4)_2(OH)_6$	Vastmanlandite,[d]	$Ce_3CaMg_2Al_2Si_5O_{19}(OH)_2F$
Fluorapatite[b,c,d]	$(Ca,REE,Na)_5(PO_4)_3(F,OH)$	Vitusite (Ce)[c]	$Na_3(Ce,La,Nd)(PO_4)_2$
Fluorite[c]	$(Ca,REE)F$	**Xenotime**[a,b,c,d]	YPO_4
Gadolinite[b,c,d]	$(REE,Y)_2Fe^{2+}Be_2Si_2O_{10}$	Yttrofluorite[c]	$(Ca,Y)F_2$
Gagarinite (Y)[c]	$NaCaY(F,Cl)_6$	Zircon[a,c,d]	$(Zr,Ln)SiO_4$

Data from [a] Dostal [55],[b] Jones et al. [107],[c] Long et al. [138], and [d] Goodenough et al. [73]
The most important and abundant ones are depicted in bold

a) b)

Fig. 4 **a** Well-developed and representative profile of Rhodoxeralf (red Mediterranean soil); **b** detail of Monazite particles that accumulate in this soil type

6.3 REEs in Soils—Anthopogenic Origin

Human activity promotes major changes in soil composition by introducing elements that are generally called potentially toxic elements. Of these, the mining and processing of REEs-containing ores (and of REEs-containing associated minerals) can be an important source of these elements in soils that lie close to corresponding activities. In the same way, any processes involving the burning or calcination of organic (coal, oil) or inorganic (metals, in metallurgy) products may release emissions containing REEs to the atmosphere, which undergo dry and/or wet deposition, and also produce the dissemination of these elements in surrounding areas, or even in remote zones, depending on the climate and meteorological conditions [101, 116, 128, 150, 236]. None of this includes the 25, which can release amounts of REEs to the environment [19, 202]. Since the 1990s, recycling of waste has become a core element of sustainable development. But in our opinion recycling can delay for some years or decades at best.

Modern agriculture very much depends on fertilizer and pesticide applications to promote the growth of crops and to protect them from pests [191, 213, 232, 239, 252]. Thus, in addition to conventional pollution sources, phosphorous fertilizers can also be a considerable source of REEs pollution because these fertilizers contain such elements [68, 77, 154, 186, 213].

Given that REEs have been extensively used in agriculture for many years and in many regions of the world, the literature about this question abounds, such as

Volokh et al. [228], Slooff et al. [202], Todorovsky et al. [214], Val'kov et al. [223], or Sabiha-Javied et al. [192], among others.

Indeed, for some authors, the most important anthropogenic source of REEs entering soil is associated with the manufacture and use of organic and mineral fertilizers [24, 97, 173, 194]. REEs enrichment in surface soils has occurred in this way [40, 65, 79, 96, 126, 146, 173, 200, 218, 254], given their characteristic persistence.

Rare earth elements (REEs) play important roles in agricultural and zootechnical applications, such as fertilizers and feed additives. For example, administration of a REE-based fertilizer was shown to increase wheat (Triticum aestivum) crop yield and REE accumulation [252] or on navel orange quality and safety [41]. The use of REE preparations as additives in livestock feed has been practiced for several decades as reviewed by Rambeck and Wehr [183] and by Redling [186]. The ultimate consequence is that they accumulate in soils, bioaccumulate in crops and finally enter the human food chain [36].

7 Geochemical Behavior of REEs in Soils

The abundance of REEs is basically controlled by their abundance in substrate rocks, although, as stated in other sections of this chaper, some anthopogenic activities may increase them; in particular, and concerning natural concentrations, Hu et al. [97] found that REEs content in regoliths very much depends on the type of parent material, which is their primary source. Mobility conditions can modify their abundance and distribution in soils. Alderton et al. [10] suggested that these elements are immobile during weathering, while Fernández-Caliani et al. [63], Pérez-Lopez et al. [177] and Edahbi et al. [60] point out relative REEs enrichment from substrate to soil. According to Zhang et al. [253], Hu et al. [97] and Cidu et al. [43], REEs mobility depends not only on parent material, but also on the geochemical and biological processes that occur in soil. Jones [106] and Cao et al. [32] describe higher mobility rates for La, Ce, Gd and Y with pH and a redox potential that are low.

All these processes can be understood by bearing in mind that during the weathering process, rainwater infiltration will hydrolize and dissolve the soluble minerals in bedrock to generate processes that enrich or deplete REEs according to their relative mobility. Thus, mineral weathering can be considered an important source of elements for soils [179]; later, REEs might still be incorporated into secondary minerals, such as clay minerals (smectite, kaolinite, etc.) or other minerals, including gypsum or anhydrite, and then these REEs remain immobilized [69, 153, 171]. Ji et al. [102] and Jin et al. [105] report the mobility of these elements during weathering and different degrees of mobility depending on climate conditions, as well as mineralogical differences and distinct organic matter contents.

The detailed study of the geochemical behaviour of REEs during rock weathering and soil formation has steadily grown in past decades [6, 11, 13, 47, 98, 118–120, 139, 142, 157, 163, 185, 217, 250, 258]. Although REEs in soils were originally assumed to be immobile or quite low [38, 255], REEs signatures have been employed to assess soil genesis because they are sufficiently mobile as to infer the intensity of some key pedogenic processes [157]. In this way, the study of these elements' distribution and behaviour in soils can considerably support the understanding of weathering and pedogenic processes, such as eluviation-illuviation, erosion, elemental depletion because of leaching, oxyhydroxide formation or haploidization, which highlight their contribution as an external source of materials (alloctonism) [97, 105, 119, 250]. Tyler and Olsson [219] indicate that the majority of REEs are 40–50% removed from the upper A and E horizons of a Swedish Haplic Podzol.

It is, therefore, concluded that REEs can be used to reveal: (1) lithologic discontinuities [6–8], (2) the presence of aeolian or anthopogenic additions [9, 160, 165], (3) estimations of weathering intensities and elemental loss rates of soils [21, 25–27, 30, 75, 85, 86, 141, 152, 158, 164, 180, 181, 218], (4) oxidation–reduction conditions in soil [3].

In short, Fig. 5 shows a scheme of the potential pathways of the edaphic cycle of REEs, in which several factors can participate, such as plant uptake, erosion, leaching, complexation, eluviation-illuviation, and the removal/adsorption of REEs by inorganic colloids (phyllosilicates and oxyhydroxides).

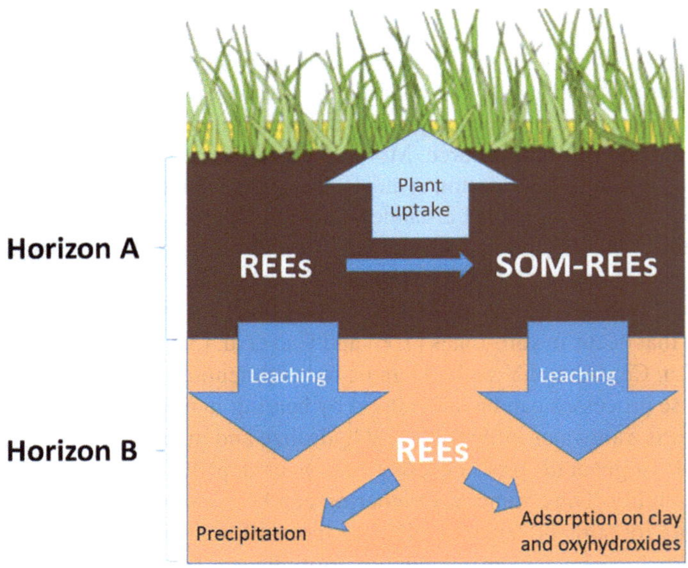

Fig. 5 Illustrative scheme of the potential pathways for REEs migration and sequestration (modified from [5])

8 REEs in the Soil–Plant System: Agricultural Applications

Ever since Drobkov [56] demonstrated in 1941 that pea plant yields increased by adding 10^{-2} g La per vessel, many agronomic studies address the effects of specific elements' plant growth processes. A review on this matter was conducted by Pang et al. [173].

Indeed, some REEs have been found to promote plant growth, development, crop production and better yields in various cultivation and application systems, but at low concentrations [205, 246]. Kabata-Pendias and Pendias [108] state that under natural conditions, plants absorb minimal amounts of REEs (1–45 mg kg^{-1} wt% in ash). Some authors [62, 166, 217] point out that certain species are capable of bioaccumulating high levels of these elements in spite of their low solubility in soil solutions.

The matter of REEs mobility and (bio)availability in soil, particularly concerning the soil-to-plant transfer, has been the subject of a number of studies, especially for assessing transfer and consequences [2, 33, 41, 53, 66, 71, 89, 90, 93, 95, 111–113, 124, 136, 176, 189, 215, 229, 247].

Wahid et al. [230] found a significant linear correlation between soil REEs content and the corresponding leaf or root REEs content in palms growing in sandy soil. Shan et al. [198] observed how the absorption capacity of REEs by plants significantly correlates with soil's cation exchange capacity. However, the amount of REEs in plants does not correlate with REEs content in the rhizosphere [49, 58, 245] due to a greater control of REEs uptake by plant properties [220].

The effects of REEs elements on plants can be divided into three groups: beneficial, inhibitory, and toxic [186]. Negative, positive or no effects of REEs on plant growth crop and tree yields in many countries around the world have been observed since 1933 [96].

Today there is limited and fragmented information on enriched REEs and the soil–plant system. The adsorption and desorption of REEs are affected by soil physico-chemical properties [49, 130, 137, 147, 230]. This means that the REEs elements uptaken by plants are not controlled only by plants themselves but depend on both their concentrations in soil and soil properties [28, 238]. For example, the uptake of REEs by plants tends to be higher at a low pH [32, 198, 217]. A positive effect of REE on crop production has been observed since 1960. El-Ramady, point out that soil pH plays a vital role in the bioavailability of various REEs and their release into soils. Tyler and Olsson [219] also studied the effects of soil pH on the uptake of metals by a grass, Agrostis capillaris L. For the lanthanoids, it was observed that root concentrations were inversely related to soil pH and positively correlated with soil concentration. When Brown et al. [29] studied the effects of REEs on various Ca-mediated biological processes in plants, they coined the term "supercalcium" to refer to La being analogous to Ca. These authors concluded that many enzymes and other functional proteins are inhibited by La. Plants with high REEs concentrations generally show a small biomass, inhibited root growth, leaf chlorosis and morphological alterations, which mostly lead to plants dying [207].

Experimental applications of REEs in the form of foliar sprays, additions of solid or liquid fertilizers, and seed treatments have been used in many species [96, 186, 192]. It is accepted that these elements can be very beneficial when used in small doses in agricultural plants [213]. D'Aquino et al. [46] report the positive effects of La (0.4–150 mg kg^{-1}) on seed germination in crops like *Oryza sativa, Solanum lycopersicum, Tritucum aestivum* or *Citrus limonia.* However, there is still controversy about the risk of uptake capacity of beneficial *vs.* toxic concentrations, and its mechanism in plants.

In light of all this, it is not surprising that REEs elements have been used to identify particulate deposition related to anthropogenic activities and the enrichment of other elements by biogenic processes, particularly through lichen, moss and four vascular plants [42]. Therefore, if these previous studies are considered, there is a considerable need for further information as to whether REEs can be used as an alternative crop nutrition option under protected and open-field agricultural conditions.

9 Environmental Consequences Linked with REEs

Soil is a component of the environment in which the anthropogenic addition of some chemical elements can pose a health risk for ecosystems and people. Today REEs element production and uses in the world represent a significant economic yield, but pose serious potential risks to health, the environment [129] and biodiversity [48]. Concerns about radioactive pollution and REEs toxicity are also voiced, as highlighted by Gwenzi et al. [83] or Ma et al. [140].

Besides natural processes, anthropogenic inputs must be taken into account because industrial, mining, and especially agricultural, activities are increasingly recognized as sources of these elements in soils [124, 258], as illustrated by Fig. 6. Therefore, despite the generally suspected low toxicity of REEs, their threshold limits and maximum permissible concentrations are still not well-known. In our opinion little is still known of the biological effects of occupational exposure to the lanthanides.

Rare earth primary production is associated with environmental impacts during mining, which depends on the type of deposit/mine and processing methods [196, 240]. According to the last authors, there may even be a major risk linked with the presence of radioactive elements (Th and U) in REEs ores.

As with other mining types, the mining and processing of REEs may generate tailings during production processes (beneficiation, separation, reduction, refining), and might change natural soil and water physico-chemical conditions, which can increase their bioavailability and migration in the environment. So, it is not surprising that high REEs concentrations have been detected in soils [135], vegetables [128] and other crops [259, 260], and even in human bodies [129] near mining and processing operations Consequently, toxicity studies of REEs have increased [83, 144, 168–170, 195, 203, 248]. Numerous examples appear in the literature to illustrate this topic. By way of example, a study by Li et al. [126] found that REEs' soil pollution linked with

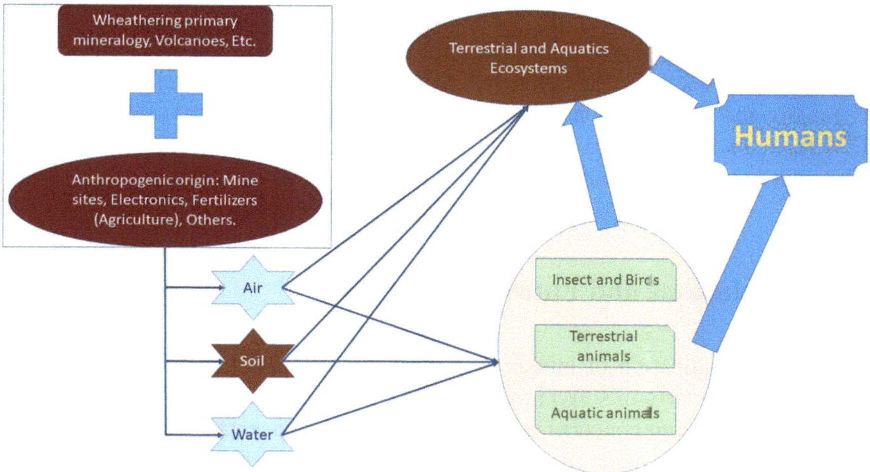

Fig. 6 Main sources of REEs entering soils and routes of impacts

tailings from an REEs processing plant in China was located approximately 7 km away from the mining area. Furthermore, Li et al. [128] study REEs concentrations in different edible plants grown on soils of a REEs mining area, in Changting County of Fujian Province (China), finding concentrations of these elements exceeding the maximum permissible concentration of REEs in vegetables in the Chinese directives (0.7 mg kg^{-1} d^{-1}), in particular in *Ipomoea aquatica* (water spinach) (0.92 mg kg^{-1} d^{-1}). The concentrations measured, in 8 different edible species, where higher in leafy vegetables than in non-leafy vegetables. They also find that REEs concentrations in local resident's blood and hair were 155.6 and 9.6 times higher than those of the general population, respectively. In another study by Slooff et al. [202], soils in polluted sites near industrial locations in the Netherlands contained high levels of REEs (at least 100-fold higher than background levels). In one of the many Chinese mining areas of these elements, Liang et al. [131] found REEs concentrations to be 200-fold higher than the Earth's crust's baseline levels for most of these. Likewise, plants bioaccumulate REEs in areas where mining activities take place [241]. Another example is that REEs also accumulate in lichens around mining areas [12, 128]. So, the presence of such excessive REEs concentrations in soils can imply serious consequences for the surrounding environment and can ultimately enter the human body via many pathways, especially by food ingestion. Hence, this concern is being increasingly voiced because, according to Pitron [178] and Carrillo [34], many deposits are located in geographical areas where environmental care is practically nonexistent.

In Campo de Montiel (south-central Spain), there are thick soils rich in monazite and, therefore, with high REEs contents (see Fig. 3). It is a unique deposit that is capable of supplying part of the European Union's needs [93]. However, as the extraction area is located in a place marked by agronomy (olive trees, vineyards, cereal crops), and with hunting reserves and protected fauna (Iberian lynx, *Lynx pardinus*, Imperial eagle, *Aquila adalberti*; red kite, *Milvus milvus*, all of which are in danger of extinction), the mining project has led to significant rejection.

Finally, the previously cited use of fertilizers in the agricultural sector [173, 192, 223] for many years could further elevate REEs soil contents [18, 103, 104, 216, 228], particularly in agricultural areas.

10 Conclusions and Future Perspectives

With the intensification of human activities, the abundance of REEs in the environment (particularly in soils) has increased dramatically in recent decades, which has raised significant concerns worldwide. Despite intense and recent research and clear progress being made in understanding various processes related to the REEs soil cycle, certain relevant aspects still need to be investigated, because they are still missing. In this context, the following specific points need to be addressed:

(i) Explore the sources of REEs in soils, especially for agricultural use, by understanding their incorporation and interactions with other soil constituents, and by considering precisely these external sources.
(ii) Investigate the safe threshold of REEs concentrations for soil, the environment and living organisms, especially concerning human health
(iii) Promote new studies about plant growth and soil–plant transfer by recognizing the interaction mechanisms and associated factors between plants and REEs
(iv) Investigate impacts for the soil environment and human and animal health because of increasing REEs utilization and their related growing presence in the environment

Given a whole series of processes related to the presence of REEs in mining and/or oil refining, the need to control or monitor them is evident due to the environmental contamination risk that these elements may entail. Other possible sources of contamination must not be ruled out, such as those deriving from the rise in discarded obsolete equipment containing REEs and from using REEs-containing phosphate fertilizers (or with slurry or WWTP sludge). However, information on the toxicity of these elements in a broad sense is scarce. Therefore, the positive and negative effects of REEs should be carefully considered.

Current research should focus on the impact that REEs use can finally have for animal and human health, because REEs can be absorbed by vegetation, may form part of mineral nutrition and finally enter the human food chain either directly with vegetals or indirectly with contaminated cattle. So, each and every lanthanide should be analyzed and assessed to clarify its comprehensive toxicity to living organisms.

Presently, the commercial recycling of rare earths is very limited. Therefore, the development of REEs recycling processes is becoming an increasingly important issue with every passing day, which is quite a challenge. In this way, the "urban mining" concept (defined as the recovery of elements and compounds from waste materials and products) is becoming more and more important.

Finally, although the current literature on REEs is not abundant, especially for their industrial and economic interest, a new impulse in their research is certainly expected to take place in the near future, particularly in relation to soils. We believe that all these studies must be placed in the ideal framework context, which would consist in linking sustainable REEs production with "innovative and clean environmental technology" based on the appropriate management of these elements.

References

1. Abbott AN, Haley BA, McManus J, Reimers CE (2015) The sedimentary flux of dissolved rare earth elements to the ocean. Geochim Cosmochim Acta 154:186–200
2. Ahmad HR, Zia-ur-Rehman M, Sohail MI, Anwar ul Ha M, Khalid H, Ayub MA, Ishaq G (2018) Effects of rare earth oxide nanoparticles on plants. Nanomater Plants Algae Microorg 1:239–275
3. Aide MT (2005) Elemental composition of soil nodules from two alfisols on an alluvial in Missouri, soil. Soil Sci 170(12):1033–1042
4. Aide M (2018) Lanthanide soil chemistry and its importance in understanding soil pathways: mobility, plant uptake, and soil health. In: Awwad NS, Mubarak AT (eds) Lanthanides. Intech Open book series. https://www.intechopen.com/chapters/62330
5. Aide MT, Aide C (2012) Rare earth elements: their importance in understanding soil genesis. ISRN Soil Sci 1–11
6. Aide M, Smith-Aide C (2003) Assessing soil genesis by rare earth elemental analysis. Soil Sci Soc Am J 5:1470–1476
7. Aide MT, Alexander L, Heberlie et al (1999) Soil genesis on felsic rocks in the St. Francois Mountains. I. The role of saprolite and its influence on soil properties. Soil Sci 164(6):428–439
8. Aide MT, Alexander L, Heberlie L, Statler P (1999) Soil genesis on felsic rocks in the St. Francois Mountains. II. The distribution of elements and their use in understanding weathering and elemental loss during genesis. Soil Sci 164(12):946–959
9. Aide MT, Aide C, Dolde J, Guffey C (2003) Geochemical indicators of external additions to soils in Big Bend National Park, Texas. Soil Sci 168(3):200–208
10. Alderton DH, Pearce M, Potts PJ (1980) Rare earth element mobility during granite alteration: evidence from southwest england. Earth Planet Sci Lett 49:149–165
11. Alfaro MR, do Nascimento CWA, Biondi CM, Silva YJAB, Accioly AM, Montero A, Ugarte OM, Estevez J (2018) Rare-earth-element geochemistry in soils developed in different geological settings of Cuba. Catena 162:317–324. https://doi.org/10.1016/j.catena.2017.10.031
12. Anawar H, Freitas MD, Canha N, Dionísio I, Dung H, Galinha C, Pacheco A (2012) Assessment of bioaccumulation of REEs by plant species in a mining area by INAA. J Radioanal Nucl Chem 294:377–381
13. Aubert D, Probst A, Stille P (2004) Distribution and origin of major and trace elements (particularly REE, U and Th) into labile and residual phases in an acid soil profile (Vosges Mountains, France). Appl Geochem 19:899–916
14. Ault T, Krahn S, Croff A (2015) Radiological impacts and regulation of rare earth elements in non-nuclear energy production. Energies 8:2066–2081
15. Balaram V (1996) Recent trends in the instrumental analysis of rare earth elements in geological and industrial materials. Trends Anal Chem 15:475–486

16. Balaram V (2016) Current advances in the miniaturization of analytical instruments—applications in cosmochemistry, geochemistry, exploration and environmental sciences. Spectroscopy 31(10):40–44
17. Balaram V (2017) Field-portable analytical instruments in mineral exploration: past, present and future. J Appl Geochem 19(4):382–399
18. Balaram V (2019) Rare earth elements: a review of applications, occurrence, exploration, analysis, recycling, and environmental impact. Geosci Front 1674–9871
19. Balassone G, Manfredi C, Vasca E, Bianco M, Boni M, Di Nunzio A, Lombardo F, Mozzillo R, Marino A, Mormone A, Mura G, Trifuoggi M, Mondillo N (2021) Recycling REEs from the waste products of Silius Mine (SE Sardinia, Italy): a preliminary study. Sustainability 13:14000
20. Bau M, Dulski P (1996) Anthropogenic origin of positive gadolinium anomalies in river waters. Earth Planet Sci Lett 143:245–255
21. Bauluz B, Mayayo MJ, Fernandez-Nieto C, Lopez JMG (2000) Geochemistry of precambrian and paleozoic siliciclastic rocks from the Iberian range (NE Spain): implications for source-area weathering, sorting, provenance, and tectonic setting. Chem Geol 168(1–2):135–150
22. Binnemans K, Jones PT, Blanpain B, Van Gerven T, Yang Y, Walton A, Buchert M (2013) Recycling of rare earths: a critical review. J Clean Prod 51:1–22. https://doi.org/10.1016/j.jclepro.2012.12.037
23. Bohn HL, McNeal BL, O'Connor GA (2001) Soil chemistry. Willey, New York
24. Borges RC, Marques LM, Mahler CF, Bernedo AVB (2018) Determination of the concentration of Ce, La, Sm and Eu in a phosphogypsum stackin Imbituba city, Santa Catarina, Brazil. Eclética Quím J 43(3):37–44. https://doi.org/10.26850/1678-4618eqj.v43.3.2018.p37-44
25. Braun JJ, Pagel M (1994) Geochemical and mineralogical behavior of REE, Th and U in the Akongo lateritic profile (SW Cameroon). Catena 21(2–3):173–177
26. Braun JJ, Pagel M, Herbilln A, Rosin C (1993) Mobilization and redistribution of REEs and thorium in a syenitic lateritic profile: a mass balance study. Geochim Cosmochim Acta 57(18):4419–4434
27. Braun JJ, Viers J, Dupre B, Polve M, Ndam J, Muller JP (1998) Solid/liquid REE fractionation in the lateritic system of Goyoum, East Cameroon: the implication for the present dynamics of the soil covers of the humid tropical regions. Geochim Cosmochim Acta 62(2):273–299
28. Brioschi L, Steinmann M, Lucot E, Pierret MC, Stille P, Prunier J, Badot PM (2013) Transfer of rare earth elements (REE) from natural soil to plant systems: implications for the environmental availability of anthropogenic REE. Plant Soil 366:143–163
29. Brown PH, Rathjen AH, Graham RD, Tribe DE (1990) Rare earth elements in biological systems. In: Gschneidner Jr KA, Eyring L (eds) Handbook on the physics and chemistry of rare earths, vol 13. Elsevier Sciences Publisher B.V., New York, pp 423–453
30. Brown DJ, Helmke PA, Clayton MK (2003) Robust geochemical indices for redox and weathering on a granitic laterite landscape in Central Uganda. Geochim Cosmochim Acta 67(15):2711–2723
31. Cao X, Wang X, Zhao G (2000) Assessment of the bioavailability of rare earth elements in soils by chemical fractionation and multiple regression analysis. Chemosphere 40:23–28
32. Cao X, Chen Y, Wang X, Deng X (2001) Effects of redox potential and pH value on the release of rare earth elements from soil. Chemosphere 44:655–661
33. Carpenter D, Boutin C, Allison JE, Parsons JL, Ellis DM (2015) Uptake and effects of six rare earth elements (REEs) on selected native and crop species growing in contaminated soils. PLoS ONE 10(6), art. no. e0129936
34. Carrillo J (2009) Extracción de tierras raras y su impacto ambiental. Asociación Geoinnova. https://geoinnova.org/blog-territorio/extraccion-tierras-raras-impacto-ambiental/
35. Chang C, Song C, Beckford HO, Wang S, Ji H (2019) Behaviors of REEs during pedogenetic processes in the karst areas of Southwest China. J Asian Earth Sci 185:104023
36. Charalampides G, Vatalis K, Karayannis V, Baklavaridis A (2016) Environmental effects and economic impact on global market of rare earth metals. IOP Conf Ser: Mater Sci Eng 161, Article 012069

37. Chen J, Yang R (2010) Analysis on REE geochemical characteristics of three types of REE-rich soil in Guizhou Province, China. J Rare Earth 28:517–522
38. Chen Y, Wang X, Peng A (1999) The research progress of fractionation among rare earth elements. Adv Environ Sci 1:10–17
39. Chen LM, Zhang GL, Jin ZD (2014) Rare earth elements of a 1000-year paddy soil chronosequence: implications for sediment provenances, parent material uniformity and pedological changes. Geoderma 230(231):274–279
40. Cheng H, Hao F, Ouyang W, Liu S, Chunye LIN, Wenjing Y (2012) Vertical distribution of rare earth elements in a wetland soil core from the Sanjiang Plain in China. J Rare Earths 30:731–738
41. Cheng J, Ding C, Li X, Zhang T, Wang X (2015) Rare earth element transfer from soil to navel orange pulp (*Citrus sinensis* Osbeck cv. Newhall) and the effects on internal fruit quality. PLoS ONE 10:e0120618
42. Chiarenzelli J, Aspler L, Dunn C, Cousens B, Ozarko D, Powis K (2001) Multi-element and rare earth element composition of lichens, mosses, and vascular plants from the central Barrenlands, Nunavut, Canada. Appl Geochem 16(2):245–270
43. Cidu R, Vittori Antisari L, Biddau R, Buscaroli A, Carbone S, Da Pelo S, Dinelli E, Vianello G, Zannoni D (2013) Dynamics of rare earth elements in water-soil systems: the case study of the Pineta San Vitale (Ravenna, Italy). Geoderma 193–194:52–67
44. Clark AM (1984) Mineralogy of the rare earth elements. In: Henderson P (ed) Rare earth element geochemistry. Elsevier, NY, pp 33–54
45. Cobelo-García A, Filella M, Croot P, Frazzoli C, Du Laing G, Ospina-Alvarez N, Rauch S, Salaun P, Schäfer J, Zimmermann S (2015) COST action TD1407: network on technology-critical elements (NOTICE)—from environmental processes to human health threats. Environ Sci Pollut Res 22:15188–15194
46. D'Aquino L, De Pinto MC, Nardi L, Morgana M, Tommasi F (2009) Effect of some light rare earth elements on seed germination, seedling growth and antioxidant metabolism in Triticum durum. Chemosphere 75:900–905
47. Da Silva YJAB, do Nascimento CWA, Miranda C, Van Straaten P, De Souza V, Agra-Becerra Y, Dos Santos CA, Trezema JC (2017) Influence of metaluminous granite mineralogy on the rare earth element geochemistry of rocks and soils along a climosequence in Brazil. Geoderma 306:28–39. https://doi.org/10.1016/j.geoderma.2017.06.031
48. Degtjarenko P, Marmor L, Randlane T (2016) Changes in bryophyte and lichen communities on Scots pines along an alkaline dust pollution gradient. Environ Sci Pollut Res 23:17413–17425
49. De Araújo ALL, De Nadai Fernandes EA, Bacchi MA, De França EJ (2012) Bioaccumulation pattern of lanthanides in pteridophytes and magnoliophytes species from Atlantic Forest. J Radioanal Nucl Chem 291:187–192
50. De Boer MA, Lammertsma K (2013) Scarcity of rare earth elements. Chem Sus Chem 6:2045–2055
51. De Vito E, Masi AN, Olsina RA (1999) Determination of trace rare earth elements by X-ray fluorescence spectrometry after preconcentration on a new chelating resin loaded with Thorin. Talanta 49:929–935. https://doi.org/10.1016/S0039-9140(99)00089-2
52. Diatloff E, Asher CJ, Smith FW (1996) Concentrations of rare earth elements in some Australian soils. Aust J Soil Res 34:735–747
53. Diatloff E, Smith FW, Asher CJ (2008) Effects of lanthanum and cerium on the growth and mineral nutrition of corn and mungbean. Ann Bot 101:971–982. PMID: 18292604
54. Donnot M, Guigues J, Lulzac YM, Magnien A, Parfenoff A, Picotet P (1973) Un nouveau type de gisement d'europium: la monazite grise à europium en nodules dans les schistes paléozoïques de Bretagne. Mineral Deposita 8(1):7–18
55. Dostal J (2017) Rare earth element deposits of alkaline igneous rocks. Resources 6(3):34
56. Drobkov AA (1941) Influence of cerium, lanthanum, and samarium on development of peas. C R Acad Sci USSR 32:669–670

57. Du X, Graedel TE (2011) Global rare earth in-use stocks in NdFeB permanent magnets. J Ind Ecol 15(6):836–843
58. Durães N, Da Silva EF, Bobos I, Ávila P (2014) Rare earth elements fractionation in native vegetation from the Moncorvo iron mines, NE Portugal. Proc Earth Planet Sci 10:376–382
59. Dutta T, Kim KH, Uchimiya M, Kwon EE, Jeon BH, Deep A, Yun ST (2016) Global demand for rare earth resources and strategies for green mining. Environ Res 150:182–190
60. Edahbi M, Plante B, Benzaazoua M, Ward M, Pelletier M (2018) Mobility of rare earth elements in mine drainage: Influence of iron oxides, carbonates, and phosphates. Chemosphere 199:647–654
61. El-Taher A (2018) Nuclear analytical techniques for detection of rare earth elements. J Radiat Nucl App 3:53–64. https://doi.org/10.18576/jrna/030107
62. Evans CH (1990) Biochemistry of lanthanides, vol 8. Pergamon Press, 444 pp
63. Fernández-Caliani JC, Barba-Brioso C, De la Rosa JD (2009) Mobility and speciation of rare earth elements in acid mine soils and geochemical implications for river waters in the southwestern Iberian margin. Geoderma 149:393–401
64. Ferreira MDS, Fontes MPF, Bellato CR, Marques Neto JDO, Lima HN, Fendorf S (2021) Geochemical signatures and natural background values of rare earth elements in soils of Brazilian Amazon. Environ Pollut 277, art. no. 116743
65. França EJ, De Nadai Fernandes EA, Bacchi MA, Tagliaferro FS (2002) Pathway of rare-earth elements in a Brazilian forestry fragment. J Alloy Compd 344:21–26
66. França EJ, De Nadai Fernandes EA, Turra C, Bacchi MA, Elias C, Tagliaferro F, Moreira C, Ribeiro A, Figueira R (2011) Survey of lanthanoids in plants from a tropical region. Int J Environ Heal 5(1–2):32–48
67. Fu S, Yan C, Wu S, Yang X (2000) Content and distribution of rare earth elements in typical soils of Guizhou province. Acta Pedol Sin 1:109–115
68. Fu F, Akagi T, Yabuki S, Iwaki M (2001) The variation of REE (rare earth elements) patterns in soil-grown plants: a new proxy for the source of rare earth elements and silicon in plants. Plant Soil 235:53–64
69. Fulignati P, Giocada A, Sbrana A (1999) Rare-earth element (REE) behaviour in the alteration facies of the active magmatic-hydrothermal system of Vulcano (Aeolian Islands, Italy). J Volcanol Geotherm Res 88:325–342
70. Gadolin J (1794) Undersökning af en svart Tung stenart ifrån ytterby stenbrott i Roslagen. Kongl Vetenskaps Academiens Nya Handlingar 15:137–155
71. Gong B, He E, Xia B, Ying R, Peijnenburg WJGM, Liu Y, Qiu H (2020) Bioavailability and phytotoxicity of rare earth metals to Triticum aestivum under various exposure scenarios. Ecotoxicol Environ Saf 205:art. no. 111346
72. González V, Vignati DA, Leyval C, Giamberini L (2014) Environmental fate and ecotoxicity of lanthanides: are they a uniform group beyond chemistry? Environ Int 71:148–157
73. Goodenough KM, Schilling J, Jonsson E, Kalvig P, Charles N, Tuduri J, Deady EA, Sadeghi M, Schiellerup H, Müller A, Bertran G, Arvanitidis N, Eliopoulos DG, Shaw RA, Thrane K, Keulen N (2016) Europe's rare earth element resource potential: an overview of REE metallogenetic provinces and their geodynamic setting. Ore Geol Rev 72(1):838–856
74. Gorbatenko AA, Revina EI (2015) A review of instrumental methods for determination of rare earth elements. Inorg Mother 51:1375–1388. https://doi.org/10.1134/S0020168515140058
75. Gouveia MA, Prudencio MI, Figueiredo MO, Pereira LJC, Waerenborgh JC, Morgado I, Pena T, Lopes A (1993) Behavior of REE and other trace and major elements during weathering of granitic rocks, Evora, Portugal. Chem Geol 107(3–4):293–296
76. Greenwood NN, Earnshaw A (1997) Chemistry of the elements. Butterworth-Heinemann, Oxford
77. Guo BS, Zhu WM, Xiong PK, Ji YJ, Liu Z, Wu ZM (1988) Rare earths in agriculture. Agricultural Scientific Technological Press, Beijing, China, p 23e208
78. Guo P, Jia X, Duan T, Zhang Y, Xu T, Chen H (2008) Special analysis of rare earth elements in soil. Chin J Anal Chem 11:1483–1487

79. Guo H, Zhang B, Wang G, Shen Z (2010) Geochemical controls on arsenic and rare earth elements approximately along a groundwater flow path in the shallow aquifer of the Hetao Basin, Inner Mongolia. Chem Geol 270:117–125
80. Gupta CK, Krishnamurthy N (2004) Extractive metallurgy of rare earths. CRC Press
81. Gupta CK, Lide DR (1997) Abundance of elements in the earth's crust and sea. In: CRC handbook of physics and chemistry, 78th edn. CRC Press, Boca Raton, p 14
82. Guyonnet D, Planchon M, Rollat A, Escalon V, Tuduri J, Charles N, Vaxelaire S, Dubois D, Fargier H (2015) Material flow analysis applied to rare earth elements in Europe. J Clean Prod 107:215–228
83. Gwenzi W, Mangori L, Danha C, Chaukura N, Dunjana N, Sanganyado E (2018) Sources, behaviour, and environmental and human health risks of high-technology rare earth elements as emerging contaminants. Sci Total Environ 636:299–313
84. Hall GEM (1992) Inductively coupled plasma mass spectrometry in geoanalysis. J Geochem Explore 44:201–249. https://doi.org/10.1016/0375-6742(92)90051-9
85. Han G, Xu Z, Yang T, Zhang G (2009) Rare earth element patterns in the Karst Terrains of Guizhou Province, China: implication for water/particle interaction. Aquat Geochem 15:457
86. Han G, Song Z, Tang Y (2017) Geochemistry of rare earth elements in soils under different land uses in a typical karst area, Guizhou Province, Southwest China. Can J Soil Sci 97:606–612
87. Haskin LA, Haskin MA, Frey FA, Wildeman TR (1968) Relative and absolute terrestrial abundances of the rare earths. In: Ahrens LH (ed) Origin and distribution of the elements. Pergamon, Oxford
88. Hayes-Labruto L, Schillebeeckx SJD, Workman M, Shah N (2013) Contrasting perspectives on China's rare earth policies: reframing the debate through a stakeholder lens. Energy Policy 63:55–68
89. He E, Qiu H (2022) Lanthanum and cerium disrupt similar biological pathways and interact synergistically in Triticum aestivum as revealed by metabolomic profiling and quantitative modeling. J Hazard Mater 426, art. no. 127831
90. He E, Gong B, Qiu H, Van Gestel CAM, Ruan J, Tang Y, Huang X, Xiao X, Li M, Qiu R (2020) Model-based rationalization of mixture toxicity and accumulation in Triticum aestivum upon concurrent exposure to yttrium, lanthanum, and cerium. J Hazard Mater 389, art. no. 121940
91. Henderson P (2013) Rare earth element geochemistry. Elsevier, Amsterdam
92. Hernández-Antonio A, Mahlknecht J, Tamez-Meléndez C, Ramos-Leal J, Ramírez-Orozco A, Parra R, Ornelas-Soto N, Eastoe CJ (2015) Groundwater flow processes and mixing in active volcanic systems: the case of Guadalajara (Mexico). Hydrol Earth Syst Sci 19:3937–3950
93. Higueras PL, Jiménez-Ballesta R, Esbrí JM, García-Giménez R, García-Noguero EM, Álvarez R, Peco JD, García-Noguero C, Campos JA (2021) Occurrence and environmental constraints of grey monazite in red soils from the Campo de Montiel area (SW Ciudad Real province, South Central Spain). Environ Sci Pollut Res 28:4573–4584
94. Hisinger W, Berzelius JJ (1804) Cerium: ein neuer metall aus einer schwedischen steinart, bastnäs tungstein gennant. Neues Allgemeines J Chem 2:397–418
95. Hu X, Ding Z, Chen Y, Wang X, Dai L (2002) Bioaccumulation of lanthanum and cerium and their effects on the growth of wheat (*Triticum aestivum* L.) seedlings. Chemosphere 48:621–629
96. Hu Z, Richter H, Sparovek G, Schnug E (2004) Physiological and biochemical effects of rare earth elements on plants and their agricultural significance: a review. J Plant Nutr 27:183–220
97. Hu ZY, Haneklaus S, Sparovek G, Schnug E (2006) Rare earth elements in soils. Commun Soil Sci Plant Anal 37:1381–1420
98. Huang C, Gong Z (2001) Geochemical implication of rare earth elements in process of soil development. J Rare Earths 19:57–62
99. IUPAC (2005) Nomenclature of inorganic chemistry. In: Connelly NG et al (eds) Recommendations 2005
100. Jackson WD, Christiansen G (1993) International strategic minerals inventory summary report: rare-earth oxides. Circular 930-N. US Geological Survey, Reston, VA, 68 pp

101. Jha AM, Singh AC (1994) Clastogenicity of lanthanides—induction of micronuclei in root tips of Vicia faba. Mutat Res Toxicol 322(3):169–172. https://doi.org/10.1016/0165-1218(94)90003-5
102. Ji H, Wang S, Ouyang Z, Zhang S, Sun C, Liu X, Zhou D (2004) Geochemistry of red residua underlying dolomites in karst terrains of Yunnan-Guizhou Plateau II. The mobility of rare earth elements during weathering. Chem Geol 203:29–50
103. Jiménez-Ballesta R, Bravo S, Pérez-De-los C, Amorós JA, García-Pradas J, Sánchez-Ormeño M, García-Navarro FJ (2021) Preliminary assessment of the occurrence of six rare earth elements in calcareous vineyard soils. Water Air Soil Pollut 232:76. https://doi.org/10.1007/s11270-021-05034-1
104. Jiménez-Ballesta R, Bravo S, Pérez-De-los-reyes C, Amorós JA, García-Pradas J, Sánchez-Ormeño M, García-Navarro FJ (2021) Exploring the presence of five rare earth elements in vineyard soils on different lithologies: Campo de Calatrava, Spain. Agronomy 11(3), art. no. 458
105. Jin L, Maa L, Dere A, White T, Mathur R, Brantley SL (2017) REE mobility and fractionation during shale weathering along a climate gradient. Chem Geol 466:352–379
106. Jones DL (1997) Trivalent metal (Cr, Y, Rh, La, Pr, Gd) sorption in two acid soils and its consequences for bioremediation. Eur J Soil Sci 697–702
107. Jones AP, Wall F, Williams CT (eds) (1996) Rare earth minerals—chemistry, origin and ore deposits, 1st edn, vol 7. Chapman & Hall, Mineralogical Society Series, London, 372 pp
108. Kabata-Pendias A, Pendias H (1984) Trace elements in soils and plants. CRC Press, Florida, p 315
109. Kabata-Pendias A, Pendias H (1989) Trace elements in soils and plants. CRC Press, Boca Raton, Fl, p 315
110. Keersemaker M (2020) Critical raw materials. In: Suriname revisited: economic potential of its mineral resources. Springer briefs in earth sciences. Springer, Cham. https://doi.org/10.1007/978-3-030-40268-6_9
111. Kotelnikova AD, Fastovets I, Rogova O, Volkov DS (2020) La, Ce and Nd in the soil-plant system in a vegetation experiment with barley (Hordeum vulgare L.). Ecotoxicol Environ Saf 206, art. no. 111193
112. Kotelnikova AD, Rogova OB, Stolbova VV (2021) Lanthanides in the soil: routes of entry, content, effect on plants, and genotoxicity (a review). Eurasian Soil Sci 54(1):117–134
113. Kovaříková M, Tomášková I, Soudek P (2019) Rare earth elements in plants. Biol Plant 63:20–32
114. Kulaksız S, Bau M (2011) Anthropogenic gadolinium as a microcontaminant in tap water used as drinking water in urban areas and megacities. Appl Geochem 26:1877–1885
115. Kümmerer K, Helmers E (2000) Hospital effluents as a source of gadolinium in the aquatic environment. Environ Sci Technol 34:573–577
116. Ladonin DV (2017) Lanthanides in soils of the Cherepovets steel mill impact zone. Eurasian Soil Sci 50:672–680. https://doi.org/10.1134/S1064229317060047
117. Lai X, Yang X (2013) Geochemical characteristics of the Bayan Obo giant REE–Nb–Fe deposit: constraints on its genesis. J S Am Earth Sci 41:99–112
118. Land M, Öhlander B, Ingri J, Thunberg J (1999) Solid speciation and fractionation of rare earth elements in a spodosol profile from northern Sweden as revealed by sequential extraction. Chem Geol 160:121–138. https://doi.org/10.1016/S0009-2541(99)00064-9
119. Laveuf C, Cornu S (2009) A review on the potentiality of rare earth elements to trace pedogenetic processes. Geoderma 154:1–12
120. Laveuf C, Cornu S, Juillot F (2008) Rare earth elements as tracers of pedogenetic processes. CR Geosci 340:523–532
121. Lawrence MG, Ort C, Keller J (2009) Detection of anthropogenic gadolinium in treated wastewater in South East Queensland, Australia. Water Res 43:3534–3540
122. Lei G, Zhang Z (1996) The distribution characteristics of REE in some soils in Guizhou. Gui Zhou Sci 3:11–17

123. Lerat-Hardy A, Coynel A, Dutruch L, Pereto C, Bossy C, Gil-Diaz T, Capdeville MJ, Blanc G, Schäfer J (2019) Rare earth element fluxes over 15 years into a major European Estuary (Garonne-Gironde, SW France): hospital effluents as a source of increasing gadolinium anomalies. Sci Total Environ 656:409–420
124. Li D, Huang S, Wang D, Wang W, Peng A (2000) Transfer characteristics of rare earth elements applied in agricultural soils. J Environ Sci Health Part A 35:1869–1881
125. Li D, Tang Y, Deng T, Chen K, Liu D (2008) Geochemistry of rare earth elements in coal—a case study from Chongqing, southwestern China. Energy Explor Exploit 26:355–362
126. Li J, Hong M, Yin X, Liu J (2010a) Effects of the accumulation of the rare earth elements on soil macrofauna community. J Rare Earth 28:957–964
127. Li QA, Li X, Zhang Q, Chen J (2010b) Effect of rare-earth element Sm on the corrosion behavior of Mg-6Al-1.2Y-0.9Nd alloy. Rare Met 29:557–560
128. Li X, Chen Z, Chen Z, Zhang Y (2013) A human health risk assessment of rare earth elements in soil and vegetables from a mining area in Fujian Province, Southeast China. Chemosphere 93(6):1240–1246
129. Li X, Chen Z, Chen Z (2014) Distribution and fractionation of rare earth elements in soil–water system and human blood and hair from a mining area in southwest Fujian Province. Environ Earth Sci 72(9):3599–3608
130. Liang T, Zhang S, Wang L, Kung HT, Wang Y, Hu A, Ding S (2005) Environmental biogeochemical behaviors of rare earth elements in soil-plant systems. Environ Geochem Health 27(4):301–311
131. Liang T, Li K, Wang L (2014) State of rare earth elements in different environmental components in mining areas of China. Environ Monit Assess 186:1499–1513
132. Lide DR (1997) Abundance of elements in the earth's crust and sea. In: CRC handbook of physics and chemistry, 78th edn. CRC Press, Boca Raton, p 14
133. Lide DR (ed) (2004) CRC handbook of chemistry and physics, 85th edn, vol 2. CRC Press, Boca Raton, 712 pp
134. Ling MX, Liu YL, Williams IS, Fang-Zhen T, Xiao-Yong Y, Xing D, Gang-Jian W, Lu-Hua X, Wen-Feng D, Wei-Dong S (2013) Formation of the world's largest REE deposit through protracted fluxing of carbonatite by subduction-derived fluids. Sci Rep 3:1776
135. Liu WS, Guo MN, Liu C, Yuan M, Chen XT, Huot H, Zhao CM, Tang YT, Morel JL, Qiu RL (2019) Water, sediment and agricultural soil contamination from an ion-adsorption rare earth mining area. Chemosphere 216:75–83
136. Liu Z, Guo C, Tai P, Sun L, Chen Z (2021) The exposure of gadolinium at environmental relevant levels induced genotoxic effects in Arabidopsis thaliana (L.). Ecotoxicol Environ Saf 215, art. no. 112138
137. Loell M, Albrecht C, Felix-Henningsen P (2011) Rare earth elements and relation between their potential bioavailability and soil properties, Nidda catchment (Central Germany). Plant Soil 349:303–317
138. Long KR, Van Gosen BS, Foley NK, Cordier D (2010) The principal rare earth elements deposits of the United States—a summary of domestic deposits and a global perspective. US Geological Survey, Scientific Investigations Report 2010–5220, 95 pp. http://pubs.usgs.gov/sir/2010/5220/. Accessed 20 Dec 2021
139. Ma JL, Wei GJ, Xu YG, Long WG, Sun W (2007) Mobilization and re-distribution of major and trace elements during extreme weathering of basalt in Hainan Island, South China. Geochim Cosmochim Acta 71:3223–3237. https://doi.org/10.1016/j.gca.2007.03.035
140. Ma L, Dang D, Wang W, Evans R, Wang W (2019) Rare earth elements in the Pearl River Delta of China: potential impacts of the REE industry on water, suspended particles and oysters. Environ Pollut 244:190–201
141. MacFarlane W, Danielson A, Holland HD, Jacobsen SB (1994) REE chemistry and Sm-Nd systematics of late Archean weathering profiles in the Fortescue Group, Western Australia. Geochim Cosmochim Acta 58(7):1777–1794
142. Mao L, Mo DW, Yang JH, Shi CX (2009) Geochemistry of trace and rare earth elements in red soils from the Dongting Lake area and its environmental significance. Pedosphere 19:615–622. https://doi.org/10.1016/S1002-0160(09)60156-9

143. Martínez RE, Pourret O, Faucon MP, Dian C (2018) Effect of rare earth elements on rice plant growth. Chem Geol 489:28–37
144. Martino C, Chianese T, Chiarelli R, Roccheri MC, Scudiero R (2022) Toxicological impact of rare earth elements (REEs) on the reproduction and development of aquatic organisms using sea urchins as biological models. Int J Mol Sci 23:2876. https://doi.org/10.3390/ijms23052876
145. McLennan SM (1989) Rare earth elements in sedimentary rocks; influence of provenance and sedimentary processes. Rev Mineral Geochem 21:169–200
146. Meryem B, Ji H, Gao Y, Ding H, Li C (2016) Distribution of rare earth elements in agricultural soil and human body (scalp hair and urine) near smelting and mining areas of Hezhang, China. J Rare Earths 34:1156–1167
147. Miao L, Ma Y, Xu R, Yan W (2011) Environmental biogeochemical characteristics of rare earth elements in soil and soil-grown plants of the Hetai goldfield, Guangdong Province, China. Environ Earth Sci 63:501–511
148. Migaszewski ZM, Gałuszka A (2015) The characteristics, occurrence, and geochemical behavior of rare earth elements in the environment: a review. Crit Rev Environ Sci Technol 45:429–471
149. Migaszewski ZM, Galuszka A, Dolegowska S (2016) Rare earth and trace element signatures for assessing an impact of rock mining and processing on the environment: Wisniowka case study, south-central Poland. Environ Sci Pollut Res Int 23:24943–24959
150. Migaszewski ZM, Gałuszka A, Dołęgowska S (2019) Extreme enrichment of arsenic and rare earth elements in acid mine drainage: case study of Wiśniówka mining area (south-central Poland). Environ Pollut 244:898–906
151. Mihajlovic J, Bauriegel A, Stärk HJ, Roßkopf N, Zeitz J, Milbert G, Rinklebe J (2019) Rare earth elements in soil profiles of various ecosystems across Germany. Appl Geochem 102:197–217
152. Minarik L, Zigova A, Bendl J, Sktivan P, Stastny M (1998) The behavior of rare-earth elements and Y during the rock weathering and soil formation in the Ricany granite massif, Central Bohemia. Sci Total Environ 215:101–111
153. Moldoveanu G, Papangelakis VG (2012) Recovery of rare earth elements adsorbed on clay minerals: I. Desorption mechanism. Hydrometallurgy 117–118:71–87
154. Naccarato A, Tassone A, Cavaliere F, Elliani R, Pirrone N, Sprovieri F, Tagarelli A, Giglio A (2020) Agrochemical treatments as a source of heavy metals and rare earth elements in agricultural soils and bioaccumulation in ground beetles. Sci Total Environ 749, art. no. 141438
155. Nakayama K, Nakamura T (2005) X-ray fluorescence analysis of rare earth elements in rocks using low dilution glass beads. Anal Sci 21:815–822
156. Navarro M, Ulbrich HHGJ, Andrade S, Janasi VA (2002) Adaptation of ICP–OES routine determination techniques for the analysis of rare earth elements by chromatographic separation in geological materials: tests with reference materials and granitic rocks. J Alloy Compd 344:40–45. https://doi.org/10.1016/S0925-8388(02)00302-X
157. Nesbitt HW (1979) Mobility and fractionation of rare earth elements during weathering of a granodiorite. Nature 279:206. https://doi.org/10.1038/279206a0
158. Nesbitt W, Markovics G (1997) Weathering of granodioritic crust, long-term storage of elements in weathering profiles, and petrogenesis of siliciclastic sediments. Geochim Cosmochim Acta 61(8):1653–1670
159. Ni Y, Hughes JM, Mariano A (1995) Crystal chemistry of the monazite and xenotime structures. Am Miner 80:21–26
160. Nikonov VV, Lukina NV, Frontas'eva MV (1999) Trace elements in Al-Fe-humus podzolic soils subjected to aerial pollution from the copper-nickel production industry in conditions of varying lithogenic background. Eurasian Soil Sci 32(3):338–349
161. Niu JX, Chen QR, Xu NX, Wei ZL (2008) Effect of combinative addition of strontium and rare earth elements on corrosion resistance of AZ91D magnesium alloy. Trans Nonferrous Metals Soc China 18:1058–1064

162. Noack CW, Dzombak DA, Karamalidis AK (2014) Rare earth element distribution and trends in natural waters with a focus on groundwater. Environ Sci Technol 48:4317–4326
163. Öhlander B, Land M, Ingri J, Widerlund A (1996) Mobility of rare earth elements during weathering of till in northern Sweden. Appl Geochem 11:93–99. https://doi.org/10.1016/0883-2927(95)00044-5
164. Öhlander B, Ingri J, Land M, Schoberg H (2000) Change of Sm-Nd isotope composition during weathering of till. Geochim Cosmochim Acta 64(5):813–820
165. Olmez I, Sholkovitz ER, Hermann D, Eganhouse RP (1991) Rare earth elements in sediments off southern California: a new anthropogenic indicator. Environ Sci Technol 25(2):310–316
166. Ozaki T, Enomoto S (2001) Uptake of rare earth elements by *Dryopteris erythrosora* (autumn fern). RIKEN Rev 35: focused on new trends in bio-trace elements research, pp 84–87
167. Pagano G (ed) (2015) Rare earth elements in human and environmental health. At the crossroads between toxicity and safety. Jenny Stanford Publishing, 280 pp
168. Pagano G, Guida M, Tommasi F, Oral R (2015) Health effects and toxicity mechanisms of rare earth elements—knowledge gaps and research prospects. Ecotoxicol Environ Saf 115:40–48
169. Pagano G, Guida M, Siciliano A, Oral R, Koçbas F, Palumbo A, Castellano I, Migliaccio O, Thomas PJ (2016) Comparative toxicities of selected rare earth elements: sea urchin embryogenesis and fertilization damage with redox and cytogenetic effects. Environ Res 147:453–560
170. Pagano G, Thomas PJ, Di Nunzio A, Trifuoggi M (2019) Human exposures to rare earth elements: present knowledge and research prospects. Environ Res 171:493–500
171. Palacios CM, Hein UF, Dulski P (1986) Behaviour of rare earth elements during hydrothermal alteration at the Buena Esperanza copper-silver deposit, northern Chile. Earth Planet Sci Lett 80:208–216
172. Pang X, Li D, Peng A (2001) Application of rare–earth elements in the agriculture of China and its environmental behavior in soil. J Soils Sediments 1:124
173. Pang X, Li D, Peng A (2002) Application of rareearth elements in the agriculture of China and its environmental behavior in soil. Environ Sci Pollut Res 9(2):143–148. https://doi.org/10.1007/BF02987462
174. Paye HS, Mello JWV, Mascarenhas GRLM, Gasparon M (2016) Distribution and fractionation of the rare earth elements in Brazilian soils. J Geoch Exp 161:27–41
175. Peng A, Zhu J (2003) Environmental chemistry and ecological effects of rare earth elements. China Environmental Press, Beijing, China, pp 196–198
176. Pepi S, Sansone L, Chicca M, Marrocchino E, Vaccaro C (2016) Distribution of rare earth elements in soil and grape berries of *Vitis vinifera* cv. "Glera" Environ Monit Assess 188(8):477
177. Pérez-López R, Delgado J, Nieto JM, Márquez-García B (2015) Rare earth element geochemistry of sulphide weathering in the São Domingos mine area (Iberian Pyrite Belt): a proxy for fluid–rock interaction and ancient mining pollution. Environ Pollut 199:139–147
178. Pitron G (2018) La guerra de los metales raros La cara oculta de la transición energética y digital. Ediciones Península Atalaya. https://static0planetadelibroscom.cdnstatics.com/libros_contenido_extra/42/41305_La_guerra_de_los_metales.pdf
179. Price DG (1995) Weathering and weathering processes. Q J Eng Geol 28:243–252
180. Price RC, Gray CM, Wilson RE, Frey FA, Taylor SR (1991) The effects of weathering on rare-earth element, Y and Ba abundances in Tertiary basalts from southeastern Australia. Chem Geol 93(3–4):245–265
181. Prudencio MI, Braga MAS, Gouveia MA (1993) REE mobilization, fractionation and precipitation during weathering of basalts. Chem Geol 107(3–4):251–254
182. Railsback LB (2003) An earth scientist's periodic table of the elements and their ions. Geology 31(9):737–740
183. Rambeck WA, Wehr U (2005) Use of rare earth elements as feed additives in pig production. Pig News Info 26:41N-47N
184. Ramos SJ, Dinali GS, Oliveira C, Martins GC, Moreira CG, Siqueira JO, Guilherme LRG (2016) Rare earth elements in the soil environment. Curr Pollut Rep 2:28–50

185. Ran Y, Liu Z (1999) Contents and distribution of rare earth elements in main types of soil in China. J Rare Earths 17:213–217
186. Redling K (2006) Rare earth elements in agriculture with emphasis on animal husbandry. PhD thesis, University of Munich, Germany, p 360
187. Rim KT (2016) Effects of rare earth elements on the environment and human health: a literature review. Toxicol Environ Health Sci 8:189–200
188. Rim KT, Koo KH, Park JS (2013) Toxicological evaluations of rare earths and their health impacts to workers: a literature review. Saf Health Work 4(1):12–26
189. Romero-Freire A, González V, Groenenberg JE, Qiu H, Auffan M, Cotelle S, Giamberini L (2021) Cytotoxicity and genotoxicity of lanthanides for Vicia faba L. are mediated by their chemical speciation in different exposure media. Sci Total Environ 790, art. no. 148223
190. Rudnick RL, Fountain DM (1995) Nature and composition of the continental crust: a lower crustal perspective. Rev Geophys 33(3):267–309
191. Saatz J, Vetterlein D, Mattusch J, Otto M, Daus B (2015) The influence of gadolinium and yttrium on biomass production and nutrient balance of maize plants. Environ Pollut 204:32–38
192. Sabiha-Javied S, Waheed N, Siddique R, Shakoor M, Tufail M (2010) Measurement of rare earths elements in Kakul phosphorite deposits of Pakistan using instrumental neutron activation analysis. Radioanal Nucl Chem 284:397–493
193. Sabot J, Maestro P (1995) Kirk-Othmer encyclopedia of chemical technology, concise, IV edn. Wiley. ISBN: 0-471-41961-3
194. Sadeghi M, Petrosino P, Ladenberger A, Albanese S, Andersson M, Morris G, Lima A, De Vivo B (2013) Ce, La and Y concentrations in agricultural and grazing-land soils of Europe. J Geochem Explor 133:202–213. https://doi.org/10.1016/j.gexplo.2012.12.007
195. Sauvé S, Desroirers M (2014) A review of what is an emerging contaminant. Chem Cent J 8(1):1
196. Schüler D, Buchert M, Liu R, Dittrich S, Merz C (2011) Study on rare earths & their recycling: final report for the Greens/EFA Group in the European Parliament, Darmstadt
197. Semhi K, Chaudhuri S, Clauer N (2009) Fractionation of rare-earth elements in plants during experimental growth in varied clay substrates. Appl Geochem 24:447–453
198. Shan XQ, Lian J, Wen B (2002) Effect of organic acids on adsorption and desorption of rare earth elements. Chemosphere 47:701–710
199. Siebert C, Rosenthal E, Möller P, Rödiger T, Meiler M (2012) The hydrochemical identification of groundwater flowing to the Bet She'an-Harod multiaquifer system (Lower Jordan Valley) by rare earth elements, yttrium, stable isotopes (H, O) and Tritium. Appl Geochem 27:703–714
200. Silva FBV, Nascimento CWA, Alvarez AM, Araújo PRM (2019) Inputs of rare earth elements in Brazilian agricultural soils via P-containing fertilizers and soil correctives. J Environ Manag 232:90–96
201. Silvestri L, Forcina A, Silvestri C, Traverso M (2021) Circularity potential of rare earths for sustainable mobility: recent developments, challenges and future prospects. J Clean Prod 292:126089. https://doi.org/10.1016/j.jclepro.2021.126089
202. Slooff W, Bont PFH, van den Hoop MAGT, Janus JA, Annema JA (1993) Exploratory report: rare earth metals and their compounds. National Institute of Public Health and Environmental Protection, Bilthoven, Report No. 710401025
203. Sneller FEC, Kalf DF, Weltje L, Van Wezel AP (2000) Maximum permissible concentrations and negligible concentrations for rare earth elements (REEs). RIVM Report 601501 011. National Institute of Public Health and the Environment, the Netherlands. http://rivm.openrepository.com/rivm/bitstream/10029/9551/1/601501011.pdf
204. Sohrabi N, Kalantari N, Amiri V, Nakhaei M (2017) Assessing the chemical behavior and spatial distribution of yttrium and rare earth elements (YREEs) in a coastal aquifer adjacent to the Urmia Hypersaline Lake, NW Iran. Environ Sci Pollut Res Int 24(25):20502–20520. https://doi.org/10.1007/s11356-017-9644-7
205. Song W, Hong F, Wan Z, Zhou Y (2002) Effects of lanthanum and europium on rooting of plantlet *Etiobotrya japonica* Lindl. in vitro. J Rare Earths 20:658–662

206. Song Z, Liu C, Han G, Wang Z, Zhu Z, Yang C (2006) Enrichment and release of rare earth elements during weathering of sedimentary rocks in Wujiang Catchments, Southwest China. J Rare Earths 24:491–496
207. Straif K, Benbrahim-Tallaa L, Baan R, Grosse Y, Secretan BA, El Ghissassi F, Bouvard VR, Guha N, Freeman C, Galichet L (2009) A review of human carcinogens—Part D: Radiation. Lancet Oncol 10(8):751–752
208. Su N, Yang S, Guo Y, Yue W, Wang X, Yin P, Huang X (2017) Revisit of rare earth element fractionation during chemical weathering and river sediment transport. Geochem Geophys Geosyst 18:935–955
209. Taylor SR (1964) Abundance of chemical elements in the continental crust: a new table. Geochim Cosmochim Acta 28(8):1273–1285
210. Taylor SR, McLennan SM (1995) The geochemical evolution of the continental crust. Rev Geophys 33:241–265
211. Taylor SR, McLennan SM, Armstrong RL, Tarney J (1981) The composition and evolution of the continental crust: rare earth element. Evidence from sedimentary rocks [and discussion]. Philos Trans R Soc Lond Ser A Math Phys Sci 301:381–399
212. Tesmer M, Möller P, Wieland S, Jahnke C, Voigt H, Pekdeger A (2007) Deep reaching fluid flow in the North East German Basin: origin and processes of groundwater salinisation. Hydrogeol J 15:1291–1306
213. Thomas PJ, Carpenter D, Boutin C, Allison JE (2014) Rare earth elements (REEs): effects on germination and growth of selected crop and native plant species. Chemosphere 96:57–66
214. Todorovsky DS, Minkova NL, Bakalova DP (1997) Effect of the application of superphosphate on rare earths' content in the soil. Sci Total Environ 203:13–16
215. Tommasi F, D'Aquino L (2016) Rare earth elements and plants. Rare earth elements in human and environmental health: at the crossroads between toxicity and safety. 107–125
216. Turra C, Fernandes EAN, Bacchi MA (2011) Evaluation on rare earth elements of Brazilian agricultural supplies. J Environ Chem Ecotoxicol 3(4):86–92
217. Tyler G (2004a) Rare earth elements in soil and plant systems—a review. Plant Soil 267:191–206. https://doi.org/10.1007/s11104-005-4888-2
218. Tyler G (2004b) Vertical distribution of major, minor, and rare elements in a Haplic Podzol. Geoderma 119(3–4):277–290
219. Tyler G, Olsson T (2002) Conditions related to solubility of rare and minor elements in forest soils. J Plant Nutr Soil Sci 165:594–601
220. Tyler G, Olsson T (2005) Rare earth elements in forest-floor herbs as related to soil conditions and mineral nutrition. Biol Trace Elem Res 106:177–191
221. United States Department of Energy (2010) Critical materials strategy. http://energy.gov/node/206101
222. Vakh EA, Vakh AS, Petukhov VI, Pavlova GY, Tarasenko IA, Zubtsova AS (2017) Study on rare-earth elements distribution in surface waters of Primorsky región. IOP Conf Ser: Earth Environ Sci 87, Article 072005
223. Val'kov A, Stepanov S, Sergievskii V, Chekmarev A (2010) Monazite raw material for the production of highly effective fertilizers. Theor Found Chem Eng 44:497–499
224. Vázquez-Ortega A, Perdrial J, Harpold A, Zapata-Ríos X, Rasmussen C, McIntosh J, Schaap M, Pelletier JD, Brooks PD, Amistadi MK, Chorover J et al (2015) Rare earth elements as reactive tracers of biogeochemical weathering in forested rhyolitic terrain. Chem Geol 391:19–32
225. Verplanck PL, Taylor HE, Nordstrom DK, Barber LB (2005) Aqueous stability of gadolinium in surface waters receiving sewage treatment plant effluent, Boulder Creek, Colorado. Environ Sci Technol 39:6923–6929
226. Verplanck PL, Furlong ET, Gray JL, Phillips PJ, Wolf RE, Esposito K (2010) Evaluating the behavior of gadolinium and other rare earth elements through large metropolitan sewage treatment plants. Environ Sci Technol 44:3876
227. Vinogradov AP (1962) Average contents of chemical elements in the main types of igneous rocks of the Earth's crust. Geokhimiya 7:555–571

228. Volokh AA, Gorbunov AV, Gundorina SF, Revich BA, Frontasyeva MV, Pal CS (1990) Phosphorous fertilizer production as a source of rare-earth elements pollution of the environment. Sci Total Environ 95:141–148
229. Vorob'ev VN, Kotov SF, Nikolenko VV, Tishin DV, Mikhailov AL (2020) Influence of light and heavy lanthanides on the physiological processes of Taraxacum hybernum. J Exp Biol Agric Sci 8(2):S298–S302
230. Wahid PA, Kamalam NV, Krishna Prabhu R, Sekhar JK, Vijayalakshmi S, Mahalingam TR, Ajith Kumar CE (2003) Rare earth element fluxes in diverse soils and their absorption by coconut Palm. J Plant Nutr 26:1427–1438
231. Walters A, Lusty P (2011) Rare earth elements—commodity profile. Br Geol Surv 2011:1–54
232. Wang Z, Zhang X, Mu Y (2008) Effects of rare-earth fertilizers on the emission of nitrous oxide from agricultural soils in China. Atmos Environ 42:3882–3887
233. Wang L, Liang T, Chong Z, Zhang C (2011) Effects of soil type on leaching and runoff transport of rare earth elements and phosphorous in laboratory experiments. Environ Sci Pollut Res 18:38–45
234. Webb GE, Nothdurft LD, Kamber BS, Kloprogge JT, Zhao JX (2009) Rare earth element geochemistry of scleractinian coral skeleton during meteoric diagenesis: a sequence through neomorphism of aragonite to calcite. Sedimentology 56:1433–1463
235. Weber R (2008) An experimental study of fractionation of the rare earth elements in poplar plants (Populous eugenei) grown in a calcium-bearing smectite soil. Master's thesis, Kansas State University, Manhattan, 50 pp
236. Weber RJ, Reisman DJ (2012) Rare earth elements: a review of production, processing, recycling, and associated environmental issues. US EPA Region 8:189–200
237. Wedepohl KH (1995) The composition of the continental crust. Geochimica et Cosmochimica Acta 59:1217–1232
238. Weltje L, Heidenreich H, Zhu W, Wolterbeek HT, Korhammer S, De Goeij JJ, Markert B (2002) Lanthanide concentrations in freshwater plants and molluscs, related to those in surface water, pore water and sediment. A case study in The Netherlands. Sci Total Environ 286:191–214
239. Wen B, Yuan DA, Shan XQ, Li FL, Zhang SZ (2001) The influence of rare earth element fertilizer application on the distribution and bioaccumulation of rare earth elements in plants under field conditions. Chem Speciat Bioavailab 13(2):39–48. https://doi.org/10.3184/095422901783726825
240. Weng ZH, Jowitt SM, Mudd GM, Haque N (2013) Assessing rare earth element mineral deposit types and links to environmental impacts. Appl Earth Sci 122(2):83–96
241. Wiche O, Heilmeier H (2016) Germanium (Ge) and rare earth element (REE) accumulation in selected energy crops cultivated on two different soils. Miner Eng 92:208–215
242. Williams-Jones AE, Migdisov AA, Samson IM (2012) Hydrothermal mobilization of the rare earth elements—a tale of "ceria" and "yttria." Elements 8:355–360
243. Wu WQ, Xu T, Hao Q, Wang Q, Zhang SJ, Zhao CY (2010) Applications of X-ray fluorescence analysis of rare earths in China. J Rare Earths 28(1):30e36
244. Wulaningsih T, Humaida H, Harijoko A, Watanabe K (2013) Major element and rare earth elements investigation of Merapi Volcano, Central Java, Indonesia. Procedia Earth Planet Sci 6:202–211
245. Wyttenbach A, Furrer V, Schleppi P, Tobler L (1998) Rare earth elements in soil and in soil-grown plants. Plant Soil 199:267–273
246. Xiong SL, Xiong ZT, Chen YC, Huang H (2006) Interactive effects of lanthanum and cadmium on plant growth and mineral element uptake in crisped-leaf mustard under hydroponic conditions. J Plant Nutr 29:1889–1902
247. Yang Q, Wang L, He J, Yang Z, Huang X (2018) Direct imaging of how lanthanides break the normal evolution of plants. J Inorg Biochem 182:158–169
248. Yin X, Martineau C, Demers I, Basiliko N, Fenton N (2021) The potential environmental risks associated with the development of rare earth element production in Canada. Environ Rev. https://doi.org/10.1139/er-2020-0115

249. Yoshida S, Muramatsu Y, Tagami K, Tagami S, Uchida S (1998) Concentrations of lanthanide elements, Th and U in 77 Japanese surface soils. Environ Int 24:275–286
250. Yusoff ZM, Ngwenya BT, Parsons I (2013) Mobility and fractionation of REEs during deep weathering of geochemically contrasting granites in a tropical setting, Malaysia. Chem Geol 349–350:71–86. ISSN 0009-2541. https://doi.org/10.1016/j.chemgeo.2013.04.016
251. Zepf V (2013) Rare earth elements: what and where they are. In: Rare earth elements. Springer, Berlin, Heidelberg, pp 11e39. Springer theses
252. Zhang S, Shan XQ (2000) Speciation of rare earth elements in soil and accumulation by wheat with rare earth fertilizer application. Environ Pollut 112(3):395–405
253. Zhang FS, Yamasaki S, Kimura K (2001) Rare earth element content in various waste ashes and the potential risk to Japanese soils. Environ Int 27:393–398
254. Zhang QH, Tong LG, Cheng L, Zhu J, Wang Q (2012) Characteristics of REE distribution in the surface soil of the farmland in wastewater irrigation area. Acta Agric Univ Jangxiensis 34:614–0618
255. Zhang Q, Han G, Liu M, Wang L (2019) Geochemical characteristics of rare earth elements in soils from Puding Karst Critical Zone Observatory, Southwest China. Sustainability 11:4963
256. Zhanheng C (2010) Outline on the development and policies of China Rare Earth Industry. U.S. Geological Survey, Reston, VA
257. Zhou W, Han G, Liu M, Song C, Li X (2020) Geochemical distribution characteristics of rare earth elements in different soil profiles in Mun River Basin, Northeast Thailand. Sustainability 12:457
258. Zhu Z, Wang Z, Li J, Li Y, Zhang Z, Zhang P (2012) Distribution of rare earth elements in sewage-irrigated soil profiles in Tianjin, China. J Rare Earths 30:609–613. https://doi.org/10.1016/S1002-0721(12)60099-4
259. Zhuang M, Zhao J, Li S, Liu D, Wang K, Xiao P et al (2016) Concentrations and health risk assessment of rare earth elements in vegetables from mining area in Shandong, China. Chemosphere 168:578–582
260. Zhuang M, Wang L, Wu G, Wang K, Jiang X, Liu T, Xiao P, Yu L, Jiang Y, Song J, Zhang J, Zhou J, Zhao J, Chu Z (2017) Health risk assessment of rare earth elements in cereals from mining area in Shandong, China. Sci Rep 7:1–6

Impact of Agricultural Wastes on Environment and Possible Management Strategies

Muhammad Numan Khan, Tanveer Ali Sial, Amjad Ali, and Fazli Wahid

Abstract The world population is growing rapidly and needs more resources exploration. The global population is expected to reach 9.7 billion by the end of 2050, which exerts a huge pressure on the agriculture sector. The agricultural activities increased to meet the need of the rapidly growing population in the past decades. Agricultural activities provide food, feed, and shelter to human. At the same time, the agricultural activities also contribute to global waste generation, environmental threats, and health hazards. At least 1/3 (1.3 billion metric tons/year) of the global food produced is lost as waste each year. Fruits and vegetables cultivation (peels, seeds, sugarcane bagasse, and woody trenches) and processing, dairy and poultry farming (urine, feaces, waste milk, bedding materials, birth tissues, bones, and blood) are the main sources of agricultural waste generation. Agricultural wastes pose threats to environment (CH_4 and CO_2), release bad smell, and influence the air and water quality, and soil health. These concerns can be addressed by proper management measures. The management practices include 3R technology (reuse, recycling, and reduction), energy recovery from wastes, rational management, bioactive compound recovery, biological application (animal feed and biofuel i.e., bioethanol, biodiesel, and biogas), and conventional food waste management (landfilling, anaerobic digestion, and composting/vermicomposting). Economic incentives can attract industrial

M. N. Khan (✉)
CAS Key Laboratory of Tropical Forest Ecology, Xishuangbanna Tropical Botanical Garden, Chinese Academy of Sciences, Menglun, Yunnan 666303, China
e-mail: muhammadnkhan@xtbg.ac.cn

Center of Plant Ecology, Core Botanical Garden, Chinese Academy of Sciences, Menglun, Yunnan 666303, China

T. A. Sial
College of Natural Resources and Environment, Northwest A&F University, Yangling 712100, Shaanxi, China

A. Ali · F. Wahid
School of Environmental and Municipal Engineering, Xi'an University of Architecture and Technology, Xi'an 710055, China

A. Ali
Department of Agriculture, University of Swabi, Swabi 23561, China

A. Núñez-Delgado (ed.), *Frontier Studies in Soil Science*, Frontier Studies in Soil Science, https://doi.org/10.1007/978-3-031-50503-4_4

sector to waste disposal and generate extra revenue. Similarly, waste can be converted to economical and precious compounds and can provide raw materials for other agro-based industries and generate high revenue and jobs market.

Keywords Crop wastes · Environmental impact · Management alternatives · Recycling

1 Introduction

Agricultural wastes are the unwanted materials either in liquid or solid forms produced from the manufacturing, processing and marketing of the agricultural products (such as crops and animals). Agriculture produces both organic and inorganic waste through a variety of farming practices, including dairy, horticulture, harvesting, and processing. According to the Agriculture Act of 1947, agriculture includes fruits and seeds cultivation, dairy farming, cattle breeding, use of grazing land, market gardens, nursery grounds, and woodland management. The agricultural activities increased to meet the need of the rapidly growing population in the past decades and it is estimated that the world population will reach 9.7 billion by the end of 2050 [107]. This rapid increase in the world population put substantial pressure on the agricultural productivity to meet the demands of future population, so the agricultural waste production will also increase. The wastes generated from the agriculture industry need proper disposal, while its inappropriate disposal causes environmental (soil and water) pollution.

Horticulture crops, including fruits and vegetables have a vital role in our daily life, hence the demand of such food stuffs is very important. According to Food and Agriculture Organization (FAO, 2017), the world produces citrus, bananas, apples, and pineapple of 124.7, 114.08, 84.63, and 25.43 million metric tons, respectively. Globally the production of some vegetables including potato, tomatoes, cabbages, brassica, carrot and turnips, cauliflower and broccoli, and peas are 3820, 171, 71.77, 38.83, 24.17, 17.42 million metric tons, respectively, (FAO, 2017). Higher production and growth of fruits and vegetables, poor handling methods and transportation make huge losses and waste of these important food stuffs being created. At least one third of the food produced globally is lost as a waste each year (1.3 billion metric tons/year). FAO; Sagar et al. [35, 88] reported that waste production in the horticulture commodities is high as compared to all other food types, about 60% of the fruits and vegetables are lost as a waste. The disposal of such waste is a critical problem all over the world because the fruit and vegetable wastes are rich sources of organic material, however, previously it has been reported that only 20% of the waste is treated before discharge [97].

Cereal and sugar producing crop also play an important role in the human life. According to Erenstein et al. [33], the worldwide production of wheat, maize, and rice was 757, 1137, and 757 million ton/year, respectively, while the annual production of sugarcane was 1.6 billion ton/year [1]. The higher production of these crops also produces waste in the form of crop residues (stem, leaves and roots). It is estimated that wheat produced 529-million-ton straw/year, of which Asia produced 43% wheat alone and an average wheat straw of 1.3–1.4 kg of wheat grain. Globally, the annual rice and maize straw production is 370–520 and 1100 million ton/year, respectively, sugarcane produces 279 million metric tons of bagasse each year [1]. Mostly the farmer burns these left-over crops residues through traditional methods in the fields. It reduces the waste volume and emits various toxic gases (carbon monoxide, carbon dioxide, particulate matter, and poly cyclic aromatic hydrocarbon) into the global environment. The burning of straw and residues also produce some dioxin during the combustion and releases chlorine content and pesticides absorbed by the crop. The Commission for Environment Co-operation [13] noticed that the dioxin was 17 time higher in Taiwan during the weeks in which intensive agriculture residues burning took place as compared to the weeks in which the burning was not done. Similarly, they compared that larger amount of dioxin (10–20%) emitted in those provinces of China which have more agricultural productions corresponds to those provinces which have less agricultural productions. This report revealed that dioxin emit during the combustion process due to the presence of chlorinated pesticides which are now a days used as agro-chemicals for the control of weeds and pests.

Globally, the growing of livestock is increasing to feed the rapid population of the world. The increase in the livestock also significantly increase the solid waste/manure of animals and poultry farms. Robbin [85] reported that a single hog produces three times more wastes and a cow produces 20 times more than human. The animal waste is a mixture of organic waste consisting of feces, urine, bedding materials, and food wastes and stored in open pit. When there is rain fall it adds to the volume of the waste (feces + urine mixture) and there may be runoff of slurry into the land and water, creating pollution in both resources. The animal waste is a rich source of nutrients like nitrate and phosphate, leach from the pit into the below ground water which causes water eutrophication. Singh et al. [100] observed that poultry is one the fast-growing industry, as crop production and dairy farming may not solve the food requirement of the humans. As the poultry farming increase, waste production also increased and about 90% of the poultry wastes spread near the poultry farming areas. The poultry industry produces huge amount of the waste including solid waste like feces, hatchery waste, saw dust, and disinfections of chicken farms. The poultry manure requires proper disposal and avoidance of these manure will create environmental pollution. The poultry manure contains high phosphorus, when it combines with mineral fertilizer have positive effect on the productivity of the crops [100]. Therefore, proper management and disposal methods for wastes needs into account to reduce the risk of water and soil pollution.

1.1 Waste

Waste is unwanted material that cannot be used by the producer and decided or required to dispose. Even this waste/substance can be reused or recycled, but legally it is considered to be waste for person/farmer who manufactured/produced it. There are different types of wastes produced globally, such as solid waste, municipal waste, hazardous waste, and agricultural waste. In this chapter we only focus on the waste that is produced during the agricultural activities either at the field or outside the field.

1.2 Agriculture Waste

Agricultural waste is produced from agricultural practices and cannot be used by farmer/producer either it is organic/inorganic and needs to discard. It can also be defined as the unwanted and unusable materials that is produced as a result of agricultural activities related to the growing of crops/farming of livestock/poultry [81, 106]. The agriculture waste mainly includes the crops residues (wheat, rice and maize straw), vegetables, and fruit wastes (the peels and seeds), sugarcane bagasse, woody trenches, and livestock residues (urine, feces, waste milk, bedding materials, birth tissues, bones, and blood from slaughter houses). The runoff of fertilizer from the field is also considered waste water and causes pollution in term of water eutrophication [81].

1.3 Main Sources and Causes of Agriculture Wastes

Farming operations are the principal source of agricultural wastes. Moreover, the agriculture waste is not restricted to the production but also include other activities on the farms and in food chain. At every stage, significant number of agricultural wastes can be produced along the agro-food cycle. Following are the wide-ranging sources of agricultural wastes:

1.3.1 Sources

Waste from Crop Production

Crop production waste refer to the waste that are produced from the agricultural activities related to crop production. Few examples of the crop production waste are crop residues (wheat, rice, and maize straw, and sugarcane bagasse).

Waste from Horticulture Crop Production

Horticulture crop waste refers to the waste generated from the maintenance and cultivation of horticulture plants and landscaping for the beautification. Examples of horticulture crops are pruning of trees, cutting of grasses, and vegetables residues (stem, leaves, and rotten vegetables).

Waste from on Farm Medications

This type of waste refers to the waste that is produced from the medicines (drug and insecticides/vaccines) used for the control of insect, pest, and for the treatment of animals (syringes, containers, and vaccines wrappers).

Waste from Animal Productions

The waste generated from the dairy framing including the feed, bedding materials, rotten milk, feaces, urine, animal carcasses, bones, and blood from the slaughterhouses.

Waste from Food and Meat Processing

The waste produced as a result of food and meat processing for the use of human being (kitchen waste, fruit peels, feathers, hoofs, rotten vegetables, and breads).

1.3.2 Causes

The main causes of the agriculture waste production are the agricultural activities at the farm during the preparation, consumption, and processing of agricultural products.

Farming Activities

These are the main source of agriculture waste production. The agricultural waste produced in the farm at every phase, such as from the clearing of land till harvesting of crops generate wastes. From preparing the feed for the live stocks till the slaughtering of the animals generated waste. Likewise, in agriculture field farmer use different insecticide and pesticides for the control of insect and pest, also generate wastes.

Poor Road Network

The transportation of agricultural products from the farm to the market or to storage halls produce huge amount of waste. As most of the farming fields are located in the villages, which results in road accidents or delay in the transportation of agricultural

Fig. 1 Agriculture product transportation **a** spoilage of food waste **b**, and kitchen and household waste **c** and **d**

commodities from the farm to the markets. Due to road accidents, most of perishable agricultural products easily get rotten. The rotten products are either thrown away on the roads or discarded once the farmer reach to the market. Figure 1a shows the transportation of agriculture product in developing countries.

Lack of Electricity

Electricity is also one of the big problems in term of storage of agricultural products. The electric power supply in the rural areas of some developing countries is not well established. If the electricity supply on the farm is not good, then farmer will have lack of cold storage facilities. Stable electric power supply enhances the cold storage and also reduces the spoilage of agricultural commodities and wastes generation.

Lack of Drying and Storage Facilities

With the right drying and storage methods, rotting of many agricultural products could be avoided. Food spoilage and agricultural waste may have been significantly reduced if farmers had access to suitable drying techniques or moisture monitoring. This would improve food security and mitigate the negative effects of agricultural wastes on human health and the environment. Many farmers depend greatly on the unreliable solar system to dry their agricultural product before it is stored. They also use the outdated and ineffective approached of conventional moisture monitoring. Aflatoxin infestation has reportedly occurred as a result of insufficient moisture content monitoring of grain prior to storage. Aflatoxin contamination of food and livestock feed can result in large amount of annual crop losses worldwide. It is both a cause and a product of food spoilage [55, 73].

Food Spoilage

The food spoilage mainly occurs due to the improper storage of food items due to the bacteria and fungi (Fig. 1b). The bacteria and fungi breakdown the food, wastes, and other product which produce foul smell and unpleasant taste. In USA alone about 40% of food is wasted annually. Li et al. [61] reported that 20, 15, and 5% of cereals, vegetable, and dairy foods are wasted annually in China, respectively.

Improper Management of Kitchen Wastes

Agricultural activities are typically carried out with the intention of producing food for domestic use. The generation of agricultural wastes is typically a byproduct of family consumption of agricultural products (Fig. 1c). Banana and orange peels, for instance, are frequently disposed as agricultural solid wastes in the houses. Moreover, the kitchen waste is also produced from the food spoilage and by adding the restaurant kitchens waste, then it becomes significantly high. Similarly, Li et al. [62] noticed that all the kitchen waste in Chinese cities are 88–94% of the total food waste generated in China annually.

2 Types of Agricultural Wastes

2.1 Fruits and Vegetables Wastes

The fruits and vegetable (FV) play a vital role in our diet. Demand of fruits and vegetables increased in the past decades due to growing human population [88]. The increase in the production and growth of FV also generated huge amount of waste because of the inappropriate handling method and poor infrastructure. The horticulture commodities (60%) in the world produce large amount of waste as compared to all other foods [88]. It has been documented by different scientist that most of waste produced in the supply chain of FV including transport, packaging, classification, grading and at home before preparation. Parfitt et al. and Plazzotta et al. [77, 79] reported that 89 million tons of the food is wasted in the European Union (EU) annually and expected that it increases by 40% in the coming 4 years. In the EU, Italy alone discarded 87% of the fruit, vegetables and cereal before reaching to the consumer [30]. Documented that the world-wide production of fruit and vegetable was 1728 million tons in 2011. Brazil is the third largest producer worldwide while in vegetable production third largest country in America. The author also reported that in Brazil 97% of the household waste composed of fruits and vegetables. Meanwhile, Sial et al. [98] observed that globally 3.5 million ton of banana peel waste produced per year in the food industries, among this waste production, China alone produced 1.3 million tons of banana peel waste, being the 2nd largest banana producing country in the world. The citrus also generate 50 to 60% waste after processing in the juice industries in the form of peel, seed, and membrane residues [113]. In fruits and vegetables approximately 25–30% of the waste generated during its processing. Apple produced 9–11% seed and pulp and 90.01% as a final product. Papaya generates 7.5–8.6% peel waste, 5.4–7.05% seeds, 31–33% unusable pulp, and 51–52% final product, while mangoes generate 9–12% peels, 11–14% seeds, and 57–59% of final product [88]. Moreover, the industries involved in canning, packaging, and freezing produced 6 million metric ton of solid waste annually. The disposal of such wastes creates both environmental and soil pollution due to its high moisture, organic compounds, and biodegradability. The FV consist of valuable nutrients, if it is not handled properly, the nutrient present in the waste is lost either through runoff or leaching. In some countries of the world the FV wastes are either burned or used as a landfilling which is also not good for the environment and release greenhouse gases (Fig. 2a) into the environment (CO_2 and CH_4). Therefore, proper management and recycling policy need to be made to reduce the loss of nutrients present in the waste and reduce the chance of greenhouse emissions.

2.2 *Cereal and Other Crops Waste*

Cereals are the most widely grown crop in the world. The three-cereals (maize, wheat, and rice) equal to 90% of the total world cereal production. Maize being the most growing cereal crop in the world followed by the wheat, rice, and barley. In cereal production, China being the largest producer of the cereals, followed by USA, India, Russia, and Brazil. China cumulatively produces more than 6×10^8 ton of straw annually, with 1.1×10^8 ton rice straw, 1.3×10^8 ton wheat straw and 2.3×10^8 ton of maize straw [50]. The main agricultural industry wastes are the wheat straw, maize straw, rice, and sugarcane bagasse [58]. Agricultural wastes compose of valuable compounds (Table 1). The straw consists of cellulose, lignin,

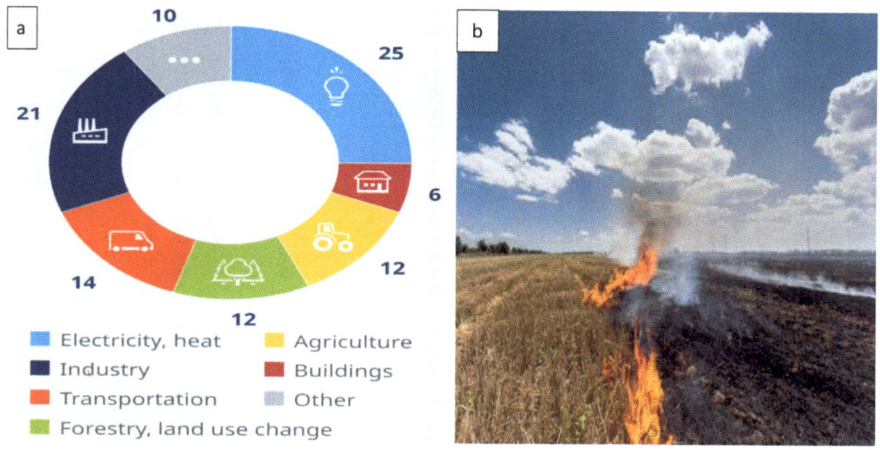

Fig. 2 Global greenhouse gas emissions by different sectors **a** straws burning in the field **b**

and hemicellulose, which when apply to the field microbes use these compounds as a source of energy (carbon) for the growth. Straw and crop residues is the most abundant and inexpensive source to reduce the environmental pollution [21, 58]. Similarly, [104] reported that Germany produced 8–13 million ton of cereal straw, whereas, UK generated 12.2 million ton of straw in 2011 from both cereals and oil seed crops. In the United Stated, the rice straw production was about 1140 million ton, in which 1.5 ton of straw was generated per ton of the rice grains [91]. In Thailand, 65 million ton of sugarcane was harvested in 2004, which generated about 18 million ton of sugarcane bagasse and 26 million ton of rice straw in the season of 2015/2016 [17, 53]. Brazil produced 279 million metric ton of sugarcane bagasse and 9.3 million ton of rice straw per year [1, 20]. The cereals and other crops straw are the major source of essential plant nutrient, but recently most of the farmer remove the straw and plant residues from the field or burn them in the field directly (Fig. 2b). The burning and removal of the crop straw from the field also reduce the nutrient concentration of the soil and burning of the residues also produce CO_2 and CH_4 which contribute to the global warming. Therefore, an efficient method is required to properly manage the crops straw or residue on the field and educate the farmers about the damages caused by the burning and removal of crop residues.

2.3 *Food Wastes*

Food waste (FW) is a serious issue for the researchers and scientists in the recent decades. The food is wasted mainly in the supply chain through different reasons, such as poor transportation, lack of storage facility, low level of farmer's education and poor management of the field. The FW losses not only threatens the world food

Table 1 Chemical composition (% dry matter) of agricultural wastes

wastes	Cellulose (%)	Hemicellulose (%)	Lignin (%)	Protein (%)	Fat (%)	Carbohydrates (%)	Moisture (%)	References
Wheat straw	31 - 33	23 - 25	12 - 14	1–5	3–5	6–8	4–5	[87, 110]
Maize straw	60 - 62	18 - 20	5 - 7	4–9	1.5–2.1	12–22	6–8	[87, 110]
Barley straw	33.5 - 34.0	21.6 - 23	12 - 14	4–7	1–3	38–44	19–23	[49, 72]
Rice straw	30 - 45	20 - 25	15 - 20	5–7	13–17	4–6	19–22	[87, 110]
Sugarcane bagasse	29 - 31	45 -55	10 - 14	0.5–0.7	2–5	50–55	6–9	[74, 87]
Cotton stalks	51–59	10–15	17–22	7–10	–	2–5	5–8	[84, 87]
Sunflower stalks	35 - 43	25 - 30	10–14	7–16	1–2	–	5–10	[70]
Sugar beet peel	18–27	15–20	1.5 - 3	6–10	2–4	70–75	8–13	[4, 31]
Potato peel	1.5- 3	4- 6	3–5	1.3–1.5	0.4–0.6	9–12	5–10	[41, 42]
Pineapple peel	15–19	25–30	1 - 2	4–10	2–4	48–78	80 - 90	[56]
Orange peel	7–11	10–12	0.5- 0.9	5–9	1.5–1.8	12–16	9–12	[64]
Banana peel	60 -64	5- 9	6 - 11	2–4	0.5–0.7	15–20	6–8	[80, 98]
Cabbage	9–12	2–4	0.5–0.6	0.5–0.9	0.04–0.06	1.6–5.01	0.3–0.5	[32, 48]
Tomato	27–33	23–27	28–31	11–15	1–13	2–4	85–95	[48, 77]
Cucumber	11–17	3–6	5–8	0.4–0.5	0.1–0.2	2–4	80–94	[48, 101]
Mango	30–38	10–14	21–28	0.7–0.9	0.5–0.7	7–10	75–85	[48, 115]
Water melon	18–20	20–24	8–11	0.8–1.0	0.3–0.4	3–5	90–94	[42, 48]

security but also have economic cost about 2.6 trillion US$, which account for 8–10% of greenhouse gas emission linked to the unconsumed food [44]. According to the report of [105], 17% of the world food production is lost as a waste, which includes 61% of waste coming from the home, 26 and 23% come from the food services and retailers, respectively. Similarly, [44] and [93] noticed that household waste is the main source of food waste. The reduction of this food wastes generation is the main concern for the world's scientists. Mainly the food waste is generated from the poor management and habits (planning, storing, eating, cooking and leftovers) [47]. Stated that the increase in the food waste quantity mainly related to the economic growth and development of the country. He further explained that most of the rural area people shifted to the industrial area, where there are a lot of job opportunities. Similarly, most of the people in developing countries shifted to more developed countries and have more job chances. Therefore, the quantity of food waste become increased. In Malaysia, the solid waste was about 2900 ton/day in 2001–2005, while it increased up to 30,000 ton/day in 2020, due the migration of citizen from different countries. Most of food waste, that is wasted before reaching to the human stomach is estimated to be 30 and 29% of the waste originated from the breads and fruits and vegetables industries, respectively [66]. In USA, the total values of the food waste was about 165.6 billion US$ in 2008, in which 26% of the total waste was fruits and vegetables [12]. This waste is a burden for mandatory waste management system, where 40% of the food produced in the US is disposed in landfills. Organic matter releases CH_4, which has 25 times more global warming potential than CO_2. The categorized food wastes are shown in Fig. 3.

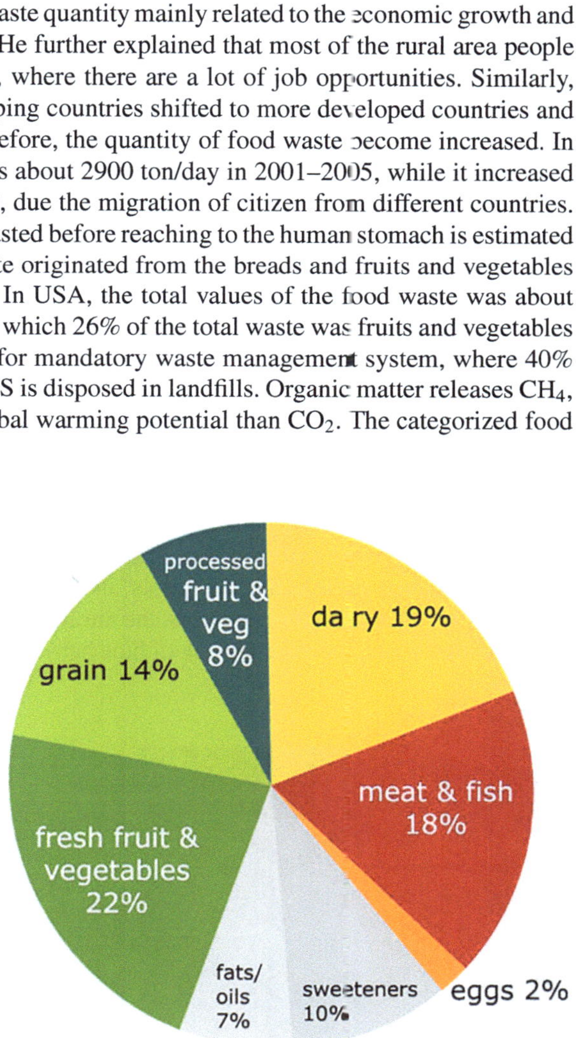

Fig. 3 Food waste by category. *Source* Internet (https://cookforgood.com/climate-change-food-waste/)

2.4 Livestock and Poultry Wastes

The livestock and poultry industries have grown very quickly because of the increase in world population. The scientists used different breeding technique to increase the animal and poultry products. The annual production of dairy and poultry are several trillion units every year in the world. Girotto and Cossu [37] reported that in Italy the average number cattle doubled between the 1990–2010, (24.1 and 45 units, respectively). Similarly, FWW (2016) stated that the US livestock and poultry generated 369 million tons of waste in 2012, which is 13 times more than the waste of US population (US population, 312 million). Globally, the main livestock unit is poultry (chicken, broilers, turkey, duck and laying hens), more than 500 billion units, the cattle and calves are 490 million units, lambs and goats are 890 million units in the world. The livestock and poultry farming produce milk, meat, eggs and yogurt, and also produce different types of waste and residues such as bones, blood, feather, bedding material, birth fluids, urine, and feaces in the farm houses, slaughtering industry, and food and meat processing companies [37, 75]. The waste produced by the livestock and poultry industry include large amount of liquid slurry and wastewater or solid, and these wastes could be either useful or harmful to the environment. When the livestock waste is used as manure in the field (Fig. 4), it is a rich source of nutrients and organic matter (OM) and increase the fertility of the soil and crop production. The animal manure contains 55–90% of nitrogen and phosphorus content in the feces and urine. If the animal manure is properly handled it have very positive effects on soil and environment, and if not then it emits dangerous gases (ammonia, hydrogen sulfide, carbon dioxide and methane). Hydrogen sulfide is very dangerous gas among other gases, it causes throat illness, headache, excessive coughing, burning eyes, and bronchitis in human. Robbin [85] reported that the farmer collects the animal waste in the "cesspools" in USA. The animal waste contains large amount of nitrogen and other nutrients, the nitrate seeping from these cesspools into below ground water, increase the risk of blue baby syndrome leading to the death of infants.

Fig. 4 Poultry waste **a** and livestock waste **b**

3 Environmental Concerns About Agricultural Wastes

3.1 *Impact of Wastes on Air Quality*

The wastes produced after the agricultural activities on the farms/field could either positively or negatively affect the world-wide air quality. Mostly for higher production and protection from insects and pests, farmers apply different inorganic fertilizers, insecticides, and pesticides. The insecticides and pesticides contain different toxic chemical (chlorofluorocarbons and aromatic hydrocarbons), which are absorbed by the crops. When these crops are burned in the open field, it releases different gases and elements into the atmosphere. It is stated that one nascent chlorine (Cl^-) destroy 100,000 ozone molecules before it is removed from the stratosphere. Similarly, the improper dumping of the agricultural waste also release CH_4 under anaerobic condition which is another toxic greenhouse gas and largely contributing to the global warming. Aneja et al. [2] noticed that the preparation of animal feed also increases the emission of organic acids and trace gases to the environment. Globally, the livestock sector emits 18% of all the greenhouse gases, which includes 65, 37, 64% anthropogenic nitrous oxide, methane, and ammonia, respectively [36]. Animal production emits hundreds of VOCs which includes acids, esters, alcohols, amides, aldehydes, aromatics, ketones, hydrocarbons, and other nitrogen and sulfur containing compounds. Some of them have very unpleasant odor such as H_2S, mainly emitted from the swine manure. The H_2S gas causes respiratory disease and also associated with death and other illness for peoples with acute exposure to gases released from the liquid manure [2, 14].

3.2 *Impact of Wastes on Water*

The agriculture waste significantly affects the quality of water. The water pollution caused by the agricultural wastes are now the major concern of the researchers and scientists. Around 80% of the world municipal wastewater discharge into the water bodies without any treatment. The processing and juice industry of fruit and vegetables discharge a lot of sewage directly into the river and stream, significantly affecting the aquatic biodiversity [65]. The water pollution caused by the agriculture wastes mainly comes from the discharge of agrochemical from the farms, drug residues and saline irrigation drainage into the underground water bodies, which significantly affect the aquatic and human ecosystem. In developed and developing countries, the pollution caused by the agriculture industry mainly on field is the leaching of different nutrient (phosphate, nitrate, and different heavy metals). The leaching of nitrate into the underground water bodies cause the eutrophication. Similarly, 38% of water bodies in the Europe Union are contaminated by the agriculture, while in the United States, agriculture sector is the major source of water pollution in streams, rivers, wetland and lake (WWAP, 2015; USEPA, 2016). The high or improper use of insecticides, pesticides, and fungicides, pollute the water bodies with carcinogens and other toxic substances. The pesticides and insecticide also affect the aquatic biodiversity by killing weeds and insects which negatively affect the food cycle of aquatic ecosystem [65, 92]. The waste from the animal processing industries are the major water pollutants. The biological oxygen demand (BOD) of pig slurry and domestic sewage was 30,000–80,000 and 200–500 mg/L, respectively, which causes hypoxia in water bodies, because the degradation of organic matter need dissolve oxygen [36].

3.3 Impact of Waste on Soil

The waste is generated from the agricultural activities on field (packaging material of pesticides and insecticide, residues of crops, animals and poultry waste) and outside field (processing and packaging industry of fruits and vegetables, feed preparation industries for the livestock and poultry, and different inorganic fertilizers industries). The sewage from these industries composed of different hazardous chemical, organic and inorganic compound, suspended particles, heavy metal, and nutrients. When these sewage sludge or waste water is applied to the agricultural land, it degrade the soil quality, diminishes the microbial activities in the soil and adversely affect the physical and chemical properties of soil. Yeilagi et al. [115] reported that the household and agriculture waste, including food, paper, plastic wastes, glass, inorganic salts, building materials, and metals are discarded by public. The decomposition and biodegradation of these wastes vary based on the physical, chemical, and biological reactions produce a liquid called "leachate". This leachate composed of different organic and inorganic compounds causing damage to both agricultural and nature ecosystem. The leachate after the biodegradation of different waste materials composed of 1000–3000 mg/L of NH_3-N in different part of the world, while in the US it has been reported that the leachate consists of 0.2–41 mg/kg Cd, 21–1500 mg/kg Cr, and 16–1300 mg/kg Ni [39, 115]. The burning of the crop residues on the farms degrades the soil quality. The straw burning reduces the volume of the residues and losses a lot of essential nutrient and remove different beneficial microbes during the burning of residues [50, 98].

4 Management

Agricultural waste such as crop straw, fruits and vegetables waste, animal manure, and poultry manure are highly biodegradable, when it properly managed, on one hand it reduced the bulk of waste and on other hand it adds to the soil fertility. Therefore, the management of wastes produced by the agricultural industry is very important, if the waste produced by the agriculture are not properly managed it create soil, water and environmental pollution and loss of essential nutrients. Due to these reasons, in the recent decades the scientist tries to develop policies for the management of waste [59]. Management of waste is the collection, recovery, transportation, and disposal of waste generated by the agricultural industries. Waste management strategies includes reduction, recycling, energy recovery, and landfilling of waste [23] can be explained in detail in the following sub topics.

4.1 Reduction of Agriculture Waste

Waste management, including reduction of agricultural waste, is considered as a top priority that mainly depends on production practices [23]. To reduce the production of agricultural waste, it is necessary to improve the quality of crops, fruits and vegetables. Similarly, Du et al. [28] investigated that 1.3 billion ton of fruits and vegetables are discarded as a waste due to its low quality standards set by the traders. To decrease the fruits and vegetable wastes produced due to low quality or substandard, different strategies are anticipated and implemented, such as the misshape fruits and vegetable are often used for the production of juice and vinegar, and reduce the price of substandard fruits and vegetable for the consumers, in order to save the fruits and vegetables to become waste [79]. Mateo-Sagasta et al. [65] reported that one quarter of the food is lost in the supply chain, and this food losses can be reduced if the food production plant is closer to the actual food demand. The other reason to generated waste on the farm is the poor storage facility. Most of the farmer lost huge number of harvested crops due to lack of storage facilities. If the storage condition is good on the farm, it reduces the chance of spoiling the crops, vegetables and fruits. We should make the storage facilities cleaned, well ventilated and disinfected before being filled with crops grain or fruits and vegetable to reduce the risk of spoiling and rotting of food stuffs.

4.2 Reuse of Agriculture Waste

The waste generates after the agricultural activities on field or outside the field, used for other purpose with or without minor changes in their properties is called the reuse of agricultural waste. The agricultural waste such as crops straw (wheat, maize and rice), sugar cane bagasse, fruit and vegetables waste are often use as a soil amendment for soil fertility. The on field agricultural waste are rich in nitrogen, phosphorus and potassium which are the main essential nutrient for the crop growth [57]. The agricultural waste can also be used for the remediation of polluted soils, by immobilizing the trace metals and metalloids and reduced the leaching of these toxic metals into the underground water [18]. Due to high fibers content the agriculture wastes such crop straw are used as a fodder for the livestock. Similarly the food waste including fruit, vegetables, meals and bakery waste are used for the preparation of poultry feed [103]. The agricultural waste such as wood, crops straw and animal dung are used for heating and cooking proposes in the rural areas, while the direct combustion of agricultural waste release toxic substance, carbon dioxide, carbon mono oxide, sulfur oxides and nitrogen [94]. To reduce the direct combustion of agricultural we should educate our rural areas peoples and arrange seminars and poster presentation to know the hazardous effect of the direct combustion of agricultural waste.

4.3 Recycling of Agricultural Waste

The agricultural waste including crop straw (wheat, maize, and rice), poultry waste, livestock waste, fruits and vegetables wastes are very important resources. The improper disposal of such agricultural waste not only treat to cause environmental pollution but also lost a lot of valuable biomass as a waste [109]. Therefore, the recycling of agricultural wastes is considered to be the important strategy to protect environmental pollution and agricultural development. Recycling is the strategies adopted to recover waste materials after a major modification in their characteristics. Recycling strategies varies, such as the whole waste could be recycled (processing to flour, composting and conversion to water) or the extraction of specific compounds from the waste. An ancient eco-friendly strategy called aerobic composting is adopted to convert organic waste into organic fertilizer. However, on the other hand anaerobic digestion is a more attractive strategy followed for the production of fertilizers from agricultural waste, because of energy recovered as bio gas [96]. Similarly, flour of agricultural waste is processed with different aims, their fibrous structure and contact surface have been used to adsorb pollutants including dyes and heavy metals from water and soil [6]. The olive seed waste used for the preparation of environmentally friendly and economical insulation material. Similarly, the olive mill waste are also used for the manufacturing of clay bricks [94]. The ash of different of agricultural waste such as wheat straw ash, wood waste ash, rice straw ash, bamboo leaves ash and sugar cane bagasse ash are used for the preparation of concrete, which is used as a cementing agent to modify the strength durability and performance of the concrete as well as reduce the cost of construction [3].

4.4 Energy Recovery from Agricultural Waste

The main of source of energy in the recent decades is the fossil fuels that are consumed by both developed and developing countries around the globe. The combustion of these fossil fuels also contributes to the global warming by emitting different toxic gases into the atmosphere [108]. Therefore, in this context agricultural waste consisting of crops straw, stalks, cobs, sugarcane bagasse, leaves, wood etc., are the carbon feasible resources for the bioenergy production and nutrient recovery [29]. Different technique has been used for the recovery of energy from agricultural waste. The methane production from the sewage sludge, animal and poultry manures, organic part of house hold materials along with energy crops, further this methane production used for the thermal and electricity [108]. Bioethanol can also be produced from the fruit, vegetables, crops straws and sugarcane bagasse, which can also be used as an energy source [8]. Briquettes is another source of energy made from the compressed wood chips, straw, charcoal, cobs, sawdust, and paper through thermochemical conversion. The briquettes are safe economical and environmentally friendly biofuel. The briquetting process usually reduce the transportation and storage cost and increase combustion quality. Briquettes made from the agricultural waste are used in power generation, straight ignition, industrial boilers, furnaces and other combustion equipment [29, 108]. The production of biohydrogen from the agro-lignocelluloses wastes to reduce the dependency on the fossil fuel. The biohydrogen can be produced through biologically decomposition or dark fermentation using livestock waste or crop residues [108].

4.5 Rational Management of Agricultural Waste

Agricultural waste is seen to be a trustworthy, affordable, and easily accessible feedstock for the possible recovery of energy, water, and useful ingredients. Integrated approaches to agricultural waste management are the dire need to maximally exploit its potentialities by developing, combining, and choosing the best strategy of reuse, recycle and energy recovery. Such strategies include waste characterization in term of amount and composition, process design including production processes of obtaining outputs, feasibility study including cost, consumer acceptability of environmental sustainability, compliance with legal standards for outputs, and appropriate processes. The conventional and emerging valorization methods are of prime importance, as the biological or chemical transformation of organic constituents normally leads to CO_2, CO, CH_4, H_2, H_2O, NH_3, and H_2S release. These technologies have the tendency to provide biogas, bio-hydrogen compost, heat and power. However, unable to recover complex components from waste. Furthermore, the environmental risk still exists due to their impact on atmospheric air pollution, greenhouse gas emissions, wastewater and solid wastes production. On the other hand, valorization methods are found effective to maintain the nutrients and other valuable compounds safe [34]. Integral valorization approaches are considered important to recover valuable components but also treat the remaining constituents conveniently. The combinations of both emerging and conventional methods would be a suitable option to promoted to reduce waste management hierarchy and avoid food waste generation [76]. Valorization technologies contribute more in reduction of final disposed by recovering and reincorporating valuable components. Strategies should be adopted to create awareness in the consumers, especially the use of green and sustainable production of food to ensure the socio-economic growth of food industry.

4.6 Bioactive Compound Present in Agricultural Waste

Globally every year huge amount of waste was produced because of the poor production chain, and estimated that 45% of fruit and vegetable is lost as waste, in which 5–25% in developed countries and 20–50% in the developing countries, which pollute soil, air and water and affecting both environment and human health [38]. Therefore, it is necessary to find a new, economical, and environmentally friendly technologies to reduce the waste and extract some bioactive compounds from the agricultural waste, a bioactive compound is a substance that has biological activity and trigger the physiological response in living organism. The bioactive compound have several beneficial effects such as anticancer, antiallergic, antioxidant, anti-inflammatory, antiproliferative and antiatherogenic [38, 40]. The agricultural waste mainly composed of lignocellulosic biomass which are the main source of sugar and phenolic compounds. The phenolic acids are composed of different groups of benzoic and cinnamic acid such as ellagic, salicylic, gallic, capsaicin, syringic, ferulic, and chlorogenic acid present in both free and bound form. Flavonoids are also bioactive compounds which two aromatic rings, furthermore the flavonoids belong to different subclasses such as flavonols, flavones, anthocyanins and isoflavones [60, 95]. Similarly, fruits and vegetables have different bioactive compounds and have different bioactivity. Apple pomace, citrus peel, seeds, mango, banana, potato, carrot, beetroot and cauliflower consist of pectin, modified citrus pectin, oligosaccharides, apigenin-glucoside, caffeic acid, ferulic acid, gallic acids and its derivatives, mangiferin, caffeic acid, sinapic, rutin, limonin, catechin, isorhamnetin, phlorizin, lutein, β-carotene, Glucoiberin Glucoerucin and ursolic acid which can be used as dietary fibers, prebiotic, antioxidant, antimicrobial, anti-inflammatory, anti-tumor, anti- diabetic, lowering blood pressure, improving blood glucose control, anti-cancer, lowering cholesterol, anti-bacteria and anti-viral [10].

4.7 Biological Approaches for Wastes Mitigation

4.7.1 Animal Feed

Agricultural wastes are used as animal feeding under controlled conditions [89]. Care should be taken while considering agricultural waste for animals feeding in order to eliminate the chances of toxin that can cause diseases or might affect dietary process [59]. Agricultural wastes are rendered safe following thermal treatments however, this process makes it more costly but still considered beneficial compared to land filling or anaerobic digestion. Agricultural waste contributes half portion of conventional feed whose production greatly influence environmental and health impacts. Countries like Japan and South Korea share food waste recycles as feed up to 35.9 and 42.5% respectively. However, such practices are illegal and strictly regulated in European countries [89].

4.7.2 Biofuels (Bioethanol, Biodiesel and Biogas)

Biofuel: The conversion of agricultural waste into biofuels are practiced in both developing and developed countries, because of energy security, environmental, and economic issues. Many countries in the world use biofuels due to its sustainability, less greenhouse gas emissions, economical, and easily degradable. Different types of biofuels include bioethanol, biodiesel, and biogas.

Bioethanol

The bioethanol can be produced from the agricultural waste (straw, bagasse, fruits and vegetables wastes) by the fermentation using different microorganism. Worldwide the annual production of bioethanol from the rice straw, corn straw, wheat straw and sugarcane bagasse was 205, 58.6, 104 and 51.3 giga liter, respectively [90]. Similarly, the bioethanol can also be produced from the vegetable and fruits wastes. Pineapple leaf waste, citrus peel, and waste citrus processing wastes give 7.12, 76, and 94% (v/v) bioethanol, respectively [99].

Biodiesel

The biodiesel can also be produced through fermentation from the agricultural waste such as animal fats, different waste cooking oils, vegetables oils and palm oil. The reduction in reservoir of natural crude oil (petrol and diesel) and high prices of the fuel forced the scientist toward the production of biodiesel. The waste palm oil, cooking oil and vegetables oil can be converted to biodiesel by different catalyst. Similarly, the pepper seed, date pits waste, palm oil, waste cooking oil and fish oil produced 94.1, 98.2, 88.15,89.6, and 97% biodiesel, respectively [99].

Biogas

Biogas can be produced from the agricultural waste (crop straw, animal manure and poultry manure) through anaerobic digestion. The anaerobic digestion of agricultural waste reduced the emissions greenhouse gases and provide sustainable energy supply. The biogas produced from the agricultural waste uses as a renewable energy, as CH_4 is the main constituent of biogas. The production of biogas from animal waste contains 65% of CH_4. The cotton stalk, cotton seed hull and cotton oil used for the biogas production in a batch reactor contain 65, 86 and 78 mL CH_4 for 1 g of cotton stalk, cotton seed hull and cotton oil cake, respectively. Similarly, the biogas production of rice straw was about 357–382 mL/day, food waste and green waste yield 358–430 and 185–245 mL/g CH_4 and cow dung yielded 47% methane content [27, 83, 99].

4.8 Conventional Food Waste Management Technologies

4.8.1 Landfilling

In the agricultural waste management, landfilling is most common and cost-effective method to handle the waste. In landfilling management practice, the waste are collect and store in a pit and then compact the waste by the compactor or bulldozers to minimize the oxygen supply. In most developing countries the agriculture waste discarded in open air landfilling site, which is very harmful for the human and create environmental pollution, while in developed countries the landfilling is engineered having the facility to collect the leachate and gases from the landfilling site, this modern landfilling technique is called sanitary landfills. The sanitary landfills also have a routine soil cover on the freshly dispose waste to reduce the odor, diseases, fire and waste scaving [43, 71]. Fruits and vegetables waste are highly rich in nutrients and are easily degradable [82]. Leachate formation, another environmental issue relates to landfilling caused by the rainwater infiltration that percolates through the waste layers and accumulates at the bottom of landfill. Leachate should be treated based on its chemical composition, municipal wastewater treatment plant under appropriate conditions. The complex matrix integrates organic matter, inorganic salts, organic trace pollutants, and heavy metals, depends on weather conditions (temperature, rainfall), waste composition, and the age, depth, design and operation of landfill [11, 68].

4.8.2 Anaerobic Digestion

Agricultural waste such animal manure, poultry manure, crop residues, fruits shell, green leaves and molasses are the potential renewable energy resources. The on field burning of wheat, rice and maize straw creates environmental pollution [24]. Therefore, there is need of economical and efficient technique which can reduces the bulk and also environmental pollution, for this purpose researchers used the anaerobic digestion. The anaerobic digesters convert the organic waste into biogas consisting of 50% methane. The anaerobic digestion convert the organic matter into methane and slurry is applied as a fertilizer in the field [67]. The gas produced in anaerobic digestion is of high quality and contain 80–90% of the atmospheric methane. The bio-methane can be used as a fuel in the vehicle, used for the cooking and for the power generation [24]. The manure and sludge produced in the treatment plant of wastewater can also be managed through anaerobic digestions. The pH, organic acid profile, alkalinity, Carbon–nitrogen ratio and the existence of heavy metals are other important parameters to be considered [117]. Compared to landfills, anaerobic digestion produces higher yields of methane due to controlled fermentation conditions in the digestor. Similarly, Wang et al. [111] reported that pig manure is one of the major animal manure and most of the manure produced cannot be well managed which causes serious threat to soil, water, air and also to livestock and poultry, the author further stated that anaerobic digestion would greatly minimize the environmental risk and also economical for their conversion into methane gas.

4.8.3 Composting and Vermicomposting

Another technology suited for the management of agricultural waste is known as composting or vermicomposting. It is environment friendly as it allows less greenhouse gas emissions and leachate generation. However, sorting of waste materials to separate organic fraction, power required for mixing and aeration, water addition and careful operating conditions should be addressed. Composting process involves the aerobic biological conversion of organic matter, results into CO_2, N_2, NH_3 and a solid recalcitrant material as compost, containing humic substances. Compost has the ability to amend soil and retain water thus reduce the needs of irrigation, precipitate heavy metals, pathogens and micro-pollutants [54]. The pH, C/N ratio, moisture content, aeration, particle size and porosity of the raw material affects composting [15].

Vermicomposting can be used as a source of nutrients and soil conditioner in agricultural applications. Vermicomposting is a recent approach for the production of bio-fertilizers, which presents a good quality nutritive vermicompost through the combine action of earthworms and microorganisms compare to conventional compost [86]. It is an efficient way of mitigating and managing environmental pollution problems. The enzymatic and microbial activities make vermicomposting a higher-quality end product compared with traditional composting [7]. Vermicomposting also results in mass reduction, reduce the processing time, contain high levels of humus, and reduce phytotoxicity of the substrate [5] Vermicomposting is considered most suitable for organic agriculture practices due to its high nutrients content and contains high quality humus, abundant plant growth hormones, enzymes, and protect crops against pests and diseases incidence [22]. Previous studies have shown the significant role of earthworms in the fruit and vegetables waste vermicomposting and concluded its beneficial role in destroying pathogenic load compare to other composts developed for soil fertilizers [101]. Composting has been promoted over landfilling by environmental legislation which share almost doubled for the developed countries (Eurostat, 2018). The average municipal solid wastes in US for the year 2018 was 292.4 million tons, of which 94 million tons were recycled and composted equivalent to 32.1% recycling and composting rate. Despite huge amount of household FW generated in Japan their share is less than 6% due to scarce implantation of separation of organic fraction as source. Like in China, almost 60% of municipal wastes are landfilled whereas less than 2% are composted which result in poor quality of compost that have high contents of heavy metals as a consequence of deficient separation practices [112].

4.8.4 Thermal Treatment

Incineration: Agriculture wastes are dumped into the landfills which are incinerated or combusted (Fig. 5). This process significantly reduces the waste volume and size. Moreover, it also extends the life of the landfill's sites. Modern incineration plants have the potential to suitably integrate air pollution issues resulting in extremely low gas emission of contaminants [46]. Nonetheless, huge investments and operation costs are required for its implementation as it need a great amount of energy due to existence of low calorific content and high moisture contents of agricultural waste. Plastic related stuff if used, not only generate harmful air pollutants (dioxins) but significantly increase atmospheric CO_2 emission [34, 54]. Natural gas being a primary source of energy could be utilized as to suitably sustain high operating temperatures. The partial recovered combustion heat is used for incineration or intended for power generation. Solid residue, ash is generated through combustion which are often land filled can be use in building material [26]. In European countries, the share of incineration among technologies utilized for municipal waste management in case of composting has almost doubled, from 15–27% within 16 years [34]. Countries like Japan and China utilized the incineration, contributed a share value of 80 and 30% in 2012, respectively [112].

Pyrolysis and gasification: Pyrolysis is the decomposition of agricultural waste (animal manure, poultry manure, carrot residues, plant materials, wood, sewage sludge, crop straw, sugarcane bagasse) in the absence or no oxygen supply resulting in the production of gas, bio-oil and biochar [51, 52]. The biochar is black carbonaceous substance which can improve the soil water holding capacity, nutrient holding capacity, increase the pH of the acidic soil, improve the soil physical and biological properties, decreased the leaching of phosphorus, nitrate and heavy metals due to high surface area of biochar and reduce greenhouse gas emission from both dryland

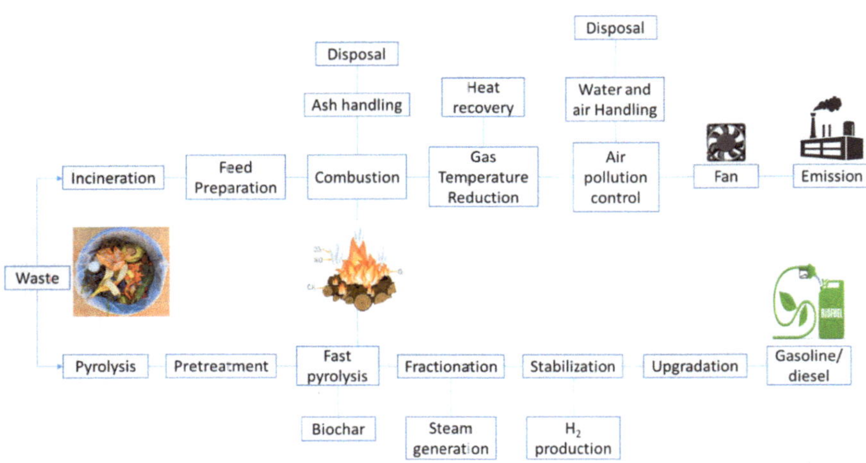

Fig. 5 Flowchart of Waste disposal through thermal treatment

and paddy soil [9, 16, 25, 98, 114]. In gasification, the wastes are decomposed in partially oxidized atmosphere for the production of gas steam called producer gas, which are used in combustion and or processed to syngas [69]. The existence of high moisture contents in food waste favors the gasification process. Nonetheless, application of high energy required to heat up and vaporize water and negatively affects the thermal efficiency of wastes treatments are the serious drawbacks in this process. Conventional techniques for managing fruit and vegetable waste has significantly improved in the last decade, only a small fraction of residues generated can be valorized by some other means [63].

5 Economic Incentives for Waste Reduction

Reducing the economic loss is an important motivating factor for changing and reducing food waste in the retail sector. Analyses of the economic cost of the in-store waste revealed that 86% of the cost derived from the cost of fresh fruit and vegetables (FFV) is based on the purchase price. The personnel cost for the working time on waste management and the waste disposal cost corresponds to 14% of the economic cost. The results show that it is essential to be aware of and included all parts when discussing the economic costs. Knowledge and awareness about the economic costs of FFV waste can create an incentive for the stores to reduce waste and the economic incentive for the stores may be stronger than by only focusing on incorporating a law concerning reduced retail wastes. Waste prevention measures will give additional advantages for the stores, such as goodwill, since awareness about food waste is becoming more well-known and trendier. The economic cost of agricultural waste is based on the purchase price, but the actual cost is even higher when the lost profit of the removed produce that will not be sold is taken into consideration. Reducing economic losses will decrease the wasted mass and climate impact at retailer level and contribute to a sustainable food supply chain. Penalizing measures to discourage the disposal of waste is mostly applied. It is mostly applied in Italy on a regional basis based on the amounts disposed waste. To exempt the environmental tax the producer must carry out treatment in a plant i.e., bio-mechanical or thermal, according to the national or regional directives. The incentives for energy production have focused on biodegradable to be composted aiming to material recovery. The biodegradables have a potential of 110–130 kg CH_4 /ton of dry matter from which we can get 220 kW h of electric energy, that corresponds to 60 €/ton [19]. The methodology to measure and calculate the economic cost of FFV waste developed in this study could be applicable for other departments in a retail store.

6 Conclusions

Agricultural activities have increased in the past decades due to increasing world population and to provide food, feed, and shelter. In addition, agricultural activities contribute to waste generation, posing environmental threats, and human health hazards. Agricultural activities like fruits and vegetables cultivation, and dairy and poultry farming generate agricultural waste, which pose threats to environment and affect the air and water quality and soil health indicators. These threats can be addressed by 3R technology, energy and bioactive compounds recovery, bioresources generation, and other conventional food waste management practices. Economic incentives can add extra benefits to wastes handling. In addition, wastes can be converted to economical compounds to provide raw materials for other agro-based industries to provide get more revenue and jobs opportunities.

Acknowledgements This work was partly supported by the National Natural Science Foundation of China (No. 32371608, 42350410444) and Yunnan Fundamental Research Projects (202101AS070010, 202201AT070673), Higher Education Commission (HEC), Islamabad, Pakistan (No. 20-LCF-84/RGM/R&D/HEC/2020). We thank the Institutional Center for Shared Technologies and Facilities of Xishuangbanna Tropical Botanical Garden, CAS for providing the required facilities.

References

1. Ajala EO, Ighalo JO, Ajala MA, Adeniyi AG, Ayanshola AM (2021) Sugarcane bagasse: a biomass sufficiently applied for improving global energy, environment and economic sustainability. Bioresources and Bioprocessing 8. https://doi.org/10.1186/s40643-021-00440-z
2. Aneja VP, Schlesinger WH, Erisman JW (2009) Effects of agriculture upon the air quality and climate: Research, policy, and regulations. Environ Sci Technol 43:4234–4240. https://doi.org/10.1021/es8024403
3. Aprianti E, Shafigh P, Bahri S, Farahani JN (2015) Supplementary cementitious materials origin from agricultural wastes—a review. Constr Build Mater 74:176–187. https://doi.org/10.1016/j.conbuildmat.2014.10.010
4. Aramrueang N, Zicari SM, Zhang R (2017) Response surface optimization of enzymatic hydrolysis of sugar beet leaves into fermentable sugars for bioethanol production. Adv Biosci Biotechnol 08:51–67. https://doi.org/10.4236/abb.2017.82004
5. Atiyeh RM, Edwards CA, Subler S, Metzger JD (2001) Pig manure vermicompost as a component of a horticultural bedding plant medium: Effects on physicochemical properties and plant growth. Biores Technol 78:11–20. https://doi.org/10.1016/S0960-8524(00)00172-3
6. Azouaou N, Sadaoui Z, Djaafri A, Mokaddem H (2010) Adsorption of cadmium from aqueous solution onto untreated coffee grounds: Equilibrium, kinetics and thermodynamics. J Hazard Mater 184:126–134. https://doi.org/10.1016/j.jhazmat.2010.08.014
7. Bajsa O, Nair J, Mathew K, Ho GE (2004) Vermiculture as a tool for domestic wastewater management. Water Sci Technol 48:125–132. https://doi.org/10.2166/wst.2004.0821
8. Balat M, Balat H (2009) Recent trends in global production and utilization of bio-ethanol fuel. Appl Energy 86:2273–2282. https://doi.org/10.1016/j.apenergy.2009.03.015

9. Bashir S, Hussain Q, Zhu J, Fu Q, Houben D, Hu H (2020) Efficiency of KOH-modified rice straw-derived biochar for reducing cadmium mobility, bioaccessibility and bioavailability risk index in red soil. Pedosphere 30:874–882. https://doi.org/10.1016/S1002-0160(20)60043-1
10. Ben-Othman S, Jõudu I, Bhat R (2020) Bioactives from agri-food wastes: Present insights and future challenges. Molecules. https://doi.org/10.3390/molecules25030510
11. Bhatt AH, Karanjekar RV, Altouqi S, Sattler ML, Hossain MDS, Chen VP (2017) Estimating landfill leachate BOD and COD based on rainfall, ambient temperature, and waste composition: Exploration of a MARS statistical approach. Environ Technol Innov 8:1–16. https://doi.org/10.1016/j.eti.2017.03.003
12. Buzby JC, Hyman J (2012) Total and per capita value of food loss in the United States. Food Policy 37:561–570. https://doi.org/10.1016/j.foodpol.2012.06.002
13. CEC (2014) Burning agricultural waste: a source of dioxins. montreal, Canada: commission for environmental cooperation 6
14. Campagna D, Kathman SJ, Pierson R, Inserra SG, Phifer BL, Middleton DC, Zarus GM, White MC (2004) Ambient hydrogen sulfide, total reduced sulfur, and hospital visits for respiratory diseases in northeast Nebraska, 1998–2000. J Expo Anal Environ Epidemiol 14:180–187. https://doi.org/10.1038/sj.jea.7500313
15. Cerda A, Artola A, Font X, Barrena R, Gea T, Sánchez A (2018) Composting of food wastes: Status and challenges. Biores Technol 248:57–67. https://doi.org/10.1016/j.biortech.2017.06.133
16. Chandra S, Medha I, Bhattacharya J (2020) Potassium-iron rice straw biochar composite for sorption of nitrate, phosphate, and ammonium ions in soil for timely and controlled release. Sci Total Environ 712:136337. https://doi.org/10.1016/j.scitotenv.2019.136337
17. Cheewaphongphan P, Junpen A, Kamnoet O, Garivait S (2018) Study on the potential of rice straws as a supplementary fuel in very small power plants in Thailand. Energies 11:1–21. https://doi.org/10.3390/en11020270
18. Clemente R, Pardo T, Madejón P, Madejón E, Bernal MP (2015) Food byproducts as amendments in trace elements contaminated soils. Food Res Int 73:176–189. https://doi.org/10.1016/j.foodres.2015.03.040
19. Cossu R, Masi S (2013) Re-thinking incentives and penalties: Economic aspects of waste management in Italy. Waste Manage 33:2541–2547. https://doi.org/10.1016/j.wasman.2013.04.011
20. Da Silva MG, Lisbôa ACL, Hoffmann R, Da Cunha Kemerich PD, De Borba WF, Fernandes GDÁ, De Souza ÉEB (2021) Greenhouse gas emissions of rice straw-to-methanol chain in Southern Brazil. J Environ Chem Eng 9. https://doi.org/10.1016/j.jece.2021.105202
21. Dang VBH, Doan HD, Dang-Vu T, Lohi A (2009) Equilibrium and kinetics of biosorption of cadmium(II) and copper(II) ions by wheat straw. Biores Technol 100:211–219. https://doi.org/10.1016/j.biortech.2008.05.031
22. Das D, Abbhishek K, Banik P, Bhattacharya P (2021) A valorisation approach in recycling of organic wastes using low-grade rock minerals and microbial culture through vermicomposting. Environ Challs 5:100225. https://doi.org/10.1016/j.envc.2021.100225
23. Demirbas A (2011) Waste management, waste resource facilities and waste conversion processes. Energy Convers Manag 52:1280–1287. https://doi.org/10.1016/j.enconman.2010.09.025
24. Demirbas A, Ozturk T (2005) Anaerobic digestion of agricultural solid residues. Int J Green Energy 1:483–494. https://doi.org/10.1081/ge-200038719
25. Demisie W, Liu Z, Zhang M (2014) Effect of biochar on carbon fractions and enzyme activity of red soil. CATENA 121:214–221. https://doi.org/10.1016/j.catena.2014.05.020
26. Dhir RK, de Brito J, Lynn CJ, Silva RV (2018) Municipal solid waste composition, incineration, processing and management of bottom ashes, sustainable. Constr Mater. https://doi.org/10.1016/b978-0-08-100997-0.00003-8
27. Dincă M, Ferdeş M, Voicu G, Paraschiv G, Moiceanu G, Muşuroi G, Ioniţă S (2014) Biogas production from mixtures of animal manure and fresh biomass with and without glucose addition

28. Du C, Abdullah JJ, Greetham D, Fu D, Yu M, Ren L, Li S, Lu D (2018) Valorization of food waste into biofertiliser and its field application. J Clean Prod 187:273–284. https://doi.org/10.1016/j.jclepro.2018.03.211
29. Du C, Li Y, Zhao X, Pei X, Yuan W, Bai F, Jiang Y (2019) The production of ethanol from lignocellulosic biomass by Kluyveromyces marxianus CICC 1727–5 and Spathaspora passalidarum ATCC MYA-4345. Appl Microbiol Biotechnol 103:2845–2855. https://doi.org/10.1007/s00253-019-09625-1
30. Edwiges T, Frare L, Mayer B, Lins L, Mi Triolo J, Flotats X, de Mendonça Costa MSS (2017) Influence of chemical composition on biochemical methane potential of fruit and vegetable waste. Waste Manage 71:618–625. https://doi.org/10.1016/j.wasman.2017.05.030
31. El-Tayeb TS, Abdelhafez AA, Ali SH, Ramadan EM (2012) Effect of acid hydrolysis and fungal biotreatment on agro-industrial wastes for obtainment of free sugars for bioethanol production. Braz J Microbiol 43:1523–1535. https://doi.org/10.1590/S1517-83822012000400037
32. Elkner K, Kosson R (2008) Dietary fibre content and its fractional composition in cabbage as affected by cultivar earliness and sauerkraut storage period. Veg Crop Res Bull 69:165–175. https://doi.org/10.2478/v10032-008-0031-2
33. Erenstein O, Jaleta M, Sonder K, Mottaleb K, Prasanna BM (2022) Global maize production, consumption and trade: trends and R&D implications. Food Secur. https://doi.org/10.1007/s12571-022-01288-7
34. Esparza I, Jiménez-Moreno N, Bimbela F, Ancín-Azpilicueta C, Gandía LM (2020) Fruit and vegetable waste management: Conventional and emerging approaches. J Environ Manage 265. https://doi.org/10.1016/j.jenvman.2020.110510
35. FAO (2014) Definitional framework of food Loss—Save food: global initiative on food loss and waste reduction. Defin Fram Food Loss—Working Pap 18
36. FAO (2006) Designing genetic impact assessment for crossbreeding with exotic beef breeds in mixed farming systems, Livestock's Long Shadow. U N, Food Agric Organ. https://doi.org/10.1177/0030727020915206
37. Girotto F, Cossu CA (2017) Animal waste and waste animal by-products generated along the livestock breeding and meat food chain. Waste Manage 70:1–2. https://doi.org/10.1016/j.wasman.2017.11.028
38. Gomes-Araújo R, Martínez-Vázquez DG, Charles-Rodríguez AV, Rangel-Ortega S, Robledo-Olivo A (2021) Bioactive compounds from agricultural residues, their obtaining techniques, and the antimicrobial effect as postharvest additives. Int J Food Sci 2021. https://doi.org/10.1155/2021/9936722
39. Gu N, Liu J, Ye J, Chang N, Li YY (2019) Bioenergy, ammonia and humic substances recovery from municipal solid waste leachate: A review and process integration. Biores Technol 293:122159. https://doi.org/10.1016/j.biortech.2019.122159
40. Guaadaoui A, Benaicha S, Elmajdoub N, Bellaoui M, Hamal A (2014) What is a bioactive compound? A combined definition for a preliminary consensus. Int J Food Sci Nutr 3, 17–179. https://doi.org/10.11648/j.ijnfs.20140303.16
41. Hijosa-Valsero M, Paniagua-García AI, Díez-Antolínez R (2018) Industrial potato peel as a feedstock for biobutanol production. New Biotechnol 46:54–60. https://doi.org/10.1016/j.nbt.2018.07.002
42. Ibrahim A, Yusof L, Beddu NS, Galasin N, Lee PY, Lee RNS, Zahrim AY (2016) Adsorption study of Ammonia Nitrogen by watermelon rind. IOP Conference Series: Earth and Environmental Science 36. https://doi.org/10.1088/1755-1315/36/1/012020
43. Iqbal N, Agrawal A, Dubey S, Kumar J (2012) Role of decomposers in agricultural waste management, biotechnological applications of biomass
44. Iranmanesh M, Ghobakhloo M, Nilashi M (2022) Impacts of the COVID-19 pandemic on household food waste behaviour: A systematic review. Appetite 176:106127. https://doi.org/10.1016/j.appet.2022.106127
45. J, K., Mbui D. N., M., M.J., B., M.F., (2020) Proximate analysis of fruits and vegetables wastes from Nairobi County, Kenya. Res J Food Sci Nutr 5:2536–7080

46. Kajiwara N, Noma Y, Sakai S (2017) Environmentally sound destruction of hexabromocyclododecanes in polystyrene insulation foam at commercial-scale industrial waste incineration plants. J Environ Chem Eng 5:3572–3580. https://doi.org/10.1016/j.jece.2017.07.006
47. Kamaruddin MA, Jantira NN, Alrozi R (2020) Food waste quantification and characterization as a measure towards effective food waste management in university. IOP Conference Series: Materials Science and Engineering 743:9. https://doi.org/10.1088/1757-899X/743/1/012041
48. Kamau JM, Mbui DN, Mwaniki MJ, Mwaura (2020) Proximate analysis of fruits and vegetables wastes from Nairobi Kenya. Res J Food Sci Nutrition 5(1):2536–7080. https://doi.org/10.31248/RJFSN2019.088. https://integrityresjournals
49. Keno MT, Tolemariam T, Demeke S, Wamatu J, Alkhtib A, Janssens GPJ (2021) Effect of barley variety on feed intake, digestibility, body weight gain and carcass characteristics in fattening lambs. Animals 11:1–9. https://doi.org/10.3390/ani11061773
50. Khan MN, Lan Z, Sial TA, Zhao Y, Haseeb A, Jianguo Z, Zhang A, Hill RL (2019) Straw and biochar effects on soil properties and tomato seedling growth under different moisture levels. Arch Agron Soil Sci 65:1704–1719. https://doi.org/10.1080/03650340.2019.1575510
51. Khan MN, Huang J, shah, A., Li, D., Daba, N.A., Han, T., Du, J. Qaswar, M., Anthonio, C.K., Sial, T.A., Haseeb, A., Zhang, L., Xu, Y., He, Z., Zhang, H., Núñez-Delgado, A., (2022) Mitigation of greenhouse gas emissions from a red acidic soil by using magnesium-modified wheat straw biochar. Environ Res 203. https://doi.org/10.1016/j.envres.2021.111879
52. Khan MN, Li D, Shah A, Huang J, Zhang L, Lan Z, Hayat S, Song Y, Bai Y, Zhang H (2022) The impact of pristine and modified rice straw biochar on the emission of greenhouse gases from a red acidic soil. Environ Res 208. https://doi.org/10.1016/j.envres.2022.112676
53. Kiatkittipong W, Wongsuchoto P, Pavasant P (2009) Life cycle assessment of bagasse waste management options. Waste Manag 29:1628–1633. https://doi.org/10.1016/j.wasman.2008.12.006
54. Kibler KM, Reinhart D, Hawkins C, Motlagh AM, Wright J (2018) Food waste and the food-energy-water nexus: A review of food waste management alternatives. Waste Manage 74:52–62. https://doi.org/10.1016/j.wasman.2018.01.014
55. Klich MA (2007) Aspergillus flavus: The major producer of aflatoxin. Mol Plant Pathol 8:713–722. https://doi.org/10.1111/j.1364-3703.2007.00436.x
56. Kodagoda K, Marapana R (2017) Development of non-alcoholic wines from the wastes of Mauritius pineapple variety and its physicochemical properties. J Pharmacogn Phytochem 6:492–497
57. Kumar A, Sengupta B, Dasgupta D, Mandal T, Datta S (2016) Recovery of value added products from rice husk ash to explore an economic way for recycle and reuse of agricultural waste. Rev Environ Sci Biotechnol 15:47–65. https://doi.org/10.1007/s11157-015-9388-0
58. Kumar AK, Sharma S (2017) Recent updates on different methods of pretreatment of lignocellulosic feedstocks: a review. Bioresour Bioprocess 4. https://doi.org/10.1186/s40643-017-0137-9
59. Laufenberg G, Kunz B, Nystroem M (2003) Transformation of vegetable waste into value added products: (A) the upgrading concept; (B) practical implementations. Biores Technol 87:167–198. https://doi.org/10.1016/S0960-8524(02)00167-0
60. Leyva-López N, Lizárraga-Velázquez CE, Hernández C, Sánchez-Gutiérrez EY (2020) Exploitation of agro-industrial waste as potential source of bioactive compounds for aquaculture. Foods 9:1–22. https://doi.org/10.3390/foods9070843
61. Li C, Bremer P, Harder MK, Lee MS, Parker K, Gaugler EC, Mirosa M (2022) A systematic review of food loss and waste in China: Quantity, impacts and mediators. J Environ Manage 303:114092. https://doi.org/10.1016/j.jenvman.2021.114092
62. Li Y, Jin Y, Li J, Chen Y, Gong Y, Li Y, Zhang J (2016) Current Situation and Development of Kitchen Waste Treatment in China. Procedia Environ Sci 31:40–49. https://doi.org/10.1016/j.proenv.2016.02.006
63. Luque R, Clark JH (2013) Valorisation of food residues: waste to wealth using green chemical technologies. Sustainable Chem Processes 1:10. https://doi.org/10.1002/ese3.9

64. Macedo ME (2018) Microfibrillated cellulose and high-value chemicals from orange peel residues. Ph.D thesis 209
65. Mateo-Sagasta J, Marjani S, Turral H, Burke J (2017) Water pollution from agriculture: a global review. FAO y IWMI 35
66. Mattsson L, Williams H, Berghel J (2018) Waste of fresh fruit and vegetables at retailers in Sweden—Measuring and calculation of mass, economic cost and climate impact. Resour Conserv Recycl 130:118–126. https://doi.org/10.1016/j.resconrec.2017.10.037
67. Merlin G, Boileau H (2013) Anaerobic digestion of agricultural waste: State of the art and future trends. Anaerobic Digestion: Types, Processes and Environmental Impact 29
68. Moody CM, Townsend TG (2017) A comparison of landfill leachates based on waste composition. Waste Manage 63:267–274. https://doi.org/10.1016/j.wasman.2016.09.020
69. Moreira R, Vaz R, Portugal A, Gil-Lalaguna N, Sánchez JL, Bimbela F (2018) Gasification of charcoal in air, oxygen, and steam mixtures over a γ-Al2O3 Fluidized Bed. Energy Fuels 32:406–415. https://doi.org/10.1021/acs.energyfuels.7b02257
70. Motte JC, Trably E, Escudié R, Hamelin J, Steyer JP, Bernet N, Delgenes JP, Dumas C (2013) Total solids content: a key parameter of metabolic pathways in dry anaerobic digestion. Biotechnol Biofuels 6:1–9. https://doi.org/10.1186/1754-6834-6-164
71. Nanda S, Berruti F (2021) Municipal solid waste management and landfilling technologies: a review. Environ Chem Lett 19:1433–1456. https://doi.org/10.1007/s10311-020-01100-y
72. Negi S, Banerjee R (2009) Optimization of extraction and purification of glucoamylase produced by Aspergillus awamori in solid-state fermentation. Biotechnol Bioprocess Eng 14:60–66. https://doi.org/10.1007/s12257-008-0107-3
73. Nierman WC, Yu J, Fedorova-Abrams ND, Losada L, Cleveland TE, Bhatnagar D, Bennett JW, Dean R, Payne GA (2015) Genome sequence of Aspergillus flavus NRRL 3357, a strain that causes aflatoxin contamination of food and feed. Genome Announc 3:4–5. https://doi.org/10.1128/genomeA.00168-15
74. Nigam JN (2000) Cultivation of Candida langeronii in sugar cane bagasse hemicellulosic hydrolyzate for the production of single cell protein. World J Microbiol Biotechnol 16:367–372. https://doi.org/10.1023/A:1008922806215
75. Ogbuewu IP, Odoemenam VU, Omede AA, Omede AA, Emenalom OO, Uchegbu MC, Okoli IC, Iloeje MU (2012) Livestock waste and its impact on the environment. Sci J Rev 1:17–32. https://doi.org/10.1177/1461444810365020
76. Papargyropoulou E, Lozano R, Steinberger K, J., Wright, N., Ujang, Z. Bin, (2014) The food waste hierarchy as a framework for the management of food surplus and food waste. J Clean Prod 76:106–115. https://doi.org/10.1016/j.jclepro.2014.04.020
77. Parfitt J, Barthel M, MacNaughton S (2010) Food waste within food supply chains: Quantification and potential for change to 2050. Philos Trans R Soc B: Biol Sci 365:3065–3081. https://doi.org/10.1098/rstb.2010.0126
78. Pirozzi A, Ferrari G, Donsì F (2022) Cellulose isolation from tomato pomace pretreated by High-Pressure homogenization. Foods 11:1–17. https://doi.org/10.3390/foods11030266
79. Plazzotta S, Manzocco L, Nicoli MC (2017) Fruit and vegetable waste management and the challenge of fresh-cut salad. Trends Food Sci Technol 63:51–59. https://doi.org/10.1016/j.tifs.2017.02.013
80. Pyar H, Peh KK (2018) Chemical compositions of banana peels (Musa sapientum) fruits cultivated in Malaysia using proximate analysis. Res J Chem Environ 22:108–113
81. Ramírez-García R, Gohil N, Singh V (2018) Recent advances, challenges, and opportunities in bioremediation of hazardous materials, phytomanagement of polluted sites: market opportunities in sustainable phytoremediation. https://doi.org/10.1016/B978-0-12-813912-7.00021-1
82. Ravindran R, Jaiswal AK (2016) Exploitation of food industry waste for High-Value products. Trends Biotechnol 34:58–69. https://doi.org/10.1016/j.tibtech.2015.10.008
83. Recebli Z, Selimli S, Ozkaymak M, Gonc O (2015) Biogas production from animal manure. J Eng Sci Technol 10:722–729. https://doi.org/10.1016/j.proeps.2015.08.144

84. Rehemujiang H, Yimamu A, Wang YL (2019) Effect of dietary cotton stalk on nitrogen and free gossypol metabolism in sheep. Asian Australas J Anim Sci 32:233–240. https://doi.org/10.5713/ajas.18.0057
85. Robbin Marks (2001) Cesspools of shame. Natural resources defense council and the clean water network
86. Rorat A, Vandenbulcke F (2019) Earthworms converting domestic and food industry wastes into biofertilizer. Emerging Concerns and Scope for Resource Recovery. Elsevier Inc., Industrial and Municipal Sludge. https://doi.org/10.1016/B978-0-12-815907-1.00005-2
87. Sadh PK, Duhan S, Duhan JS (2018) Agro-industrial wastes and their utilization using solid state fermentation: a review. Bioresources and Bioprocessing 5:1–15. https://doi.org/10.1186/s40643-017-0187-z
88. Sagar NA, Pareek S, Sharma S, Yahia EM, Lobo MG (2018) Fruit and vegetable waste: bioactive compounds, their extraction, and possible utilization. Compr Rev Food Sci Food Saf 17:512–531. https://doi.org/10.1111/1541-4337.12330
89. Salemdeeb R, zu Ermgassen, E.K.H.J., Kim, M.H., Balmford, A., Al-Tabbaa, A., (2017) Environmental and health impacts of using food waste as animal feed: a comparative analysis of food waste management options. J Clean Prod 140:871–880. https://doi.org/10.1016/j.jclepro.2016.05.049
90. Sarkar N, Ghosh SK, Bannerjee S, Aikat K (2012) Bioethanol production from agricultural wastes: an overview. Renewable Energy 37:19–27. https://doi.org/10.1016/j.renene.2011.06.045
91. Satlewal A, Agrawal R, Bhagia S, Das P, Ragauskas AJ (2018) Rice straw as a feedstock for biofuels: availability, recalcitrance, and chemical properties. Biofuels, Bioprod Biorefin 12:83–107. https://doi.org/10.1002/bbb.1818
92. Schreinemachers P, Tipraqsa P (2012) Agricultural pesticides and land use intensification in high, middle and low income countries. Food Policy 37:616–626. https://doi.org/10.1016/j.foodpol.2012.06.003
93. Setti M, Banchelli F, Falasconi L, Segrè A, Vittuari M (2018) Consumers' food cycle and household waste. When behaviors matter. J Clean Prod 185:694–706. https://doi.org/10.1016/j.jclepro.2018.03.024
94. Shaaban S, Nasr M (2020) Toward Three R's Agricultural Waste in MENA: Reduce. Reuse, and Recycle, in: Chapter 17:337–353. https://doi.org/10.1007/978-3-030-18350-9_17
95. Shahidi F, Ambigaipalan P (2015) Phenolics and polyphenolics in foods, beverages and spices: antioxidant activity and health effects—a review. J Funct Foods 18:820–897. https://doi.org/10.1016/j.jff.2015.06.018
96. Sharma VK, Testa C, Lastella G, Cornacchia G, Comparato MP (2000) Inclined-plug-flow type reactor for anaerobic digestion of semi-solid waste. Appl Energy 65:173–185. https://doi.org/10.1016/S0306-2619(99)00084-7
97. Shen F, Yuan H, Pang Y, Chen S, Zhu B, Zou D, Liu Y, Ma J, Yu L, Li X (2013) Performances of anaerobic co-digestion of fruit & vegetable waste (FVW) and food waste (FW): single-phase vs. two-phase. Biores Technol 144:80–85. https://doi.org/10.1016/j.biortech.2013.06.099
98. Sial TA, Khan MN, Lan Z, Kumbhar F, Zhao Y, Zhang J, Sun D, Xiu L (2018) Contrasting effects of banana peels waste and its biochar on greenhouse gas emissions and soil biochemical properties. Process Saf Environ Prot 122:366–377. https://doi.org/10.1016/j.psep.2018.10.030
99. Singh AP, Mustafi NN, Sharma YC, Agarwal AK (2020) Introduction to alternative fuels and their utilization strategies in internal combustion engines. Energy, Environ, Sustain. https://doi.org/10.1007/978-981-15-0418-1_1
100. Singh P, Mondal T, Sharma R, Mahalakshmi N, Gupta M (2018) Poultry waste management. Int J Curr Microbiol Appl Sci 7:2319–7706. https://doi.org/10.20546/ijcmas.2018.708.077
101. Soobhany N (2018) Preliminary evaluation of pathogenic bacteria loading on organic Municipal Solid Waste compost and vermicompost. J Environ Manage 206:763–767. https://doi.org/10.1016/j.jenvman.2017.11.029

102. Szymanska-Chargot M, Chylinska M, Gdula K, Koziol A, Zdunek A (2017) Isolation and characterization of cellulose from different fruit and vegetable pomaces. Polymers 9. https://doi.org/10.3390/polym9100495
103. Truong L, Morash D, Liu Y, King A (2019) Food waste in animal feed with a focus on use for broilers. Int J Recycl Org Waste Agric 8:417–429. https://doi.org/10.1007/s40093-019-0276-4
104. Tufail T, Saeed F, Afzaal M, Ain HBU, Gilani SA, Hussain M, Anjum FM (2021) Wheat straw: A natural remedy against different maladies. Food Sci Nutr 9:2335–2344. https://doi.org/10.1002/fsn3.2030
105. UNEP (2021) UN Environment Program's (UNEP) Food Waste Index Report
106. Ugwuoke CU, Monwuba N, Onu FM, Shimave AG, Okonkwo EN, Opurum CC (2018) Impact of agricultural waste on sustainable environment and health of rural women. Civ Environ Res 10:1–9
107. United Nation (2019) World population prospects 2019, United Nation
108. Vaish B, Srivastava V, Singh PK, Singh P, Singh RP (2020) Energy and nutrient recovery from agro-wastes: rethinking their potential possibilities. Environ Eng Res 25:623–637. https://doi.org/10.4491/eer.2019.269
109. Wang B, Dong F, Chen M, Zhu J, Tan J, Fu X, Wang Y, Chen S (2016) Advances in recycling and utilization of agricultural wastes in china: based on environmental risk, crucial pathways, influencing factors, policy mechanism. Procedia Environ Sci 31:12–17. https://doi.org/10.1016/j.proenv.2016.02.002
110. Wang B, Sun H, Wang D, Liu H, Liu J (2022) Constraints on the utilization of cereal straw in lactating dairy cows: a review from the perspective of systems biology. Animal Nutrition 9:240–248. https://doi.org/10.1016/j.aninu.2022.01.002
111. Wang Y, Zhang J, Sun Y, Yu J, Zheng Z, Li S, Cui Z, Hao J, Li G (2020) Effects of intermittent mixing mode on solid state anaerobic digestion of agricultural wastes. Chemosphere 248:126055. https://doi.org/10.1016/j.chemosphere.2020.126055
112. Wei Y, Li J, Shi D, Liu G, Zhao Y, Shimaoka T (2017) Environmental challenges impeding the composting of biodegradable municipal solid waste: a critical review. Resour Conserv Recycl 122:51–65. https://doi.org/10.1016/j.resconrec.2017.01.024
113. Wilkins MR, Suryawati L, Maness NO, Chrz D (2007) Ethanol production by *Saccharomyces cerevisiae* and *Kluyveromyces marxianus* in the presence of orange-peel oil. World J Microbiol Biotechnol 23:1161–1168. https://doi.org/10.1007/s11274-007-9346-2
114. Yang X, Liu J, McGrouther K, Huang H, Lu K, Guo X, He L, Lin X, Che L, Ye Z, Wang H (2016) Effect of biochar on the extractability of heavy metals (Cd, Cu, Pb, and Zn) and enzyme activity in soil. Environ Sci Pollut Res 23:974–984. https://doi.org/10.1007/s11356-015-4233-0
115. Yeilagi S, Rezapour S, Asadzadeh F (2021) Degradation of soil quality by the waste leachate in a Mediterranean semi-arid ecosystem. Sci Rep 11:1–12. https://doi.org/10.1038/s41598-021-90699-1
116. Yingkamhaeng N, Sukyai P (2014) The potential of mango peel utilization for cellulose extraction by hydrothermal pretreatment Naiyasit Yingkamhaeng, Prakit Sukyai*. The Potential of Mango Peel Utilization for Cellulose Extraction by Hydrothermal Pretreatment
117. Zhang C, Su H, Baeyens J, Tan T (2014) Reviewing the anaerobic digestion of food waste for biogas production. Renew Sustain Energy Rev 38:383–392. https://doi.org/10.1016/j.rser.2014.05.038

Organic Carbon Management and the Relations with Climate Change

Abderraouf Benslama, Fouzi Benbrahim, Jose Navarro-Pedreño, Ignacio Gómez Lucas, Manuel Miguel Jordán Vidal, and María Belén Almendro-Candel

Abstract Soil organic carbon (SOC) plays a key role in global warming mitigation. The type of SOC and the decomposition rate determine the organic matter storage in the soils. The environmental conditions, soil properties and land use and management are the main factors determining the ability of soil to store carbon. However, there are no common rules that can be applied for all the conditions, and it is important to understand the local and regional environmental processes to achieve the better strategy to increase SOC. Moreover, it is important to improve the methods to easily know the type of organic carbon stored and determine if SOC is increased by the increment of the stabilised and long-time resistant organic matter. More research is needed if we want to stablish a common framework of actions that can be used under different situations to reduce greenhouse emissions from agriculture and ensure the sequestration and storage of carbon in soils. Even more, all would be under a general SOC methodology to estimate the carbon content and their resilience to changes. The only generalization that can be made is that farm soils have lower SOC levels than soils with natural vegetation under similar environmental factors and this is the key to look for new strategies to combat climate change.

Keywords Global warming · Land management · Land use · Organic matter · Resilience

A. Benslama (✉)
Laboratoire de valorisation et conservation des écosystèmes arides (LVCEA), Faculté des Sciences de la Nature et de la Vie et Science de la Terre, Université de Ghardaïa, Ghardaïa, Algeria
e-mail: bensrofa@yahoo.fr

F. Benbrahim
École Normale Supérieure de Ouargla, BP 398, Haï Ennasr, 30000 Ouargla, Algeria

A. Benslama · J. Navarro-Pedreño (✉) · I. G. Lucas · M. M. J. Vidal · M. B. Almendro-Candel
Department of Agrochemistry and Environment, University Miguel Hernández of Elche, Avda. de la Universidad S/N, 03202 Elche, Spain
e-mail: jonavar@umh.es

A. Núñez-Delgado (ed.), *Frontier Studies in Soil Science*, Frontier Studies in Soil Science, https://doi.org/10.1007/978-3-031-50503-4_5

1 Introduction

Soil is an important resource, essential for filtering water, maintaining biodiversity, and storing carbon from the atmosphere. As a major carbon sink, soils play an important role in combating rising atmospheric greenhouse gas (GHG) concentrations [94]. This is why their use is at the heart of many sustainable development issues and objectives [2], like those proposed by United Nations for 2030.

Soils are dynamic living ecosystems that are much more complex than they appear. Insects, bacteria and fungi are essential organisms for soil carbon, i.e., they decompose organic matter (carbon-based plant and animal remains) and release carbon into the soil.

In some ecosystems, this carbon can remain in the soil for a long time, but the quality of carbon that can be stored in soils depends on local ecosystems, land use and land cover [94]. Regardless of its source or state of decomposition, soil organic carbon (SOC) is a measurement of the total amount of organic carbon in the soil. Its presence in soils is not only important for reducing global warming, but also a key of the carbon cycling and, it varies spatially depending on the environmental factors.

Interest in SOC is common among soil scientists and relevant practitioners because of its importance for key physical, chemical, biological, and ecological functions and because SOC is a universal indicator of soil quality [131].

C sequestration in soils has also been promoted as a strategy to mitigate the effects of increasing GHG to the atmosphere [58, 67, 72]. The creation of solutions to increase the stock of SOC and thus partially offset the increase in atmospheric carbon dioxide depends on carbon sequestration [89]. Sequestration and storing are closely related because the major source of organic matter into the soils are plants.

The United Nations Sustainable Development Goals, including Zero Hunger (SDG 2), Climate Action (SDG 13) and Life on Earth (SDG 15), recognise the importance of maintaining soil carbon stocks to reduce greenhouse gas emissions and ensure sustainable food security by 2030 [39]. Regarding global warming and food security, soils—carbon stock in the form of soil organic matter (SOM)—represent a crucial component [116, 155].

Although we know that the dynamics and storage of carbon can mitigate the negative effects of climate change, we still do not know well how all the processes work and how to measure the SOC along these processes, at any stage. This review gives a general overview of additional aspects, such as soil characteristics and land management, that affect how biological systems work and can be improved by adopting deliberate land use policies in order to address the reduction of atmospheric CO_2. This study will favour the increase of our knowledge of carbon stock variability and land use management methods across various land use types, guaranteeing sustainable land use for mitigation and adaptation strategies to climate change. Additionally, it will offer concise and current information on the primary causes of land use change to agricultural land managers, environmentalists, and legislators. To acquire accurate estimations of carbon sequestration potential, we must understand the effects of various land management techniques and land use. It has been estimated that the cumulative net CO_2 emissions from land use changes between 1750 and 2011 are estimated at approximately180 ± 80 PgC [20].

2 Importance of Knowledge of Soil Organic Carbon Content as the Main Terrestrial Carbon Reservoir

Soil organic carbon plays a major role in maintaining soil properties and is an important carbon reservoir sensitive to anthropogenic disturbances, including changes in land use or management [53]. It is critical the absence of basic rules and legislation related to the soil management changes in most of the countries. At the same time, the absence of a common framework to determine the effects of the land management changes in soil quality.

In terms of the environment, increased carbon storage in the form of soil organic matter could be crucial in averting climate change by limiting the emission of greenhouse gases into the atmosphere [9, 90]. The aims are to reduce SOC stock losses and explore SOC sequestration potential [63] in order to increase the presence of stabilised organic matter in the soils. On the other hand, soils constitute a significant source of carbon stored as soil organic matter (SOM), making them potentially extremely vulnerable to climate changes [1]. Soils can help to mitigate negative effects, but they are also subjected to them.

Organic carbon accounts for 50% of soil organic matter overall. It is a mixture of materials, mainly from living organisms that are more or less recognisable and mixed with inorganic fraction of the soil. The stock of organic carbon is constantly being renewed in the soil and determines many soil properties: soil fertility, its structure, and the biodiversity that inhabits soil. Depending on the climate, land use and management, the organic carbon supply may also be a source of greenhouse gases due the decomposition processes and the metabolism of the biota. Understanding the storage and destocking mechanisms of soil organic matter is therefore a crucial issue both for the sustainability of cropping systems and for environmental objectives [98].

Agricultural land undergoes SOC changes over time that are characterized by dynamic exchange processes that are influenced by environmental factors like soil texture, temperature, and precipitation as well as management techniques like cropping systems, fertilization, residue disposal, and tillage regimes [6, 27, 31, 63, 68, 138]. Therefore, carbon sequestration in soils, in the form of organic matter, is beneficial for the fertility of agricultural and forest soils, and also reduces CO_2 in the atmosphere, making them a major player in ongoing climate change and its mitigation.

Research on how climate change affects carbon stocks in agricultural and forest soils is an issue that is constantly expanding. As a result of the rising temperature, the data point to two opposing effects: an acceleration of carbon mineralization, or its conversion to CO_2, and an increase in biomass with carbon being recycled into the soil as organic matter. Contrarily, the process of releasing CO_2 into a land unit's

Fig. 1 Organic carbon storage process: humification

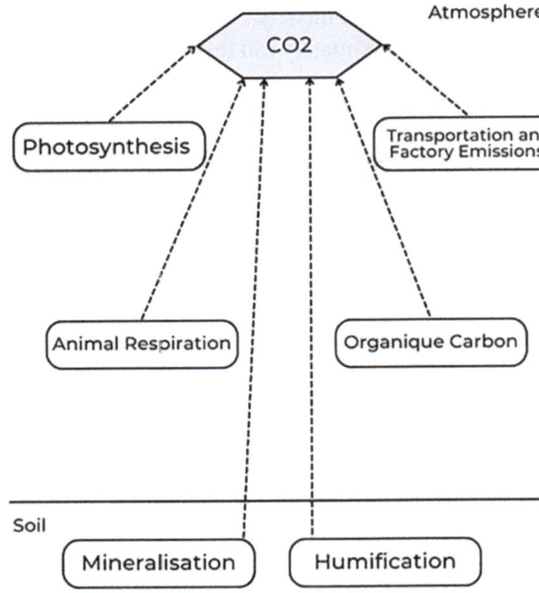

soil through plants, plant waste, and other organic solids helps the land unit retain or store carbon as a component of the soil organic matter (humus). The key to this process, as shown in Fig. 1, is to obtain stabilised organic matter (humified organic matter), which develops complex materials that can be kept in the soil for an extended period of time. The promotion of the humification seems to be the naturally strategy to increase SOC.

The retention time of organic carbon sequestered and stored in the soil (terrestrial reservoir) can range from short-term storage (but not immediately released to the atmosphere) to long-term storage (millennia), so the process of SOC sequestration is expected to increase net SOC storage [99, 100] if it is followed by adequate humification. In these circumstances, our society must understand and forecast the fate of organic carbon in soils in order to comprehend not only the current but also the projected worldwide changes [87].

2.1 *The Need for SOC Measurement Protocols*

The reduction potential in the agricultural sector is less precisely defined than in other sectors due to the largely diffuse nature of emissions, the complexity of the underlying biophysical and behavioural processes, and the vast diversity of production systems. However, it is crucial to fairly assess and contrast the costs of the several available levers for purposes of policymaking [105].

It is necessary to establish a soil sampling protocol to certify changes in organic carbon stock in mineral soils [131] and in organic soils (peatlands), and guidelines for measuring and modelling SOC stocks and stock changes in livestock production systems [40].

The United Nations Framework Convention on Climate Change (UNFCCC) [135] and its embodiments in the Kyoto Protocol [136] and the Paris Agreement [137] require reporting countries to monitor and report changes in SOC stock.

Regulations, procedures, and reporting requirements have been created as a result, including Decision 529/2013, Regulation 2018/841, and Regulation (EU) 525/2013 [35] on greenhouse gas emissions and removals from land use, land use change, and forestry (LULUCF) activities [36, 37].

The historical Paris Agreement, which was recently signed in 2015, intends to keep global warming well below 2 °C, ideally below 1.5 °C. Emission reductions must be combined with swiftly implemented mitigation measures in order to achieve this [92].

The role of United Nations and in this case, Food and Agricultural organization, is critical to establish and spread a protocol that can be useful to measure the changes of SOC in soils and account how these changes occur. This is a new step that should be applied after the establishment in 2017 the Global Soil Laboratory Network (GLOSOLAN) to build and strengthen the capacity of laboratories in soil analysis and to respond to the need for harmonizing soil analytical data.

2.2 *Agriculture and Mitigation Measures*

Reducing N_2O, CH_4 and CO_2 emissions, storing more carbon in soils and biomass, and creating bioenergy (biofuels, biogas) to replace fossil fuels can all be used by agriculture to help achieve national and international GHG reduction targets [104]. We know what to do but the way to do is subjected to many circumstances, including environmental, technical and socioeconomic ones. Additionally it is widely acknowledged that land use in agriculture may play the most important role in lowering and mitigating emissions because of the large surface occupied by cropping systems, close to 40% of land area [41]. Land use emissions are increasingly included in nationally determined contributions, NDCs, officially INDCs [92].

Numerous mitigation strategies have been put forth on a global or for individual countries, with agricultural sectors such as the International 4/1000 initiative (which calls for an increase in soil organic matter of 0.4% annually) launched at the 2015 United Nations Climate Change Conference in Paris highlighting the potential role of soils in mitigation activities. According to [71], a 0.4% increase in soil organic carbon (SOC) at a depth of 30 cm might reduce anthropogenic greenhouse gas (GHG) emissions by about 25% annually (using data from 2014 emissions) but also at different levels: at the global level [126, 127], at the scale of the continent [3], at the national level [44, 113, 123], for the livestock sector [93, 122], for soil organic carbon storage [129], for agroforestry [158].

Researchers have discovered, for instance, that the topsoil SOC in an agricultural system is lower than that of the nearby original forest soil because the farmland ecosystem exports a significant quantity of organic C, resulting in a limited supply of SOC [48]. For example, optimised agricultural practices with high organic inputs, permanent vegetation cover and reduced tillage can play a key role in soil carbon sequestration, defined in contrast to a reference cropping system (i.e., in ref. 84), and thus in climate change mitigation [8, 46, 109, 141].

The effects of conservation agriculture (CA) on soil carbon dynamics directly translate into an increase in carbon in the soil fraction because CA helps to mitigate climate change. By reducing mechanical soil manipulation, this can significantly slow down the oxidation of organic carbon [54] and a consequence, soil organic compounds could be more time inalterable in the soils.

To discover soil carbon sequestration/storing mechanisms, information on the consequences of intense cultivation with various agricultural systems and associated management approaches is required. Once discovered, this knowledge can help to create agricultural practices modified that have a great chance of improving SOC stocks, which would reduce the amount of CO_2 in the atmosphere and slow down global warming [85]. The three most crucial factors influencing SOC dynamics are the quantity of carbon intake, total SOC content, and environmental conditions (precipitation and temperature). These factors could account for the majority of the variation in SOC change rates [80].

Annual carbon losses from agricultural soils are thought to make up about 12% of all human-caused emissions on a worldwide scale [17]. The conversion of primary forests to agricultural land is therefore regarded as SOC depletion, whereas afforestation is seen as a way to replenish SOC supplies [26, 28, 50, 55, 83, 147]. Soils have lost huge amounts of organic carbon due to agriculture (Table 1), due to the conversion of natural ecosystems to cropland and, more recently, due to agricultural intensification in the face of demographic change [17]. On the other hand, The SOC pool is vulnerable to human intervention, particularly in the form of land use and land cover changes [147].

Table 1 Practices that increase or decrease the SOC

SOC	Increase SOC	Decrease SOC
Practices	– High organic inputs such as: manure, unharvested and unburned residues – Mineral nitrogen content – Irrigation – Fertilisation – Reduced tillage – Conservation agriculture – Afforestation	– Agricultural intensification – Human interference – Change in land use and land cover – Conversion of Natural Ecosystems to cropland – Intercropping

2.3 *Importance of Carbon Sequestration*

A critical component of keeping global warming to 2 °C is carbon sequestration [150]. Concerns about climate change's effects on world SOC stocks have been raised [147]. Therefore, SOC plays a key role in the C cycle and in mitigating global warming [152, 154]. In this cycle, sequestration by living organisms has the main role as a source of organic matter for soils.

One of the main reasons for the prominence of soil carbon sequestration in national GHG reduction programmes is the assumed "win–win" situation of mitigating GHG while improving food security through improved soil health and cropland fertility [97, 120, 132]. About one-third of the increase in atmospheric CO_2 is attributable to soil carbon loss due to changes in land use [101, 111]. When forecasting future carbon cycling and potential sequestration, specific consideration of soil temperature and texture is necessary [154] as both have a key role: formation of aggregates and soil structure, and temperature control biological activity and organic matter decomposition.

It is important to cope the physiological process of carbon sequestration by living organisms with the SOC accumulation and the mitigation of climate change, emphasizing on the essentiality of tackling adverse impacts of the ongoing process of climate change and engaging scientific and academic community in the depth-exploration of causes and outcomes of climate change impacts [65]. Carbon sequestration is linked to living organisms and those to environmental conditions and specially temperature.

3 The Relationship Between Soil Organic Carbon and Climate Change

Changes in soil organic matter can have a considerable impact on the GHG responsible for climate change since soil organic matter is recognized as a fundamental component of the global carbon cycle [15, 47, 49]. Soil is not only a sink but also a source of GHG. In this sense, we should be able to ensure conditions in cropping systems to increase the SOC stock in front of the emissions of GHG from soils. This is, in our opinion, one of the most important lines for research in Soil Science.

The distribution, composition, structure, and function of terrestrial ecosystems are significantly impacted by small changes in the soil's organic carbon storage, which in turn affects the availability of nutrients to terrestrial vegetation [76, 148]. Although there is a consensus that spatial variation in SOC is an important input to models used to understand the current and future carbon cycle and to predict global climate change, neither how to quantify the relative contribution of factors such as topography or canopy composition on SOC variation nor how the intensity of soil sampling affects the estimated fraction are completely clear [151].

Future climate and ecosystem quality will be impacted by how soil carbon dynamics respond to climate and land use change. The majority of the soil carbon pool is made up of deep soil carbon (deeper than 20 cm), yet little is known about its dynamics [87]. [87] suggested that additional analysis revealed that climate and agricultural practices had the greatest impact on the age of topsoil carbon. The latter demonstrated the high dependence of soil carbon dynamics on other soil properties, such as clay content and mineralogy, by confirming that the age of deep soil carbon was more affected by soil taxa than by climate. As a result, the future balance of soil carbon, or the balance between ecosystem production of dead organic matter and heterotrophic respiration, will have a significant impact on climate change [59]. Additionally, the SOC increment strikes a balance between the biota of the soil and the soil's inorganic composition, which would aid in the storage of SOC in the profile.

Not only temperature, but the importance of soil moisture was confirmed by [143] in their study on the distribution of SOC and N stocks in Bavaria based on agricultural regions. The Tertiary hills and loess regions, which had significant areas with potentially high soil moisture content in existing floodplains, had the highest SOC and N stocks for cropland. Due to low temperatures, heavy precipitation, and high soil moisture content in regions of glacial denudation, grassland soils demonstrated the largest accumulation of SOC and N in the Alps and Pre-Alps. The study done by [147], on the other hand, demonstrated that the rate of SOC sequestration was not only influenced by land use and land cover, but also by climatic conditions that interacted with land use and land cover. Higher rainfall seemed to slow the rate of SOC sequestration in the topsoil, but a warmer environment tended to speed up SOC accumulation. To better understand how SOC reacts to climate change, it is necessary to investigate how land use/land cover and climate interact [147]. One of the major concerns nowadays is the impact of climate change in boreal soils because the increment of temperature facilitates the decomposition of osil organic matter by the microorganisms [5].

Several researches have provided various insights into the geographical environment; [60] revealed that SOC increases with rising temperature and precipitation. Environmental factors have a bigger impact on SOC when the temperature is lower. Instead, the more substantial the effect of soil texture on SOC, the higher the temperature [154].

SOC is frequently predicted using climatic variables, such as precipitation and air temperature, as well as other covariables. As was previously demonstrated, some studies indicate that precipitation has a strong positive link with SOC, while air temperature has a strong negative correlation [7, 13, 38, 101]. It has been suggested that changes in precipitation patterns have an even greater impact on ecosystem dynamics than the singular or combined effects of increasing atmospheric CO_2 concentration and temperature [140]. Therefore, understanding ecosystem carbon cycle responses to changes in precipitation is of paramount importance for accurately predicting the rate and extent of climate change [57, 157]. The interconnections between soil, climate, vegetation cover, and land management determine how projected future climate change scenarios would affect the SOC and its dynamics [70, 78].

All of these findings suggested that there are not universal laws that apply everywhere, and it is important to know the local conditions, environmental factors, and

Table 2 Some examples of the expected positive and negative effects on SOC storage

Effect	Positive	Negative	References
LULC	Grassland	Cropland	[147]
Humidity	High	Low	[143]
Precipitation	High	Low	[60]
Temperature	Low	High	[154]
Texture	Clay soils	Sandy soils	[117]

soil type to understand and decide the best method to enhance SOC. Table 2 presents some of the more prevalent trends, nevertheless.

Changes in precipitation may have an impact on interactions with other global change factors by altering ecosystem structure and function and terrestrial carbon cycles [14], for instance, CO_2 increment in the atmosphere and global warming. Global climate models anticipate that future precipitation patterns would alter significantly [86] due to the intensification of the global water cycle as a result of global warming, according to [57]. This might have a significant impact on ecosystem carbon dynamics. Although several controlled experiments have been carried out to investigate how ecosystem carbon processes respond to changed precipitation, it is still unknown how soil carbon storage responds to both drought and irrigation throughout the world's biomes [157]. In fact, it is thought that enhanced forest and agricultural land management has a great potential for reducing CO_2 on a worldwide scale through the storage of atmospheric carbon in soils and irrigation would be a desirable factor to be used for organic carbon store in cropping systems.

Pastures are particularly significant in this context as they account for about 25% of the potential carbon sequestration in the world's soils [45], yet they have so far been largely under-resourced [107]. In addition, the relatively high sequestration rate combined with the large spatial extent of potential cropping areas leads to the conclusion that cover crop cultivation is a sustainable and effective measure to mitigate climate change [92]. In sum, agriculture, forestry, and other land uses are responsible for 24% (10–12 Pg CO_2 per year) of global greenhouse gas emissions, as the example of Spain, agriculture also accounts for 10% of GHG emissions [104].

4 Organic Carbon Storage Dynamics

Numerous studies have revealed that recent climate change has caused changes in the distribution of forest vegetation [73, 74, 112, 133], leading to changes in the carbon sequestration potential [11, 34, 149] and affecting the carbon cycle [19, 42, 156]. Forests land globally absorb roughly $2.5 + 1.3$ PgC yr^{-1}, or roughly one-third of our fossil emissions [20]. By accelerating the rate at which greenhouse gases are accumulating in the atmosphere, a decrease in the size of these sinks could hasten climate change [34].

Planning for climate change mitigation and adaptive management must take all of these factors into account because the consequences of climate change on forests differ depending on species, site circumstances, management, and fire regime [34]. In Florida (USA), sugarcane and wetlands had the largest SOC stocks, followed by improved pastures, urban areas, mesic upland forests, rangelands, and pine forests, while crops, citrus, and xeric upland forests had the lowest. These differences were reported in the study done by [147]. Temperate forests offer many opportunities to increase carbon sinks, however, the risk of negative effects on climate change and poor management decisions can limit these opportunities [34]. Therefore, forest soils are an important component of the global carbon cycle as they store large amounts of organic carbon [110].

4.1 The Effect of Temperature on Soil Organic Matter (SOM)

A significant amount of research has been done on the effects of temperature on microbial decomposition of SOM in the context of global climate change. For example, increasing temperature and decreasing precipitation together can increase the carbon storage capability of forest vegetation [43, 52, 62]. Numerous studies on the impact of temperature on the microbial degradation of SOM have been conducted in relation to global climate change (Kirschbaum 1995) [43, 52, 156]. According to certain research, temperature increases can hasten the microbial decomposition of SOM, which causes a net loss of carbon to the atmosphere [10, 25, 32]. However, other research showed that raising temperature can enhance soil organic carbon (SOC) by encouraging biomass input that outpaces an increase in breakdown [12, 95]. The decomposition process can also be affected by many factors, such as the type of organic matter and soil texture [75]. For example, the DNDC (DeNitrification-DeComposition) model can accurately simulate SOC and its dynamic change under specific climates ($R^2 = 0.96$) and can also make a long-term estimate [64, 153].

This suggests that the adoption of established technologies can sequester/store carbon at a rate of 5–10 kg/ha year on dry areas, 50–500 kg/ha year in pastures, 500–1000 kg/ha year in croplands, and 500–1000 kg/ha year in forests [150]. Soil management should therefore be prioritized for implementation if it is to be used to combat the issue of global warming [125]. Opposite to those results, [143], on the other hand, claim that climate effects (mean annual temperature and precipitation) had little impact on agricultural soils since management choices partially offset them, especially in cultivated soils. However, the results of [91] showed that SOC content is strongly correlated with precipitation and temperature under cropland and with texture and drainage under grassland, but not only that the combined effect of increasing temperature and decreasing precipitation can increase the SOC of forest vegetation [156].

4.2 Effect of Cropping Practices: Intercropping

As it seems, cropland has a huge potential global carbon sink combined with global food production needs, increasing due to the growing world population and the wealth of emerging economies. In comparison to soils with wild vegetation and under the same environmental conditions, farmland soils typically have lower SOC concentrations. Finding practical ways to boost SOC stocks while enhancing and sustaining high agricultural output is therefore essential [107].

Indeed, intercropping systems have a greater potential to reduce atmospheric carbon dioxide concentration compared to single cropping systems [103]. As well as [107] suggested, cover cropping is a sustainable and successful method to combat climate change based on the relatively high sequestration rate and the vast spatial breadth of viable cropping regions. Even more, cover crops and mulching systems can be useful to keep soil moisture at higher levels than conventional agriculture [114], and this effect can be an additional factor join to the application of organic wastes as a soil cover that can promote the increment of SOC.

Models must account for geographical and temporal variations in these agricultural practices in order to assess long-term changes in soil carbon, for instance on a European scale [21]. Agricultural practices, both for CO_2 and N_2O fluxes, as well as the consequences of climate change, all affect the GHG balance of croplands in Europe [51]. The following aspects of agricultural practices are affected: (i soil carbon supply, such as from manure and unharvested, unburned residues,(ii soil carbon decomposition, such as from timing and intensity of tillage, soil mineral nitrogen content, and irrigation; and (iii soil N_2O emissions from fertilization practices. According to [80], the variation in rates of SOC change observed in their study showed that the age of topsoil carbon was mainly affected by climate and cultivation (environment and land management).

4.3 Effect and the Role of Microbial Activity

According to estimates, the Earth's terrestrial carbon storage is made up of around two-thirds of soil organic carbon [130]. To find pathways for carbon sequestration in soils and to keep SOC at a level necessary for preserving soil health and also for controlling global warming, it is necessary to understand the dynamics of the carbon stock in soils as influenced by management practices [50] and the effects on soil microbiota. A lot of research has been done on the function and significance of soil biota, particularly microbial activity. The need of energy and nutrients to maintain the biological activity of soils is one of the most important bottlenecks that limits the increment of SOC. The other one is the temperature and the opposite effects from cold to warm temperature commented previously.

Environmental conditions control the microbial activity as far as the SOC, and in most cases at higher temperatures favour a faster degradation of organic matter in the soil due to a thermal boost to microbial activity [66]. Microbial activity and decomposition of soil organic matter depend on the amount of SOM, structure, texture, composition of biotic material input into soils, and environmental conditions, i.e. temperature, soil moisture, and nutrient status.

Recent research demonstrates that the amount of soil organic carbon protection (the clay-humus complex) and, subsequently, the accessibility of substrates for microbial activity, impact the rate of organic carbon breakdown [142]. Therefore, there are two paths that can be opposed. On the one hand, the biological activity of the soils is necessary for the transformation of fresh organic matter and residues used as amendments. On the other hand, the maintenance of this activity consumes and decomposes organic matter that can release GHG into the atmosphere. Reaching the balance and knowing what it is for each type of soil and environmental conditions, continues to be a major challenge for researchers.

5 Increasing SOC Stocks

Any plan to increase SOC is extremely difficult due to the ephemeral nature of carbon stored in soils and the lack of a thorough understanding of SOM stabilization mechanisms [70]. Increasing soil carbon stocks can have many advantages in addition to the potential for mitigation, such as better water infiltration and storage, improved nutrient cycling, increased land productivity, and increased below-ground and possibly above-ground biodiversity, all of which can improve livelihoods [92]. Again, the measures proposed relate to nitrogen management, livestock diets, and on-farm energy production and consumption reached great importance [104], avoiding possible nitrogen pollution, especially for water.

In fact, given the climate and land use/land cover, soils are projected to constitute a significant carbon sink [147]. Additionally, SOC stocks are often regulated by C outputs [121] like SOM mineralization, soil erosion, and export (including dissolved organic matter) as well as C inputs like residues, secretions, and exudates from plants, animals, and microorganisms [115, 154]. In this regard, increases in SOC concentration in the near-surface soil due to conservation agriculture led to improvements in soil physical conditions,these should contribute to increased sustainability and adaptation to climate change, even if they do not necessarily lead to a steady increase in agricultural yields [107]. In particular, wetland soils are likely to provide a substantial carbon sink [147], although they can constitute, mainly in younger stages and associated to the depth, an atmospheric GHG source [4, 124, 146]. Therefore, it is better to increase organic matter stocks through labile than stable inputs [107]. However, this should be studied in depth because not all the organic amendments increase carbon storage and some of them can also increase the CO_2 emissions, depending on environmental conditions and soil management [94].

Cover crops are crops that are sown in place of bare fallow during the winter and are ploughed in as green manure before planting the following major crop. They are also known as intercrops or catch crops [22]. The potential of cover crops to increase

SOC stocks and thus mitigate climate change has been demonstrated in very few studies [69] but the number of them is increasing in order to understand and enhance soil organic matter [119, 139]. Additionally, cover crops can help decrease fertilizer leaching and improve nutrient efficiency, decrease wind and water erosion, and manage pests, making them both environmentally and commercially advantageous in the long run [107].

The use of straw has been widely recommended as an environmentally friendly practice to manage carbon sequestration/storage in agricultural ecosystems. A general trend and magnitude of changes in soil C in response to the return of straw remains un-certain [77]. Active fractions of SOC [18, 84] and the emission of three major biogenic GHG, i.e., CO_2, CH_4, and N_2O, can all be increased by straw incorporation [33, 82, 159].

Therefore, raising soil organic matter through management approaches considers difficulties with soil carbon as well as agronomic, environmental, sociological, economic, and ethical aspects [77]. Despite extensive research on the subject, the impacts of environmental factors and farm management techniques on SOC are still unclear. For instance, some studies [23, 61, 134], found that higher temperatures increased SOC content on cultivated land, while other studies (finding that higher temperatures lowered it) discovered the opposite [96, 106].

From a management standpoint, conservation agriculture techniques including converting from monoculture to varied rotational cropping, minimizing soil disturbance through no-till techniques, and improving primary production through fertilization are typically thought to improve SOC content [29, 139, 141]. By lowering decomposition, raising sequestration, or doing both, these strategies raise SOC. The advantages of these agricultural methods heavily depend on the soil and climate conditions for certain cropping systems [81, 128]. According to Davidson and Janssens [24] environmental restrictions and the naturally diverse character of SOC have disguised the reactions of SOC dynamics to increased temperatures [155]. Other results show that SOC content is strongly correlated with precipitation and temperature under cropland and with texture and drainage under grassland [91].

On the other hand, mitigation of GHG emissions by increasing SOC stocks should not be seen as a way to compensate for CO_2 emissions from fossil fuel combustion or N_2O or CH_4 emissions from agricultural activities and thus allow business as usual to continue, but rather as an additional lever in the portfolio of options that countries can consider to reduce their agricultural GHG emissions [145]. The amount of CO_2 stored in soils must not be equal to CO_2 emissions due human activity; in such a way it is not possible to reduce CO_2 presented in the atmosphere.

Diverse inputs (fertilizers, organic matter, chemicals, water, tillage, etc.) are utilized throughout cultivation, mostly to boost yields. As a result, there are more crop wastes, which must be returned into the soil. It is typical to anticipate that intense plant cultivation using these inputs will result in a build-up of carbon in the soil. Contrarily, numerous researches demonstrate that with intensive farming employing contemporary inputs, the carbon content in soils has declined [85].

Soil de-sealing and creating Technosols maybe a good strategy to increase the carbon storage as a form of organic carbon recovering soil functions. Moreover, the pressure on agricultural and natural soils is growing due to the trend of increasing world population and the number of urban settlements [116], which may have a negative effect on the likelihood of using soils as a carbon storage to mitigate climate change. Anthropogenic mand-made soils can help in urban and peri-urban environments to increment the possibilities of SOC storage. Those including the use of green roofs in urban areas [118].

6 Conclusions

A lot of knowledge about dynamics and storage of carbon, including how climate change affects carbon stocks either directly or indirectly, have been accumulated mainly about laboratory, local and regional experiments. Indeed, most research has focused on biological aspects, some authors have shown that many soil properties modify the functions of ecological systems and are crucial for long-term storage, including the soil texture and the interactions between soil organic matter and inorganic particles. As a result, much remains to be known about the effects associated with these soil properties and carbon storage.

Increasing soil organic carbon stocks has additional benefits in terms of improving soil fertility and agricultural productivity. Improving sustainable cropland management could contribute significantly to CO_2 mitigation. It is important to note that the benefits associated with carbon sequestration go beyond CO_2 mitigation, as increased SOM is associated with improved soil fertility, soil structure, water retention capacity and thus higher productivity [144]. Sustainable cropland management should imply the increment of SOC because in other way, this management would not be considered as sustainable under the climate change scenery that we are living.

Incorporation of manure and crop residues, conservation tillage, fallowing, organic farming, bioenergy, perennial crops, improved management of harvested peat soils, improved pasture and livestock management, and conversion of cropland to grassland are promising agricultural practices to increase organic carbon stocks [30, 46, 102]. Thus, it is true that alternative cropping systems have the potential to sequester organic carbon under temperate climate conditions, through higher carbon input rather than through the effect of reduced tillage [8]. When residues are placed on the surface, as in some conservation agriculture practices, they are more susceptible to environmental factors (such as temperature and moisture) than when they are integrated into the soil profile [56].

Short-scale soil incubation experiments (days to year) have shown that the rate of soil organic carbon mineralisation could be increased by higher inputs of fresh organic matter [16]. A meta-analysis of 257 studies shows that nitrogen addition significantly increased above-ground, belowground and litter carbon pools by 35.7%, 23.0% and 20.9% respectively [79].

The incorporation of cover crops into cropping systems is a viable strategy for sequestering carbon in agricultural soils. Cover crops have an advantage over other management techniques for increasing soil organic carbon (SOC) since they don't diminish yields like extensification of farming or transfer carbon to other systems like organic manure treatments do [107]. Straw integration and extensification through arable and barley rotations, organic manure or sewage sludge supplements, and more recently, the growth of winter cover crops, have all been recommended methods for enhancing carbon inputs [88].

Numerous conservation agriculture techniques, such as zero tillage, boost the amount of SOC in the soil's surface layers, improving soil quality. However, these techniques do not always raise the stock of SOC, which is required to reduce climate change. Additionally, it is incorrect to assume that any management strategy that increases SOC stock mitigates climate change; instead, interpretation must take into account the mechanisms causing the increase in SOC and whether there is a net transfer of carbon from the atmosphere to the soil in comparison to the standard strategy [109, 129, 130]. The greatest amounts of SOC sequestration are achieved by taking agricultural land out of production and returning it to native grasslands or forests [127], but this land-use change is in conflict with food security objectives [108].

More research is needed if we want to stablish a common framework of actions that can be used under different situations to reduce greenhouse emissions and ensure the sequestration and storage of carbon in soils. Even more, all would be under a general SOC methodology to estimate the carbon content and their resilience to changes in all the countries as we need to know the global balance between soils and atmosphere. The only generalization that can be made is that farm soils have lower SOC levels than soils with natural vegetation under similar environmental factors, and to change this, is the challenge for researchers.

Acknowledgements This work has been possible in part due to the participation of corresponding author in the EU project LIFE CLIMAMED CCM/GR/000087.

References

1. Abdalla M, Hastings A, Chadwick DR, Jones DL, Evans CD, Jones MB, Rees RM, Smith P (2018) Critical review of the impacts of grazing intensity on soil organic carbon storage and other soil quality indicators in extensively managed grasslands. Agr Ecosyst Environ 253:62–81. https://doi.org/10.1016/j.agee.2017.10.023
2. Abera W, Tamene L, Abegaz A, Hailu H, Piikki K, Söderström M, Givrvetz E, Sommer R (2021) Estimating spatially distributed SOC sequestration potentials of sustainable land management practices in Ethiopia. J Environ Manage 286:112191. https://doi.org/10.1016/j.jenvman.2021.112191
3. Aertsens J, De Nocker L, Gobin A (2013) Valuing the carbon sequestration potential for European agriculture. Land Use Policy 31:584–594. https://doi.org/10.1016/j.landusepol.2012.09.003
4. Ahn C, Jones S (2013) Assessing organic matter and organic carbon contents in soils of created mitigation Wetlands in Virginia. Environ Eng Res 18(3):151–156. https://doi.org/10.4491/eer.2013.18.3.151

5. Allison SD, Treseder KK (2011) Climate change feedbacks to microbial decomposition in boreal soils. Fungal Ecol 4(6):362–374. https://doi.org/10.1016/j.funeco.2011.01.003
6. Alston JM, Beddow JM, Pardey PG (2009) Agricultural research, productivity, and food prices in the long run. Science 325:1209–1210. https://doi.org/10.1126/science.11704
7. Alvarez R, Lavado RS (1998) Climate, organic matter and clay content relationships in the Pampa and Chaco soils Argentina. Geoderma 83(1–2):127–141. https://doi.org/10.1016/S0016-7061(97)00141-9
8. Autret B, Mary B, Chenu C, Balabane M, Girardin C, Bertrand M, Grandeau G, Beaudoin N (2016) Alternative arable cropping systems: a key to increase soil organic carbon storage? Results from a 16 year field experiment. Agr Ecosyst Environ 232:150–164. https://doi.org/10.1016/j.agee.2016.07.008
9. Banwart S, Noellemeyer E, Milne E (2014) Soil Carbon: science, management and policy for multiple benefits, volume 71 of Scope. CABI, Wallingford
10. Bellamy PH, Loveland PJ, Bradley RI, Lark RM, Kirk GJD (2005) Carbon losses from all soils across England and Wales 1978–2003. Nature 437:245–248. https://doi.org/10.1038/nature04038
11. Boisvenue C, Running SW (2010) Impacts of climate change on natural forest productivity-evidence since the middle of the 20th century. Glob Change Biol 12:862–882. https://doi.org/10.1111/j.1365-2486.2006.01134.x
12. Bond-Lamberty B, Thomson A (2010) Temperature-associated increases in the global soil respiration record. Nature 464:579–582. https://doi.org/10.1038/nature08930
13. Burke IC, Yonker CM, Parton WJ, Cole CV, Flach K, Schimel DS (1989) Texture, climate, and cultivation effects on soil organic matter content in U.S. Grassland Soils. Soil Sci Soc Am J 53(3):800–805. https://doi.org/10.2136/sssaj1989.03615995005300030029x
14. Cable JM, Ogle K, Williams DG, Weltzin JF, Huxman TE (2008) Soil texture drives responses of soil respiration to precipitation pulses in the Sonoran Desert: implications for climate change. Ecosystems 11:961–979. https://doi.org/10.1007/s10021-008-9172-x
15. Caddeo A, Marras S, Sallustio L, Spano D, Sirca C (2019) Soil organic carbon in Italian forests and agroecosystems: estimating current stock and future changes with a spatial modelling approach. Agric For Meteorol 278:107654. https://doi.org/10.1016/j.agrformet.2019.107654
16. Cardinael R, Eglin T, Guenet B, Neill C, Houot S, Chenu C (2015) Is priming effect a significant process for long-term SOC dynamics? Analysis of a 52-years old experiment. Biogeochemistry 123(1):203–219. https://doi.org/10.1007/s10533-014-0063-2
17. Chaplot V (2021) Evidences of plants' impact on land degradation and climate change: an urgent call for new multi-disciplinary research. Geoderma 392:114984. https://doi.org/10.1016/j.geoderma.2021.114984
18. Chen H, Hou R, Gong Y, Gong Y, Li H, Fan M, Kuzyakov Y (2009) Effects of 11 years of conservation tillage on soil organic matter fractions in wheat monoculture in Loess Plateau of China. Soil Tillage Res 106:85–94. https://doi.org/10.1016/j.still.2009.09.009
19. Chiang J, Iverson L, Prasad A, Brown KJ (2008) Effects of climate change and shifts in forest composition on forest net primary production. J Integr Plant Biol 50:1426–1439. https://doi.org/10.1111/j.1744-7909.2008.00749.x
20. Ciais P, Sabine C, Bala G, Bopp L, Broykin V, Canadell J, Chahbra A, DeFries R, Galloway J, Heimann M, Jones C, Le Quéré C, Myneni RB, Piao S, Thornton P (2013) Carbon and other biogeochemical cycles supplementary material. In: Stocker TF, Qin D, Plattner G-K, Tignor M, Allen SK, Boschung J, Nauels A, Xia Y, Bex V, Midgley PM (eds) Climate change 2013: the physical science basis. Contribution of working group I to the fifth assessment report of the intergovernmental panel on climate change. www.climatechange2013.org and www.ipcc.ch
21. Ciais P, Wattenbach M, Vuichard N, Smith PC, Piao SL, Don A, Luyssaert§ S, Janssens§ IA, Bondeau A, Dechow KR, Leip A, Smith PC, CARBOEUROPE Synthesis Team (2010) The European carbon balance. Part 2: croplands. Glob Chang Biol 16(5):1409–1428. https://doi.org/10.1111/j.1365-2486.2009.02055.x

22. Dabney SM, Delgado JA, Reeves DW (2001) Using winter cover crops to improve soil and water quality. Commun Soil Sci Plant Anal 32:1221–1250. https://doi.org/10.1081/CSS-100104110
23. Dalias P, Anderson JM, Bottner P, Couteaux MM (2001) Long-term effects of temperature on carbon mineralisation processes. Soil biology and Biochemistry 33:1049–1057. https://doi.org/10.1016/S0038-0717(01)00009-8
24. Davidson A, Janssens IA (2006) Temperature sensitivity of soil carbon decomposition and feedbacks to climate change. Nature 440:165–173. https://doi.org/10.1038/nature04514
25. Davidson A, Trumbore E, Amundson R (2000) Soil warming and organic carbon content. Nature 408:789–790. https://doi.org/10.1038/35048672
26. Dawson JJC, Smith P (2007) Carbon losses from soil and its consequences for land-use management. Sci Total Environ 382(2–3):165–190. https://doi.org/10.1016/j.scitotenv.2007.03.023
27. De Gryze S, Lee J, Ogle S, Paustian K, Six J (2011) Assessing the potential for greenhouse gas mitigation in intensively managed annual cropping systems at the regional scale. Agr Ecosyst Environ 144:150–158. https://doi.org/10.1016/j.agee.2011.05.023
28. De Gryze S, Six J, Paustian K, Morris SJ, Paul EA, Merckx R (2004) Soil organic carbon pool changes following land-use conversions. Glob Change Biol 10(7):1120–1132. https://doi.org/10.1111/j.1365-2486.2004.00786.x
29. Deen W, Kataki PK (2003) Carbon sequestration in a long-term conventional versus conservation tillage experiment. Soil Tillage Res 74:143–150. https://doi.org/10.1016/S0167-1987(03)00162-4
30. Dendoncker N, Van Wesemael B, Rounsevell M, Roelandt C, Lettens S (2004) Belgium's CO_2 mitigation potential under improved cropland management. Agr Ecosyst Environ 103:101–116. https://doi.org/10.1016/j.agee.2003.10.010
31. Dolan S, Clapp C, Allmaras R, Baker J, Molina J (2006) Soil organic carbon and nitrogen in a Minnesota soil as related to tillage, residue and nitrogen management. Soil Tillage Res 89:221–231. https://doi.org/10.1016/j.still.2005.07.015
32. Dorrepaal E, Toet S, van Logtestijn R, Swart E, van de Weg M, Callaghan V, Aerts R (2009) Carbon respiration from subsur-face peat accelerated by climate warming in the subarctic. Nature 460:616–619. https://doi.org/10.1038/nature08216
33. Duiker S, Lal R (2000) Carbon budget study using CO_2 flux measurements from a no till system in central Ohio. Soil Tillage Res 54:21–30. https://doi.org/10.1016/S0167-1987(99)00101-4
34. Dymond CC, Beukema S, Nitschke CR, Coates KD, Scheller RM (2016) Carbon sequestration in managed temperate coniferous forests under climate change. Biogeosciences 13(6):1933–1947. https://doi.org/10.5194/bg-13-1933-2016
35. European Union (2013a) Decision no 529/2013/eu of the European parliament and of the council of 21 may 2013 on ac-counting rules on greenhouse gas emissions and removals resulting from activities relating to land use, land-use change and forestry and on information concerning actions relating to those activities
36. European Union (2013b) Regulation (eu) no 525/2013 of the European parliament and of the council of 21 may 2013 on a mechanism for monitoring and reporting greenhouse gas emissions and for reporting other information at national and union level relevant to climate change and repealing decision no 280/2004/ec
37. European Union (2018) Regulation (eu) 2018/841 of the European parliament and of the council of 30 May 2018 on the inclusion of greenhouse gas emissions and removals from land use, land use change and forestry in the 2030 climate and energy framework, and amending regulation (eu) no 525/2013 and decision no 529/2013/eu
38. Evans JR, Von Cammerer S (2013) Temperature response of carbon isotope discrimination and mesophyll conduct-ance in tobacco. Plant, Cell Environ 36:745–756. https://doi.org/10.1111/j.1365-3040.2012.02591.x
39. FAO (2018) FAO and the 17 Sustainable Developmnet Goals. Food and Agriculture Organsation of the United Nations, Rome. (Accessed 3 May 2018). http://www.fao.org/3/a-i4997e.pdf

40. FAO (2018) Measuring and modelling soil carbon stocks and stock changes in livestock production systems. In: Guide-lines for assessment (Draft for public review). Livestock Environmental Assessment and Performance (LEAP) Partner-ship. FAO, Rome, Italy
41. FAOSTAT (2023) FAO Statistical Database. In: https://www.fao.org/faostat/en/#home
42. Fei XH, Song QH, Zhang YP, Liu YT, Sha LQ, Yu GR, Zhang L, Duan C, Deng Y, Wu C, Lu Z, Luo K, Chen A, Xu K, Liu W, Huang H, Jin Y, Zhou R, Li J, Lin Y, Zhou L, Fu Y, Bai X, Tang X, Gao J, Zhou W, Grace J (2018) Carbon ex-changes and their responses to temperature and precipitation in forest ecosystems in Yunnan, Southwest China. Sci Total Environ 616:824–840. https://doi.org/10.1016/j.scitotenv.2017.10.239
43. Fierer N, Craine JM, McLauchlan K, Schimel JP (2005) Litter quality and the temperature sensitivity to decomposition. Ecology 86(2):320–326. https://www.jstor.org/stable/3450950
44. Fitton N, Ejerenwa CP, Bhogal A, Edgington P, Black H, Lilly A, Barraclough D, Worrall F, Hillier J, Smith P (2011) Greenhouse gas mitigation potential of agricultural land in Great Britain. Soil Use Manage 27:491–501. https://doi.org/10.1111/j.1475-2743.2011.00365.x
45. Follett RF, Reed DA (2010) Soil carbon sequestration in grazing lands: societal benefits and policy implications. Rangel Ecol Manage 63(1):4–15. https://doi.org/10.2111/08-225.1
46. Freibauer A, Rounsevell MDA, Smith P, Verhagen J (2004) Carbon sequestration in the agricultural soils of Eu-rope. Geoderma 122:1–23. https://doi.org/10.1016/j.geoderma.2004.01.021
47. Friedlingstein P, O'Sullivan M, Jones MW, Andrew RM, Hauck J, Olsen A, Peters GP, Peters W, Pongratz J, Sitch S, Le Quéré C, Canadell JG, Ciais P, Jackson RB, Alin S, Aragão LEOC, Arneth A, Arora V, Bates NR, Becker M, Benoit-Cattin A, Bittig HC, Bopp L, Bultan S, Chandra N, Chevallier F, Chini LP, Evans W, Florentie L, Forster PM, Gasser T, Gehlen M, Gilfillan D, Gkritzalis T, Gregor L, Gruber N, Harris I, Hartung K, Haverd V, Houghton RA, Ilyina T, Jain AK, Joetzjer E, Kadono K, Kato E, Kitidis V, Korsbakken JI, Landschützer P, Lefèvre N, Lenton A, Lienert S, Liu Z, Lombardozzi D, Marland G, Metzl N, Munro DR, Nabel JEMS, Nakaoka S-I, Niwa Y, O'Brien K, Ono T, Palmer PI, Pierrot D, Poulter B, Resplandy L, Robertson E, Rödenbeck C, Schwinger J, Séférian R, Skjelvan I, Smith AJP, Sutton AJ, Tanhua T, Tans PP, Tian H, Tilbrook B, van der Werf G, Vuichard N, Walker AP, Wanninkhof R, Watson AJ, Willis D, Wiltshire AJ, Yuan W, Yue X, Zaehle S (2020) Global carbon budget. Earth Syst Sci Data 12:3269–3340. https://doi.org/10.5194/essd-12-3269-2020
48. Gao X, Meng T, Zhao X (2017) Variations of soil organic carbon following land use change on deep-loess hillslopes in China. Land Degrad Dev 28(7):1902–1912. https://doi.org/10.1002/ldr.2693
49. Garnier J, Billen G, Tournebize J, Barré P, Mary B, Baudin F (2022) Storage or loss of soil active carbon in cropland soils: the effect of agricultural practices and hydrology. Geoderma 407:115538. https://doi.org/10.1016/j.geoderma.2021.115538
50. Genxu W, Ju Q, Guodong C, Yuanmin L (2002) Soil organic carbon pool of grassland soils on the Qinghai-Tibetan Plateau and its global implication. Sci Total Environ 291(1–3):207–217. https://doi.org/10.1016/s0048-9697(01)01100-7
51. Gervois S, Ciais P, Noblet-Ducoudre N, Brisson N, Vuichard N, Viovy N (2008) The carbon and water balance of European croplands throughout the 20th Century. Glob Biogeochem Cycles 22, GB2022. https://doi.org/10.1029/2007GB003018
52. Giardina CP, Ryan MG (2000) Evidence that decomposition rates of organic carbon in mineral soil do not vary with temperature. Nature 404:858–861. https://doi.org/10.1038/35009076
53. Gong H, Li Y, Li S (2021) Effects of the interaction between biochar and nutrients on soil organic carbon sequestration in soda saline-alkali grassland: a review. Glob Ecol Conserv 26:e01449. https://doi.org/10.1016/j.gecco.2020.e01449
54. González-Sánchez EJ, Ordóñez-Fernández R, Carbonell-Bojollo R, Veroz-González O, Gil-Ribes JA (2012) Meta-analysis on atmospheric carbon capture in Spain through the use of conservation agriculture. Soil Tillage Res 122:52–60. https://doi.org/10.1016/j.still.2012.03.001

55. Guo LB, Gifford RM (2002) Soil carbon stocks and land use change: a meta analysis. Glob Change Biol 8(4):345–360. https://doi.org/10.1046/j.1354-1013.2002.00486.x
56. Helgason BL, Gegorich EG, Janzen HH, Ellert BH, Lorenz N, Dick RP (2014) Long-term microbial retention of residue is site-specific and depends on residue placement. Soil Biol Biochem 68:231–240. https://doi.org/10.1016/j.soilbio.2013.10.002
57. Houghton R (2007) Balancing the global carbon budget. Annual Review of Earth and Planetary Sciences 35, 313–347. ttps://doi.org/https://doi.org/10.1146/annurev.earth.35.031306.140057.
58. Janzen HH (2004) Carbon cycling in earth systems - a soil science perspective. Agr Ecosyst Environ 104:399–417. https://doi.org/10.1016/j.agee.2004.01.040
59. Jones C, Cox P, Huntingford C (2003) Uncertainty in climate-carbon-cycle projections associated with the sensitivity of soil respiration to temperature. Tellus B 55:642–648. https://doi.org/10.3402/tellusb.v55i2.16760
60. Kang X, Li Y, Wang J, Yan L, Zhang X, Wu H, Yan Z, Zhang K, Hao Y (2020) Precipitation and temperature regulate the carbon allocation process in alpine wetlands: quantitative simulation. J Soils Sediments 20(9):3300–3315. https://doi.org/10.1007/s11368-020-02643-x
61. Keatterer T, Reichstein M, Andren O, Lomander A (1998) Temperature dependence of organic matter decomposition: a critical review using literature data analyzed with different models. Biol Fertil Soils 27:258–262. https://doi.org/10.1007/s003740050430
62. Kirschbaum MUF (1995) The temperature dependence of soil organic matter decomposition, and the effect of global warming on soil organic C storage. Soil Biol Biochem 27(6):753–60. https://doi.org/10.1016/0038-0717(94)00242-S
63. Knotters M, Teuling K, Reijneveld A, Lesschen JP, Kuikman P (2022) Changes in organic matter contents and carbon stocks in Dutch soils, 1998–2018. Geoderma 414:115751. https://doi.org/10.1016/j.geoderma.2022.115751
64. Ku H-H, Ryu J-H, Bae H-S, Jeong C, Lee S-E (2019) Modelling a long-term effect of rice straw incorporation on SOC content and grain yield in rice field. Arch Agron Soil Sci 65(14):1941–1954. https://doi.org/10.1080/03650340.2019.1583330
65. Kumar A (2022) Coping with climate change. In: Kumar A (ed) Ecosystem-based adaptation. Elsevier, Amsterdam, The Netherlands, pp 143–233. https://doi.org/10.1016/B978-0-12-815025-2.00004-6
66. Kutsch W, Bahn M, Heinemeyer A (2010) Soil carbon relations: an overview. Soil carbon dynamics: an integrated methodology. Cambridge University Press, Cambridge, UK, pp 1–15. https://doi.org/10.1017/CBO9780511711794.002
67. Lal R (2001) World cropland soils as a source or sink for atmospheric carbon. Adv Agron 71:145–191. https://doi.org/10.1016/S0065-2113(01)71014-0
68. Lal R (2004) Soil carbon sequestration impacts on global climate change and food security. Science 304:1623–1627. https://doi.org/10.1126/science.1097396
69. Lal R (2004) Soil carbon sequestration to mitigate climate change. Geoderma 123:1–22. https://doi.org/10.1016/j.geoderma.2004.01.032
70. Lal R (2013) Soil carbon management and climate change. Carbon Manag 4(4):439–462. https://doi.org/10.4155/cmt.13.31
71. Lal R (2015) Cover cropping and the "4 per Thousand" proposal. J Soil Water Conserv 70(6):141A. https://doi.org/10.2489/jswc.70.6.141A
72. Lal R, Kimble JM, Follett RF, Stewart BA (Eds) (1998) Soil processes and the carbon cycle. Boca Raton, USA: CRC Press. 609 pp.
73. Ledig FT, Rehfeldt GE, Sáenz-Romero C, Flores-López C (2010) Projections of suitable habitat for rare species under global warming scenarios. Am. J. Bot. 97:970–987. https://www.jstor.org/stable/20700429
74. Lenihan JM, Drapek R, Bachelet D, Neilson RP (2003) Climate change effects on vegetation distribution, carbon, and fire in California. Ecol Appl 13:1667–1681. https://www.jstor.org/stable/4134769
75. Li S, Li J, Li C, Huang S, Li X, Li S, Ma Y (2016) Testing the RothC and DNDC models against long-term dynamics of soil organic carbon stock observed at cropping field soils in

North China. Soil and Tillage Research. 163:290–297. https://doi.org/10.1016/j.still.2016.07.001
76. Liao S, Tan S, Peng Y, Wang D, Ni X, Yue K, Wu F, Yang Y (2020) Increased microbial sequestration of soil organic carbon under nitrogen deposition over China's terrestrial ecosystems. Ecol Process 9(1):52. https://doi.org/10.1186/s13717-020-00260-7
77. Liu C, Lu M, Cui J, Li B, Fang C (2014) Effects of straw carbon input on carbon dynamics in agricultural soils: a meta-analysis. Glob Change Biol 20(5):1366–1381. https://doi.org/10.1111/gcb.12517
78. Loehman RA, Keane RE, Holsinger LM (2020) Simulation modelling of complex climate, wildfire, and vegetation dynamics to address wicked problems in land management. Frontiers in Forests and Global Change 3:3. https://doi.org/10.3389/ffgc.2020.00003
79. Lu M, Zhou X, Luo Y, Yang Y, Fang C, Chen J, Li B (2011) Minor stimulation of soil carbon storage by nitrogen addition: a meta-analysis. Agr Ecosyst Environ 140(1–2):234–244. https://doi.org/10.1016/j.agee.2010.12.010
80. Luo Z, Feng W, Luo Y, Baldock J, Wang E (2017) Soil organic carbon dynamics jointly controlled by climate, carbon inputs, soil properties and soil carbon fractions. Glob Change Biol 23(10):4430–4439. https://doi.org/10.1111/gcb.13767
81. Luo Z, Wang E, Sun OJ (2010) Can no-tillage stimulate carbon sequestration in agricultural soils? A meta-analysis of paired experiments. Agr Ecosyst Environ 139:224–231. https://doi.org/10.1016/j.agee.2010.08.006
82. Ma J, Xu H, Yagi K, Cai Z (2008) Methane emission from paddy soils as affected by wheat straw returning mode. Plant Soil 313:167–174. https://doi.org/10.1007/s11104-008-9689-y
83. Maia SMF, Ogle SM, Cerri CEP, Cerri CC (2010) Soil organic carbon stock change due to land use activity along the agricultural frontier of the southwestern Amazon, Brazil, between 1970 and 2002. Glob Change Biol 16(10):2775–2788. https://doi.org/10.1111/j.1365-2486.2009.02105.x
84. Malhi SS, Nyborg M, Goddard T, Puurveen D (2011) Long-term tillage, straw and N rate effects on quantity and quality of organic C and N in a Gray Luvisol soil. Nutr Cycl Agroecosyst 90(1):1–20. https://doi.org/10.1007/s10705-010-9399-8
85. Mandal B, Majumder B, Bandyopadhyay PK, Hazra GC, Gangopadhyay A, Samantaray RN, Mishra AK, Chaudhury J, Saha MN, Kundu S (2007) The potential of cropping systems and soil amendments for carbon sequestration in soils under long-term experiments in subtropical India. Glob Change Biol 13(2):357–369. https://doi.org/10.1111/j.1365-2486.2006.01309.x
86. Marco-Dos Santos G, Meléndez-pastor I, Navarro-Pedreño J, Koch M (2019) Assessing water availability in Mediterranean regions affected by water conflicts through MODIS data time series analysis. Remote Sens 11:1355. https://doi.org/10.3390/rs11111355
87. Mathieu JA, Hatté C, Balesdent J, Parent É (2015) Deep soil carbon dynamics are driven more by soil type than by climate: a worldwide meta-analysis of radiocarbon profiles. Glob Change Biol 21(11):4278–4292. https://doi.org/10.1111/gcb.13012
88. Mazzoncini M, Sapkota TB, Bàrberi P, Antichi D, Risaliti R (2011) Long-term effect of tillage, nitrogen fertilization and cover crops on soil organic carbon and total nitrogen content. Soil Tillage Res 114:165–174. https://doi.org/10.1016/j.still.2011.05.001
89. McNally SR, Beare MH, Curtin D, Meenken ED, Kelliher FM, Calvelo Pereira R, Shen Q, Baldock J (2017) Soil carbon sequestration potential of permanent pasture and continuous cropping soils in New Zealand. Glob Change Biol 23(11):4544–4555. https://doi.org/10.1111/gcb.13720
90. McSherry ME, Ritchie ME (2013) Effects of grazing on grassland soil carbon: a global review. Glob Change Biol 19(5):1347–1357. https://doi.org/10.1111/gcb.12144
91. Meersmans J, van Wesemael B, Goidts E, Van Molle M, De Baets S, De Ridder F (2011) Spatial analysis of soil organic carbon evolution in Belgian croplands and grasslands, 1960–2006. Glob Change Biol 17(1):466–479. https://doi.org/10.1111/j.1365-2486.2010.02183.x
92. Milne E, Aynekulu E, Bationo A, Batjes NH, Boone R, Conant R, Davies J, Hanan N, Hoag D, Herrick E, Knaussenberger J, Neely W, Njoka C, Ngugi J, Patron B, Paustian K, Reid R, Said M, Shepherd M, Swift D, Thornton P, Wiliams S, Miller S, Nkonya E (2016) Grazing

lands in Sub-Saharan Africa and their potential role in climate change mitigation: what we do and don't know. Environ Dev 19:70–74
93. Monteny GJ, Bannink A, Chadwick D (2006) Greenhouse gas abatement strategies for animal husbandry. Agr Ecosyst Environ 112:163–170. https://doi.org/10.1016/j.agee.2005.08.015
94. Navarro-Pedreño J, Almendro-Candel MB, Zorpas AA (2021) The increase of soil organic matter reduces global warming, myth or reality? Sci 3(1):18. https://doi.org/10.3390/sci3010018
95. Nemani RR, Keeling CD, Hashimoto H, Jolly WM, Piper SC, Tucker CJ, Myneni RB, Running SW (2003) Climate-driven increases in global terrestrial net primary production from 1982 to 1999. Science 300(5625):1560–1563. https://doi.org/10.1126/science.1082750
96. Oechel WC, Vourlitis GL, Hastings SJ, Zulueta RC, Hinzman L, Kane D (2000) Acclimation of ecosystem CO_2 exchange in the Alaskan Arctic in response to decadal climate warming. Nature 406:978–981. https://doi.org/10.1038/35023137
97. Oldfield EE, Wood SA, Palm CA, Bradford MA (2015) How much SOM is needed for sustainable agriculture? Front Ecol Environ 13:527–527. https://doi.org/10.1890/1540-9295-13.10.527
98. Olorunfemi IE, Olufayo AA, Fasinmirin JT, Komolafe AA (2021) Dynamics of land use land cover and its impact on carbon stocks in Sub-Saharan Africa: an overview. Environ Dev Sustain 24:40–76. https://doi.org/10.1007/s10668-021-01484-z
99. Olson KR, Al-Kaisi MM, Lal R, Lowery B (2014) Experimental consideration, treatments, and methods in determining soil organic carbon sequestration rates. Soil Sci Soc Am J 78(2):348–360. https://doi.org/10.2136/sssaj2013.09.0412
100. Olson KR (2013) Soil organic carbon sequestration, storage, retention and loss in U.S. croplands: Issues paper for protocol development. Geoderma 195–196:201–206. https://doi.org/10.1016/j.geoderma.2012.12.004
101. Paramesh V, Kumar P, Nath AJ, Francaviglia R, Mishra G, Arunachalam V, Toraskar S (2022) Simulating soil organic carbon stock under different climate change scenarios: a RothC model application to typical land-use systems of Goa India. Catena 213:106129. https://doi.org/10.1016/j.catena.2022.106129
102. Paustian K, Andren O, Janzen HH, Lal R, Smith P, Tian G, Tiessen H, Van Noordwijk M, Woomer PL (1997) Agricultural soils as a sink to mitigate CO_2 emissions. Soil Use Manage 13:230–244. https://doi.org/10.1111/j.1475-2743.1997.tb00594.x
103. Peichl M, Thevathasan NV, Gordon AM, Huss J, Abohassan RA (2006) Carbon sequestration potentials in temperate tree-based intercropping systems, southern Ontario Canada. Agrofor Syst 66(3):243–257. https://doi.org/10.1007/s10457-005-0361-8
104. Pellerin S, Bamière L, Angers D, Béline F, Benoît M, Butault JP, Chenu C, Colnenne-David C, de Cara S, Delame N, Doreau M, Dupraz P, Faverdin P, Garcia-Launay F, Hassouna M, Hénault C, Jeuffroy MH, Klumpp K, Metay A, Moran D, Recous S, Samson E, Savini I, Pardon L (2013) How can French agriculture contribute to reducing greenhouse gas emissions? Abatement potential and cost of ten technical measures. [Technical Report] 2013. Hal-INRAE. https://hal.inrae.fr/hal-02809908
105. Pellerin S, Bamière L, Angers D, Béline F, Benoit M, Butault JP, Chenu C, Colnenne-David C, De Cara S, Delame N, Doreau M, Dupraz P, Faverdin P, Garcia-Launay F, Hassouna M, Hélault C, Jeuffroy M-H, Klumpp K, Metay A, Moran D, Recous S, Samson E, Savini I, Pardon L, Chemineau P (2017) Identifying cost-competitive greenhouse gas mitigation potential of French agriculture. Environ Sci Policy 77:130–139. https://doi.org/10.1016/j.envsci.2017.08.003
106. Peterjohn WT, Melillo JM, Steudler PA, Newkirk KM, Bowles FP, Aber JD (1994) Responses of trace gas fluxes and N availability to experimentally elevated soil temperatures. Ecol Appl 4:617–625. https://doi.org/10.2307/1941962
107. Poeplau C, Don A (2015) Carbon sequestration in agricultural soils via cultivation of cover crops-A meta-analysis. Agr Ecosyst Environ 200:33–41. https://doi.org/10.1016/j.agee.2014.10.024

108. Powlson DS, Stirling CM, Thierfelder C, White RP, Jat ML (2016) Does conservation agriculture deliver climate change mitigation through soil carbon sequestration in tropical agro-ecosystems? Agr Ecosyst Environ 220:164–174. https://doi.org/10.1016/j.agee.2016.01.005
109. Powlson DS, Glendining MJ, Coleman K, Whitmore AP (2011) Implications for soil properties of removing cereal straw: results from long-term studies. Agron J 103:279. https://doi.org/10.2134/agronj2010.0146s
110. Prietzel J, Christophel D (2014) Organic carbon stocks in forest soils of the German Alps. Geoderma 221:28–39. https://doi.org/10.1016/j.geoderma.2014.01.021
111. Pugh TAM, Arneth A, Olin S, Ahlström A, Bayer AD, Klein Goldewijk K, Lindeskog M, Schurgers G (2015) Simulated carbon emissions from land-use change are substantially enhanced by accounting for agricultural management. Environ Res Lett 10:12–124008. https://doi.org/10.1088/1748-9326/10/12/124008
112. Rachael H, Davidb R, Janek H, Richard F, Chrisd T (2010) The distributions of a wide range of taxonomic groups are ex-panding polewards. Glob Change Biol 12:450–455. https://doi.org/10.1111/j.1365-2486.2006.01116.x
113. Rees RM, Baddeley JA, Bhogal A, Ball BC, Chadwick DR, Macleod M, Lilly A, Pappa VA, Thorman RE, Watson CA, Williams JR (2013) Nitrous oxide mitigation in UK agriculture. Soil Sci Plant Nutr 59:3–15. https://doi.org/10.1080/00380768.2012.733869
114. Rico JR, Navarro-Pedreño J, Gómez I (2016) Evaluation of plant waste used as mulch on soil moisture retention. Span J Soil Sci 6(2):133–144. https://doi.org/10.3232/SJSS.2016.V6.N2.05
115. Robinson JM, Barker SLL, Arcus VL, McNally SR, Schipper LA (2020) Contrasting temperature responses of soil respiration derived from soil organic matter and added plant litter. Biogeochemistry 150(1):45–59. https://doi.org/10.1007/s10533-020-00686-3
116. Rodríguez-Espinosa T, Navarro-Pedreño J, Gómez-Lucas I, Jordán-Vidal MM, Bech-Borras J, Zorpas AA (2021) Urban areas, human health and technosols for the green deal. Environ Geochem Health 43(12):5065–5086. https://doi.org/10.1007/s10653-021-00953-8
117. Routhier M, Lafleur B, Bélanger N (2014) Accumulation des stocks de carbone dans les sols sous des cultures bioénergétiques de Populus spp., Salix spp. et Panicum Virgatum. VertigO - la revue électronique en sciences de l'environnement 14(2). https://doi.org/10.4000/vertigo.15076
118. Shafique M, Xue X, Luo X (2020) An overview of carbon sequestration of green roofs in urban areas. Urban For & Urban Green 47:126515. https://doi.org/10.1016/j.ufug.2019.126515
119. Sánchez-Navarro V, Martínez-Martínez S, Acosta JA, Almagro M, Martínez-Mena M, Boix-Fayos C, Díaz-Pereira E, Temnani A, Berrios P, Pérez-Pastor A, Zornoza R (2023) Soil greenhouse gas emissions and crop production with implementation of alley cropping in a Mediterranean citrus orchard. Eur J Agron 142:126684. https://doi.org/10.1016/j.eja.2022.126684
120. Sanderman J, Creamer C, Baisden WT, Farrell M, Fallon S (2017) Greater soil carbon stocks and faster turnover rates with increasing agricultural productivity. Soil 3(1):1–16. https://doi.org/10.5194/soil-3-1-2017
121. Sasmito SD, Kuzyakov Y, Lubis AA, Murdiyarso D, Hutley LB, Bachri S, Friess DA, Martius C, Borchard N (2020) Organic carbon burial and sources in soils of coastal mudflat and mangrove ecosystems. CATENA 187:104414. https://doi.org/10.1016/j.catena.2019.104414
122. Schils RLM, Eriksen J, Ledgard SF, Vellinga TV, Kuikman PJ, Luo J, Petersen SO, Velthof GL (2013) Strategies to mitigate nitrous oxide emissions from herbivore production systems. Animal 7:29–40. https://doi.org/10.1017/s175173111100187x
123. Schneider UA, McCarl BA, Schmid E (2007) Agricultural sector analysis on greenhouse gas mitigation in US agriculture and forestry. Agric Syst 94:128–140. https://doi.org/10.1016/j.agsy.2006.08.001
124. Sirivedhin T, Gray KA (2006) Factors affecting denitrification rates in experimental wetlands: field and laboratory studies. Ecol Eng 26(2):167–181. https://doi.org/10.1016/j.ecoleng.2005.09.001

125. Smith P (2008) Land use change and soil organic carbon dynamics. Nutr Cycl Agroecosyst 81(2):169–178. https://doi.org/10.1007/s10705-007-9138-y
126. Smith P, Haberl H, Popp A, Erb KH, Lauk C, Harper R, Tubiello FN, Pinto AD, Jafari M, Sohi S, Masera O, Bottcher H, Berndes G, Bustamante M, Ahammad H, Clark H, Dong HM, Elsiddig EA, Mbow C, Ravin-dranath NH, Rice CW, Abad CR, Romanovskaya A, Sperling F, Herrero M, House JI, Rose S (2013) How much land-based greenhouse gas mitigation can be achieved without compromising food security and environmental goals? Glob Change Biol 19:2285–2302. https://doi.org/10.1111/gcb.12160
127. Smith P, Martino D, Cai Z, Gwary D, Janzen H, Kumar P, McCarl B, Ogle S, O'Mara F, Rice C, Scholes B, Si-rotenko O, Howden M, McAllister T, Pan G, Romanenkov V, Schneider U, Towprayoon S, Wattenbach M, Smith J (2008) Greenhouse gas mitigation in agriculture. Philos Trans R Soc B: Biol Sci 363:789–813. https://doi.org/10.1098/rstb.2007.2184
128. Smith P, Powlson DS, Glendining MJ, Smith JU (1997) Potential for carbon sequestration in European soils: preliminary estimates for five scenarios using results from long-term experiments. Glob Change Biol 3:67–79. https://doi.org/10.1046/j.1365-2486.1997.00055.x
129. Sommer R, Bossio D (2014) Dynamics and climate change mitigation potential of soil organic carbon sequestration. J Environ Manage 144:83–87. https://doi.org/10.1016/j.jenvman.2014.05.017
130. Stockmann U, Adams MA, Crawford JW, Field DJ, Henakaarchchi N, Jenkins M, Minasny B, McBratney A, Remy de Courcelles V, Singh K, Wheeler I, Abbott L, Angers DA, Baldock J, Bird M, Brookes PC, Chenu C, Jastrow JD, Lal R, Lehman J, O'Donnell AG, Parton WJ, Whitehead D, Zimmermann M (2013) The knowns, known unknowns and unknowns of sequestration of soil organic carbon. Agr Ecosyst Environ 164:80–99. https://doi.org/10.1016/j.agee.2012.10.001
131. Stolbovoy V, Montanarella L, Filippi N, Jones A, Gallego J, Grassi G (2007) Soil sampling protocol to certify the changes of organic carbon stock in mineral soil of the European Union. techreport EUR 21576 EN/2. European Commission, Joint Research Centre
132. Tiessen H, Cuevas E, Chacon P (1994) The role of soil organic matter in sustaining soil fertility. Nature 371:783–785. https://doi.org/10.1038/371783a0
133. Trindade W, Santos MH, Artoni RF (2020) Climate change shifts the distribution of vegetation types in south Brazilian hotspots. Reg Environ Change 20:90. https://doi.org/10.1007/s10113-020-01686-7
134. Trumbore SE, Chadwick OA, Amundson R (1996) Rapid exchange between soil carbon and atmospheric carbon dioxide driven by temperature change. Science 272:393–396. https://doi.org/10.1126/science.272.5260.393
135. United Nations (1992) United nations framework convention on climate change. FCCC/INFORMAL/84. GE.05-62220 (E) 200705
136. United Nations (1998) Kyoto protocol to the United Nations Framework Convention on Climate Change. In: https://unfccc.int/resource/docs/convkp/kpeng.pdf
137. United Nations (2015) Paris agreement. In: https://unfccc.int/process-and-meetings/the-paris-agreement/the-paris-agreement
138. Van Wesemael B, Paustian K, Meersmans J, Goidts E, Barancikova G, Easter M (2010) Agricultural management ex-plains historic changes in regional soil carbon stocks. Proc Natl Acad Sci 107:14926–14930. https://doi.org/10.1073/pnas.1002592107
139. Vanino S, Di Bene C, Piccini C, Fila G, Pennelli B, Zornoza R, Sanchez-Navarro V, Álvaro-Fuentes J, Hüppi R, Six J, Farina R (2022) A comprehensive assessment of diversified cropping systems on agro-environmental sustainability in three Mediterranean long-term field experiments. Eur J Agron 140:126598. https://doi.org/10.1016/j.eja.2022.126598
140. Weltzin JF, Loik ME, Schwinning S, Williams DG, Fay PA, Haddad BM, Harte J, Huxman TE, Knapp AK, Lin G (2003) Assessing the response of terrestrial ecosystems to potential changes in precipitation. Bioscience 53:941–952. https://doi.org/10.1641/0006-3568(2003)053[0941:ATROTE]2.0.CO;2
141. West TO, Post WM (2002) Soil organic carbon sequestration rates by tillage and crop rotation. Soil Sci Soc Am J 66:1930–1946. https://doi.org/10.2136/sssaj2002.1930

142. Whitehead D, Schipper LA, Pronger J, Moinet GY, Mudge PL, Pereira RC, Kirschbaum MUF, McNally SR, Beare MH, Camps-Arbestain M (2018) Management practices to reduce losses or increase soil carbon stocks in temperate grazed grasslands: New Zealand as a case study. Agr Ecosyst Environ 265:432–443. https://doi.org/10.1016/j.agee.2018.06.022
143. Wiesmeier M, Hübner R, Barthold F, Spörlein P, Geuß U, Hangen E, Reischl A, Schilling B, von Lützow M, Kögel-Knabner I (2013) Amount, distribution and driving factors of soil organic carbon and nitrogen in cropland and grassland soils of southeast Germany (Bavaria). Agr Ecosyst Environ 176:39–52. https://doi.org/10.1016/j.agee.2013.05.012
144. Wiesmeier M, Hübner R, Spörlein P, Geuß U, Hangen E, Reischl A, Schilling B, von Lützow M, Kögel-Knabner I (2014) Carbon sequestration potential of soils in southeast Germany derived from stable soil organic carbon saturation. Glob Change Biol 20:653–665. https://doi.org/10.1111/gcb.12384
145. Wollenberg E, Richards M, Smith P, Havlík P, Obersteiner M, Tubiello FN, Herold M, Gerber P, Carter S, Reisinger A, van Vuuren DP, Dickie A, Neufeldt H, Sander BO, Wassmann R, Sommer R, Amonette JE, Fal-cucci A, Herrero M, Opio C, Roman-Cuesta RM, Stehfest E, Westhoek H, Ortiz- Monasterio I, Sapkota T, Rufi-no MC, Thornton PK, Verchot L, West PC, Soussana JF, Baedeker T, Sadler M, Vermeulen S, Campbell BM (2016) Reducing emissions from agriculture to meet the 2°C target. Glob Change Biol 22:3859–3864. https://doi.org/10.1111/gcb.13340
146. Wu H, Zhang J, Ngo HH, Guo W, Hu Z, Liang S, Fan J, Liu H (2015) A review on the sustainability of constructed wetlands for wastewater treatment: design and operation. Biores Technol 175:594–601. https://doi.org/10.1016/j.biortech.2014.10.068
147. Xiong X, Grunwald S, Myers DB, Ross CW, Harris WG, Comerford NB (2014) Interaction effects of climate and land use/land cover change on soil organic carbon sequestration. Sci Total Environ 493:974–982. https://doi.org/10.1016/j.scitotenv.2014.06.088
148. Xu Z, Yu G, Zhang X, He N, Wang Q, Wang S, Xu X, Wang R, Zhao N (2018) Biogeographical patterns of soil microbial community as influenced by soil characteristics and climate across Chinese forest biomes. Appl Soil Ecol 124:298–305. https://doi.org/10.1016/j.apsoil.2017.11.019
149. Yao YT, Piao SL, Wang T (2018) Future biomass carbon sequestration capacity of Chinese forests. Sci Bull 63:1108–1117. https://doi.org/10.1016/j.scib.2018.07.015
150. Yin S, Zhang X, Lyu J, Zhi Y, Chen F, Wang L, Chunjiang L, Zhou S (2020) Carbon sequestration and emissions mitigation in paddy fields based on the DNDC model: a review. Artif Intell Agric 4:140–149. https://doi.org/10.1016/j.aiia.2020.07.002
151. Yuan Z, Gazol A, Lin F, Ye J, Shi S, Wang X, Wang M, Hao Z (2013) Soil organic carbon in an old-growth temperate forest: spatial pattern, determinants and bias in its quantification. Geoderma 195:48–55. https://doi.org/10.1016/j.geoderma.2012.11.008
152. Zhang Y, Li P, Liu X, Xiao L, Shi P, Zhao B (2019) Effects of farmland conversion on the stoichiometry of carbon, nitro-gen, and phosphorus in soil aggregates on the Loess Plateau of China. Geoderma 351:188–196. https://doi.org/10.1016/j.geoderma.2019.05.037
153. Zhang L, Shao J (2017) Dynamic simulation of farmland SOC in parallel ridge valley area in eastern Sichuan in the next 30 years. Chin J Eco-Agric 12:1848–1857
154. Zhang Y, Li P, Liu X, Xiao L, Li T, Wang D (2022) The response of soil organic carbon to climate and soil texture in China. Front Earth Sci:1–11.https://doi.org/10.1007/s11707-021-0940-7
155. Zhao G, Bryan BA, King D, Luo Z, Wang E, Song X, Yu Q (2013) Impact of agricultural management practices on soil organic carbon: simulation of Australian wheat systems. Glob Change Biol 19(5):1585–1597. https://doi.org/10.1111/gcb.12145
156. Zhou R, Zhang Y, Peng M, Jin Y, Song Q (2022) Effects of climate change on the carbon sequestration potential of forest vegetation in Yunnan Province, southwest China. Forests 13(2):306. https://doi.org/10.3390/f13020306
157. Zhou X, Zhou L, Nie Y, Fu Y, Du Z, Shao J, Zheng Z, Wang X (2016) Similar responses of soil carbon storage to drought and irrigation in terrestrial ecosystems but with contrasting mechanisms: A meta-analysis. Agr Ecosyst Environ 228:70–81. https://doi.org/10.1016/j.agee.2016.04.030

158. Zomer RJ, Neufeldt H, Xu JC, Ahrends A, Bossio D, Trabucco A, van Noordwijk M, Wang MC (2016) Global tree cover and biomass carbon on agricultural land: the contribution of agroforestry to global and national carbon budgets. Sci Rep 6:12. https://doi.org/10.1038/srep29987
159. Zou JW, Huang Y, Lu YY, Zheng X, Wang Y (2005) Direct emission factor for N_2O from rice-winter wheat rotation systems in southeast China. Atmos Environ 39:4755–4765. https://doi.org/10.1016/j.atmosenv.2005.04.028

Agricultural Soil Management Impacts on Soil Microbial Communities

Lyudmyla Symochko and Paulo Pereira

Abstract Soil microbial communities are significantly affected by agricultural soil management. Inappropriate management can be a cause of land degradation and ecosystem services loss. This chapter aims to assess the impacts of agricultural soil management on soil microbiome. In large amounts, fertilizers and pesticides can affect the soil microbial structure and communities and change ground interactions between different ecological-trophic groups of microorganisms. Soil microbiological activity directly impacts plant biomass production and determines the ecological state of soils. The excessive use of mineral fertilizers affects the mineralization process and reduces microbiological diversity and soil organic matter. Organic fertilizers can positively affect soil properties (e.g., soil structure, fertility) and soil microbiome (diversity, microbial biomass, activity). Nevertheless, before using it, it is essential to identify the presence of antibiotic residues not to increase soil pollution and prevent it from entering the food chain. Control of contamination by antimicrobial substances is crucial for the elimination of spreading antibiotic-resistant bacteria in the environment. This is an additional tool for the prevention formation of soil resistome.

Keywords Soil microbiome · Agriculture · Mineral fertilizers · Organic fertilizers · Resistome

L. Symochko
University of Coimbra, Coimbra, Portugal

Uzhhorod National University, Uzhhorod, Ukraine

P. Pereira (✉)
Environmental Management Laboratory, Mykolas Romeris University, Vilnius, Lithuania
e-mail: pereiraub@gmail.com

A. Núñez-Delgado (ed.), *Frontier Studies in Soil Science*, Frontier Studies in Soil Science, https://doi.org/10.1007/978-3-031-50503-4_6

1 Introduction

Different agriculture practices can significantly alter the soil's biogeochemical, biological and microbiological properties. Soil microbiome plays a vital role in forming soil fertility and productivity of agroecosystems. It is one of the most critical factors in forming a healthy environment [23, 47, 52, 58]. Soil microbial biomass, community structure, and microbial diversity are sensitive indicators of soil quality. The structure of microbial communities and microbial activity can impact plant growth and crop yield [6, 35, 63, 65]. Changes in soil microbiome functional activity and integrity reduce soil quality and agroecosystems productivity. In particular, soil microbial biomass, activity and diversity are indicators of soil fertility and ecosystem productivity. For this reason, they are used as indicators of soil quality and health. The ecological status of soil depends on qualitative and quantitative changes in the microbiome structure, ratio, and several certain ecological trophic groups of microorganisms [50, 56]. Any microbiome consists of microorganisms of different functional and taxonomic groups, which differ in their requirements for nutritional conditions and energy sources. There are specific ecological niches in the soil with properties inherent only to them. These are caused by the combined action of several environmental factors, such as temperature, moisture, the reaction of the environment (pH) and anthropogenic activity. The objective of this book chapter is to assess the impacts of agriculture practices on microbial communities.

2 Effect of Mineral Fertilizers (Nitrogen-Phosphorous-Potassium) on Soil Microbial Communities and Their Activity

The impact of agro-technologies on soil microbiomes can be either direct or indirect. They can affect the physical and chemical parameters of the microclimate and energy sources (e.g., carbon), the structural reorganization of microbial communities and their functional activity [2, 31]. Microbial activity and diversity are related to soil ecosystem multifunctionality, plant productivity, nitrate, ammonium, phosphorus availability and nitrogen mineralization rates [11, 40]. Mineral (phosphate, potash, nitrogen or their complex) and organic manure can significantly affect soil microbiome [12, 21, 48]. The main nitrogen sources available to plants are gaseous nitrogen from the atmosphere and nitrate and ammonium nitrogen from the soil. Nitrogen nutrition is vital for the growth of all crops. With a lack of nitrogen growth, plants deteriorate sharply. Compared to other essential elements, it contributes to the growth of vegetative organs, the formation of the ovary, and increased harvest and quality. Nitrogen plays a vital role in increasing the energy of photosynthesis. Nitrogen also promotes the active growth of the roots of cultivated plants. Both lack and excess nitrogen weaken root growth. Nitrogen excess fertilizers block the supply of available copper, which leads to a decreased nutritional value of protein in feed, and negatively affects the shelf life of vegetables and their quality for canning. Excessive nitrogen nutrition can cause the accumulation of nitrates in plants that are harmful to humans and animals. The use of nitrogen fertilizers regulates the nitrogen nutrition of plants [18, 36]. The application of mineral nitrogen fertilizers affects the soil microbiome [19] and activates microorganisms (bacteria, algae, fungi and actinomycetes) and microfauna (protozoa and nematodes) [38, 64, 65]. Plants use up to 40% of nitrogen fertilizers. Approximately 30% can be lost by leaching, and 30% is fixed in organic form. One of the main problems of increasing soil fertility is providing them with nitrogen,with intensive use of soils, assimilated nitrogen is at a minimum. Nitrogen fertilizers applied in large doses change the properties of the soil: migration along the soil profile increases fulvic acids and humic acids, calcium and magnesium cations, as well as acidifies the soil, which negatively affects the root system of plants. A single application of high doses of nitrogen fertilizers increases the loss of nutrients. At the same time, not only the nitrogen from fertilizers but also from the soil is washed out because the mobility of soil nutrients increases under the influence of fertilizers [17, 60]. Therefore, using appropriate doses and forms of nitrogen fertilizers is one of the main factors in improving agricultural production. The use of nitrogen fertilizers in the world is increasing. Therefore it is key to know their behavior on the soil microbiological activities. High concentrations of ammonia can suppress soil microbiome vital activity. This leads to the inhibition of processes of nitrification or ammonification. Long-term studies have shown that double doses of nitrogen fertilizers decrease the number of ammonifiers and nitrogen-fixing microbiota in the soil. Instead, the content of soil micromycetes increases. Nitrogen fertilizers, directly and indirectly, influence the diversity of soil bacterial, fungal, and protist communities. The indirect effect on the soil microbiome is due to soil physicochemical properties changing. The alteration of inter-kingdom interactions within soil microbiomes is due to competition and predator–prey interactions between the different microbial groups [66].

Phosphorus is one of the most critical elements of plant nutrition because it is part of proteins. Nitrogen in the soil can be replenished by biological fixation from the air, phosphorus only by phosphates application to the soil in the form of fertilizers. The main phosphorus sources are phosphorites, apatite, vivianite, and waste from the meat and metallurgical industries. Enriched apatite are the best raw materials for the production of industrial phosphors [5]. When phosphorite flour is applied together with phosphorus, uranium, radium and other radioactive elements enter the soil, albeit in small quantities. Different types of superphosphates contain different amounts of heavy metal impurities. So, depending on the type of fertilizer, the cadmium content ranges from 1 to 170 mg/kg. With the systematic application of increased doses of superphosphate containing about 1.5% of water-soluble fluorine, the latter accumulates in large quantities and pollutes the soil and plants. Increased amounts of fluorine inhibit the action of enzymes in the plant, inhibit photosynthesis and disrupt protein metabolism [3, 39]. Fluoride also negatively affects the activity of microbiological processes in the soil. The processes of mineralization and humus formation are slow down. Potassium plays a significant role in the life of plants. It is most abundant in young growing organs, the cell sap of plants, as it helps the rapid accumulation of carbohydrates. Potassium enhances ammonia nitrogen's use during synthesizing amino acids and protein. Under the influence of potassium, the winter resistance of plants increases. To obtain high yields of crops, exchangeable potassium accumulated in the soil is not enough, so it is necessary to apply potash fertilizers [30, 62]. Potash fertilizers have a significant drawback. Most potash fertilizers contain a significant amount of chlorine, which hurts the growth and development of plants [1]. Applying large doses of potassium fertilizers can lead to increased chloride ions, displace calcium and magnesium from the soil absorption complex, and increase their migration along the soil profile [9, 16]. Sodium in potassium salt and sylvinite pollutes the soil. It deteriorates the physical and chemical properties of many soils, especially chernozem, chestnut and saline soils, as it makes them more salting. Finding potassium in feed in excessive amounts can cause animal poisoning. An excess of chlorine also negatively affects microbial communities' taxonomic and functional structure. The number of "agronomically useful" microbiota, particularly nitrogen-fixing and phosphate-mobilizing bacteria, is decreasing. In microbial communities, the number of oligotrophic and pedotrophic microbiota is increasing, which indicates the deterioration of the ecological status of soils [52]. Complex fertilizers (Nitrogen-Phosphorous-Potassium) have a high concentration of nutrients, so their use significantly reduces farm expenses for their transportation, mixing, storage and introduction. Among the disadvantages of complex fertilizers is that the proportions in the Nitrogen-Phosphorous-Potassium content vary within narrow limits. Therefore, when applying, for example, the required amount of nitrogen, other nutrients are applied less or more than required [32, 46]. Mineral fertilizers and the main nutrients (nitrogen, phosphorus, potassium, calcium, magnesium, sulfur) contain various impurities and ballast substances: salts of heavy metals, radioactive substances, and organic compounds. These impurities are transferred from the raw fertilizer materials. Their content can be up to 5%. Most of the impurities belong to toxic substances. Various industrial wastes are often used as micro-fertilizers in enterprises. For example, pyrite nitrite is used as a copper fertilizer, although it contains lead, arsenic, and other harmful substances. Ballast substances (e.g., chlorine, sodium) that enter the soil with fertilizers negatively affect its properties, cause changes in physiological and biochemical processes in plants, affect soil microbial communities that worsen the quality of agricultural products and contaminate groundwater [4, 29].

3 Organic Fertilizers in Agroecosystems, Impact on Soil Microbiome

In traditional agriculture, due to active use of mineral fertilizers and chemical plant protection agents may cause side effects: a change in the structure of the soil microbiome, a decrease in its diversity, and a violation of functional parameters. In the last century, agricultural intensification caused significant biodiversity loss in most agroecosystems. Biodiversity and ecosystem services might be protected with agri-environmental schemes and agro-technologies [26, 28, 42].

Organic mechanisms of pathogen control were used in agriculture plant organisms through competition or antagonism using measures which contributed to the improvement of the functional structure of the soil microbiome and increasing its species diversity. The application of organic fertilizers in agriculture improves soil fertility [33]. It contributes to increasing crop yields. However, it can be a source of the spread of antibiotic-resistant microorganisms, which pose a danger not only to the environment but is a risk factor to human health [51]. This is because antibiotics are widely used on farms for therapeutic purposes and prophylactically, as well as growth factors. Getting into the body of animals, they are excreted together with feces and urine, characterized by a long half-life. Together with organic fertilizers, they get into the environment. High levels of reproduction and significant biomass and adaptation possibilities characterize soil microorganisms. The biggest danger is that these microorganisms are vectors of horizontal and vertical gene transfer of antibiotic resistance. Therefore, an essential aspect of studying the soil microbiome is not only functional parameters and taxonomic characteristics but also verification of the presence of antibiotic-resistant microorganisms [53, 54]. The basis of sustainable and productive ecosystems is healthy soil. For that, it is necessary to implement good agricultural practices with healthy principles such as minimal soil disturbance, crop rotation, and organic manure-enhancing soil microbiome [41, 57]. Good Agricultural Practices is a collection of principles to apply for on-farm production and postproduction processes, resulting in safe and healthy food and non-food agriculture products while taking into account economic, social and environmental sustainability (FAO 2016). Implementation of these principles in agriculture has some difficulties. In recent decades, obtaining high yields was achieved due to using intensive agricultural technologies with high doses of mineral fertilizers and pesticides [12, 43].

4 Antibiotics in Agriculture—Soil Resistome

Soil is an important reservoir of antibiotic resistance genes (ARGs). However, their potential risk in different ecosystems and their response to anthropogenic land use change are still unknown [45]. Environmental contamination by antibiotics is linked to the development of antimicrobial resistance in non-target species of bacteria. Antibiotic substances and their metabolites excreted by food animals in primary production enter the environment (soil, water) and are a selector for the emergence and evolution of resistances. Selection for resistance already occurs at low concentrations of antibiotics in nature [24]. Using antibiotics in animal husbandry and their faeces as organic fertilizer is problematic for agriculture. Antibiotic residues affect soil microbiome and resistome [37, 55]. The environment, particularly soil, has been identified as a huge reservoir of resistance genes, the so-called "environmental resistome" [10]. There is also a risk that antimicrobial resistance genes can be mobilized and transferred to human pathogens [61]. However, the transfer routes of resistance genes from the environmental resistome to humans could be clearer. Whether the bacteria perform critical ecosystem services, pose a health threat as pathogens, or have incompletely understood functions, the development of antimicrobial resistance because of human activities is problematic. Incorporating genetic information encoding can alter susceptibility characteristics of microorganisms for resistance or by a mutation in their DNA. Antibiotic resistance genes are recognized as important environmental contaminants [44, 61]. It is now accepted that resistance is a natural property of all bacteria. The term resistome is used to describe the framework that encompasses all forms of resistance and precursor elements. Resistance genes originating from the guts of animals can, for example, reach soil with manure and spread via horizontal gene transfer, which is promoted by the simultaneous presence of antibiotics (and other pollutants) in manure and soil [15, 59]. The bi-directional nature of resistance gene transfer between the food chain and the environment is important because it allows the environment to act as a repository or reservoir that intercepts and re-releases resistance genes [7]. Recognizing that human health, animal health, and ecosystem health are inextricably linked, the "One Health" [14, 34]. Taken together in a complex picture (Fig. 1), the transfer of resistance genes between pathogenic and non-pathogenic bacteria that make up the microbiome of humans, animals, and free-living bacteria in the air, soil and water is revealed. The antimicrobial agents used in humans and animals act as selectors within the intestinal mucosa, but also their degradation products, which can be detected as residues in food or get via the manure into the soil, groundwater and drinking water [20].

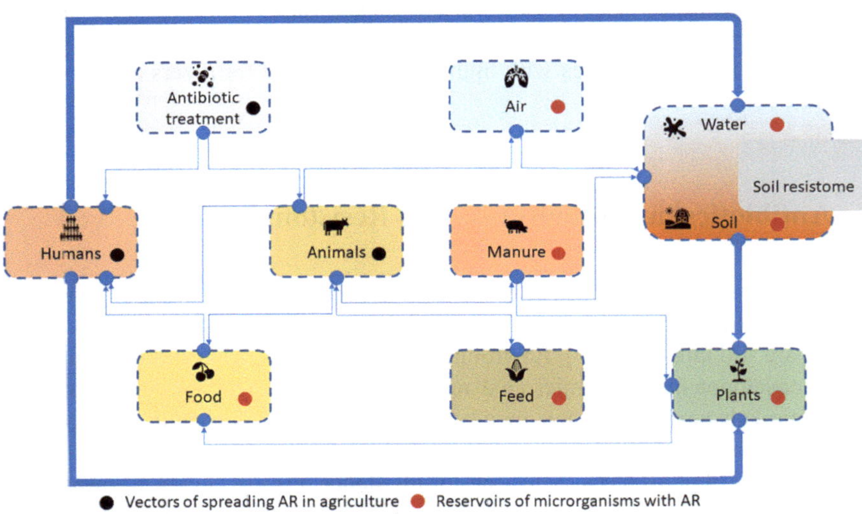

Fig. 1 Vectors of spreading antibiotic resistant (AR) bacteria and formation of soil resistome

Data supports that the reservoirs for resistant bacteria present in humans, animals and the environment are linked, identical bacterial clones [8], and resistance genes can be found in all three reservoirs.

Studies of the agricultural environment and the food chain have revealed that pathogen reservoirs exist outside their animal host, which presents a potential infection reservoir. It has been shown that gene flow in the environment is extensive for bacterial populations and is likely to be promoted by the availability of nutrients. This gene flow also implies an increased transfer of drug-resistance genes encoded by plasmids and gene cassettes in biofilms and slurries [22].

Additional research has focused on the ecological roles of antibiotics, and evidence has been reported on selective conditions in soil for expression of antibiotic-resistant gene clusters using reporter gene fusions and by detecting mRNA extracted from soil. Soil can be a major reservoir of microbial pathogens. In situ analysis of different environments revealed that pollution with antibiotics can drive the selection for adaptive mutations or uptake of resistance genes similar to those already rapidly evolving in bacteria found in hospitals and farms due to the frequent use of biocides and antibiotics [27, 49].

A large fraction of the antibiotics administered to animals is excreted and reaches agricultural soils with animal manure fertilizers [8, 15]. It has been shown that the spreading of manure from pigs in the soil treated with sulfadiazine (SDZ) resulted in a higher abundance of sulfonamide. Several antibiotics, such as fluoroquinolones, tetracyclines and sulfonamides, persist in soil for weeks to months. Soil bacterial genomes consist of large parts of mobile genetic elements. Plasmid-mediated genetic variation allows bacteria to respond rapidly with adaptive responses to challenges such as irregular antibiotic or metal concentrations or opportunities such as utilizing xenobiotic compounds. Cultivation-independent detection methods such as qPCR and capture of plasmids from soil bacteria into Escherichia coli, and complete sequencing have provided new insights into the role and ecology of plasmids [22]. It was shown that mobile genetic elements from soil bacterial communities could be captured into food chain-relevant organisms such as *E. coli* [59]. This shows that the environmental antibiotic-resistant bacteria pool is a potential source of AR bacteria selected for food production animals. The soil microbiome and resistome depend on anthropogenic impact [55]. A significant number of aerobic microorganisms with multiple antibiotic resistance present in the agroecosystem, part of them are the causative agents of foodborne infections. Soil particles carrying viable bacteria can be transported over long distances and might contribute to the spreading of antibiotic resistant bacteria over wide geographic ranges [13, 25].

5 Conclusion

The use of inappropriate farming practices: high doses of chemical fertilizers and pesticides and frequent changes in land use may cause variability in soil microbial communities, which can have important effects on soil fertility and productivity.

Applying chemical fertilizers impacts populations of microbes such as bacteria, micromycetes and streptomycetes. They play different functions in the soil, such as maintenance of soil structure, suppressing effects of toxins and pathogens and indirectly helping in sustainable agricultural productivity. The use of high doses of mineral fertilizers leads to a decrease in microbial biodiversity and a violation of the integration of microbial coenosis. The number of oligotrophic microbiota increases and the soil's mineralization processes are disturbed. This can cause losses of soil organic matter and, as a result, reduce soil fertility. Inorganic and organic fertilizers have a vital role in increasing agricultural production, but the use of mineral fertilizers is constantly growing. There is increasing concern regarding the adverse environmental effects of chemical fertilizers on soil microbiota. Applying organic fertilizers improves soil structure and is responsible for a more balanced and stable nutrient supply, which can sustain a more diverse microbial community than mineral fertilizers. Moreover, organic fertilization positively affects microbial activity taxonomic structure of soil microbiome and improves the chemical and physical properties of soil better than inorganic fertilization. However, despite organic fertilizers' positive effects, some must be controlled. Organic fertilizers should be checked for the presence of antibiotic residues and antibiotic-resistant microorganisms to prevent soil contamination since it plays a leading role in forming soil resistome and environmental resistance in general. The soil microbiome plays a vital role in agroecosystems' ecological status and in the development and spread of antibiotic resistance in humans and the environment. That is why soil microbial communities' structure, diversity, and functions should be better managed through good agricultural practices.

References

1. Ashraf MY, Gul A, Ashraf M, Hussain F, Ebert G (2010) Improvement in yield and quality of kinnow (Citrus deliciosa × Citrus nobilis) by potassium fertilisation. J Plant Nutr 33:1625–1637. https://doi.org/10.1080/01904167.2010.496887
2. Auffret MD, Karhu K, Khachane A, Dungait JAJ, Fraser F, Hopkins DW, Wookey PA, Singh BK, Freitag TE, Hartley IP, Prosser JI (2016) The role of microbial community composition in controlling soil respiration responses to temperature. PLoS One 11(10):e0165448. https://doi.org/10.1371/journal.pone.0165448
3. Azzi V, Kazpard V, Lartiges B, Kobeissi A, Kanso A, El Samrani AG (2017) Trace metals in phosphate fertilizers Used in Eastern Mediterranean Countries. Clean Soil Air Water 45. https://doi.org/10.1002/clen.201500988
4. Bai YC, Chang YY, Hussain M, Lu B, Zhang JP, Song XB, Lei XS, Pei D (2020) Soil chemical and microbiological properties are changed by long-term chemical fertilizers that limit ecosystem functioning. Microorganisms 8(5):694. https://doi.org/10.3390/microorganisms8050694
5. Bakalár T, Pavolová H, Šimková Z, Bednárová L (2022) Phosphorus management in Slovakia—a case study. Sustainability 14(16):10374. https://doi.org/10.3390/su141610374
6. Bender S, Van der Heijden M (2015) Soil biota enhance agricultural sustainability by improving crop yield, nutrient uptake and reducing nitrogen leaching losses. J Appl Ecol 52:228–239. https://doi.org/10.1111/1365-2664.12351

7. Bengtsson-Palme J, Kristiansson E, Larsson DGJ (2018) Environmental factors influencing the development and spread of antibiotic resistance. FEMS Microbiol Rev 2018 Jan 1;42(1):fux053. https://doi.org/10.1093/femsre/fux053
8. Bulteel AJB, Larson EL, Getahun H (2020) Identifying global research gaps to mitigate antimicrobial resistance: a scoping review. Am J Infect Contr 49:818–824. https://doi.org/10.1016/j.ajic.2020.11.024
9. Chen W, He ZL, Yang XE, Mishra S, Stoffella PJ (2010) Chlorine nutrition of higher plants: progress and perspectives. J Plant Nutr 33:943–952. https://doi.org/10.1080/01904160903242417
10. Costa VM, McGrann KM, Hughes DW, Wright GD (2006) Sampling the antibiotic resistome. Science, 20, 311(5759):374–377. https://doi.org/10.1126/science.1120800. PMID: 16424339
11. Delgado-Baquerizo M et al (2016) Microbial diversity drives multifunctionality in terrestrial ecosystems. Nat Commun 7:10541. https://doi.org/10.1038/ncomms10541
12. Demyanyuk O, Symochko L, Mostoviak I (2020) Soil microbial diversity and activity in different climatic zones of Ukraine. Regul Mech Biosyst 11(2):338–343 https://doi.org/10.15421/022051
13. Ding GC, Radl V, Schloter-Hai B, Jechalke S, Heuer H, Smalla K, Schloter M (2014) Dynamics of soil bacterial communities in response to repeated application of manure containing sulfadiazine. PLoS One 26;9(3):e92958. https://doi.org/10.1371/journal.pone.0092958
14. Garg S, Banerjee B (2021) One world, one health. Indian J Community Med 46(4):581–583. https://doi.org/10.4103/ijcm.ijcm_1230_21
15. Gaze WH, Zhang L, Abdouslam NA, Hawkey PM, Calvo-Bado L, Royle J, Brown H, Davis S, Kay P, Boxall AB, Wellington EM (2011) Impacts of anthropogenic activity on the ecology of class 1 integrons and integron-associated genes in the environment. ISME J 5(8):1253–61. https://doi.org/10.1038/ismej.2011.15. Epub 2011 Mar 3. PMID: 21368907; PMCID: PMC3146270
16. Geilfus CM (2018) Review on the significance of chlorine for crop yield and quality. Plant Sci 270:114–122. https://doi.org/10.1016/j.plantsci.2018.02.014
17. Gheysari M, Mirlatifi SM, Homaee M, Asadi ME, Hoogenboom G (2009) Nitrate leaching in a silage maise field under different irrigation and nitrogen fertilizer rates. Agri Water Manag 96:946–954. https://doi.org/10.1016/j.agwat.2009.01.005
18. Hachiya T, Sakakibara H (2017) Interactions between nitrate and ammonium in their uptake, allocation, assimilation, and signaling in plants. J Exp Bot. 1;68(10):2501–2512. https://doi.org/10.1093/jxb/erw449
19. Hammad HM, Chawla MS, Jawad R, Alhuqail A, Bakhat HF, Farhad W, Khan F, Mubeen M, Shah AN, Liu K, Harrison MT, Saud S, Fahad S (2022) Evaluating the impact of nitrogen application on growth and productivity of maize under control conditions. Front Plant Sci 13:885479. https://doi.org/10.3389/fpls.2022.885479
20. Hao H, Cheng G, Iqbal Z, Ai X, Hussain HI, Huang L, Dai M, Wang Y, Liu Z, Yuan Z (2014) Benefits and risks of antimicrobial use in food-producing animals. Front Microbiol 12(5):288. https://doi.org/10.3389/fmicb.2014.00288
21. Hartmann M, Frey B, Mayer J, Mader P, Widmer F (2015) Distinct soil microbial diversity under long-term organic and conventional farming. Multidisc J Microbial Ecol 9:1177–1194. https://doi.org/10.1038/ismej.2014.210
22. Heuer H, Smalla K (2012) Plasmids foster diversification and adaptation of bacterial populations in soil. FEMS Microbiol Rev 36(6):1083–1104. https://doi.org/10.1111/j.1574-6976.2012.00337.x
23. Ichihashi Y et al (2020) Multi-omics analysis on an agroecosystem reveals thesignificant role of organic nitrogen to increaseagricultural crop yield. PNAS 117(25):14552–14560. https://doi.org/10.1073/pnas.1917259117
24. Irfan M, Almotiri A, AlZeyadi ZA (2022) Antibiotics (Basel), 5, 11(10):1362. https://doi.org/10.3390/antibiotics11101362.PMID: 36290020
25. Kaviani Rad A, Astaykina A, Streletskii R, Afsharyzad Y, Etesami H, Zarei M, Balasundram SK (2022) An overview of antibiotic resistance and abiotic stresses affecting antimicrobial

resistance in agricultural soils. Int J Environ Res Pub Health 12;19 (8):4666. https://doi.org/10.3390/ijerph19084666
26. Kleijn D, Baquero RA, Clough Y, Díaz M, De Esteban J, Fernández F, Gabriel D, Herzog F, Holzschuh A, Jöhl R, Knop E, Kruess A, Marshall EJ, Steffan-Dewenter I, Tscharntke T, Verhulst J, West TM, Yela JL (2006) Mixed biodiversity benefits of agri-environment schemes in five European countries. Ecol Lett Mar; 9(3):243–254; discussion 254–257. https://doi.org/10.1111/j.1461-0248.2005.00869.x
27. Koch N, Islam NF, Sonowal S, Prasad R, Sarma H (2021) Environmental antibiotics and resistance genes as emerging contaminants: methods of detection and bioremediation. Curr Res Microb Sci 14(2):100027. https://doi.org/10.1016/j.crmicr.2021.100027
28. Krauss J, Gallenberger I, Steffan-Dewenter I (2011) Decreased functional diversity and biological pest control in conventional compared to organic crop fields. PLoS One 6(5):e19502. https://doi.org/10.1371/journal.pone.0019502
29. Kubier A, Wilkin RT, Pichler T (2019) Cadmium in soils and groundwater: a review. Appl Geochem 1(108):1–16. https://doi.org/10.1016/j.apgeochem.2019.104388
30. Larramendy M, Soloneski S (2021) Soil contamination—threats and sustainable solutions. IntechOpen, London. https://doi.org/10.5772/intechopen.87652
31. Lauber CL, Ramirez KS, Aanderud Z, Lennon J, Fierer N (2013) Temporal variability in soil microbial communities across land-use types. Multidisc J Microbial Ecol 7:1641–1650. https://doi.org/10.1038/ismej.2013.50
32. Liu X, Liu X, Liu W, Tan Q, Hu C, Li J (2021) Nutritional status of different citrus trees and the recommended dosages of N, P and K for citrus production in China. J Plant Nutr 27:565–574. https://doi.org/10.11674/zwyf.20457
33. Lupatini M, Korthals GW, de Hollander M, Janssens TK, Kuramae EE (2017) Soil microbiome is more heterogeneous in organic than in conventional farming system. Front Microbiol 4(7):2064. https://doi.org/10.3389/fmicb.2016.02064
34. Mackenzie JS, McKinnon M, Jeggo M (2014) One health: from concept to practice. Confronting Emerg Zoonoses 19:163–189. https://doi.org/10.1007/978-4-431-55120-1_8
35. Martin EA, Feit B, Requier F, Friberg H (2019) Assessing the resilience of biodiversity-driven functions in agroecosystems under environmental change. Adv Ecol Res 60:59–123. https://doi.org/10.1016/bs.aecr.2019.02.003
36. Muratore C, Espen L, Prinsi B (2021) Nitrogen uptake in plants: the plasma membrane root transport systems from a physiological and proteomic perspective. Plants (Basel). Apr 1; 10(4):681. https://doi.org/10.3390/plants10040681
37. Pal C, Bengtsson-Palme J, Kristiansson E, Larsson DG (2016) The structure and diversity of human, animal and environmental resistomes. Microbiome, 7, 4(1):54. https://doi.org/10.1186/s40168-016-0199-5.
38. Patyka VP, Symochko LY (2013) Soil microbiological monitoring of natural and transformed ecosystems in the trans-Carpathian region of Ukraine. Microbiol J 75(2):21–31
39. Pawuła A (2022) Radioactive elements in phosphorous fertilizer-basalt flour-recommended mineral fertilizer. J Geosci Environ Protect 10:15–32. https://doi.org/10.4236/gep.2022.104002
40. Pereira P (2020) Ecosystem services in a changing environment. Sci Total Environ 702:135008
41. Pereira P, Barcelo D, Panagos P (2020) Soil and water threats in a changing environment. Environ Res 186:109501
42. Pereira P, Bogunovic I, Munoz-Rojas M, Brevik E (2018) Soil ecosystem services, sustainability, valuation and management. Curr Opin Environ Sci Health 5:7–13
43. Popp J, Pető K, Nagy J (2013) Pesticide productivity and food security. A review. Agron Sustain Dev 33:243–255. https://doi.org/10.1007/s13593-012-0105-x
44. Pruden A, Pei R, Storteboom H, Carlson K (2006) Antibiotic resistance genes as emerging contaminants: studies in northern Colorado. Environ Sci Technol 40(23):7445–7450. https://doi.org/10.1021/es0604131
45. Qian X, Gunturu S, Guo J (2021) Metagenomic analysis reveals the shared and distinct features of the soil resistome across tundra, temperate prairie, and tropical ecosystems. Microbiome 9:108. https://doi.org/10.1186/s40168-021-01047-4

46. Quaggio JA, Mattos D, Cantarella H, Almeida ELE, Cardoso SAB (2002) Lemon yield and fruit quality affected by NPK fertilisation. Sci Horticult 96:151–162. https://doi.org/10.1016/S0304-4238(02)00121-8
47. Santos LF, Olivares FL (2021) Plant microbiome structure and benefits for sustainable agriculture. Curr Plant Biol 26:100198. https://doi.org/10.1016/j.cpb.2021.100198
48. Schmidt R, Gravuer K, Bossange AV, Mitchell J, Scow K (2018) Long-term use of cover crops and no-till shift soil microbial community life strategies in agricultural soil. PLoS One 13:e0192953. https://doi.org/10.1371/journal.pone.0192953
49. Sengupta S, Chattopadhyay MK, Grossart HP (2013) The multifaceted roles of antibiotics and antibiotic resistance in nature. Front Microbiol 12(4):47. https://doi.org/10.3389/fmicb.2013.00047
50. Sharma SK, Ramesh A, Sharma MP, Joshi OP, Govaerts B, Steenwerth KL, Karlen DL (2010) Microbial community structure and diversity as indicators for evaluating soil quality. biodiversity, biofuels, agroforestry and conservation agriculture. Sustainable agriculture reviews. Springer, Dordrecht, pp 317–358. https://doi.org/10.1007/978-90-481-9513-8_1
51. Singh B, Trivedi P (2017) Microbiome and the future for food and nutrient security. Microb Biotechnol 10:50–53. https://doi.org/10.1111/1751-7915.12592
52. Symochko L (2020) Soil microbiome: diversity, activity, functional and structural successions. Int J Ecosyst Ecol Sci (IJEES) 10(2):277–284. https://doi.org/10.31407/ijees10.206
53. Symochko L, Bugyna L, Hafiiyak O (2021) Ecological aspects of biosecurity in modern agroecosystems. Int J Ecosyst Ecol Sci (IJEES) 11(1):181–186.https://doi.org/10.31407/ijees11.124
54. Symochko L, Meleshko T, Symochko V, Boyko N (2018) Microbiological control of soil-borne antibiotic resistance human pathogens in agroecosystems. Int J Ecosyst Ecol Sci (IJEES) 8(3):591–598. https://doi.org/10.31407/ijees8320
55. Symochko L, Hamuda H, Demyanyuk O, Symochko V, Patyka V (2019) Soil microbial diversity and antibiotic resistance in natural and transformed ecosystems. Int J Ecosyst Ecol Sci (IJEES) 9(3):581–590. https://doi.org/10.31407/ijees
56. Symochko L.Yu., Kalinichenko A.V. (2018) Soil Microbiome of Primeval Forest Ecosystems in Transcarpathia. Mikrobiolohichnyi Zhurnal, 80 (3), 3–14. doi: https://doi.org/10.15407/microbiolj80.03.003
57. Trivedi P et al (2021) Enabling sustainable agriculture through understanding and enhancement of microbiomes. New Phytol 230(6), 2129-2147. https://doi.org/10.1111/nph.17319
58. Wakelin SA (2018) Managing soil microbiology: realising opportunities for the productive land-based sectors. N Z J Agric Res 61:358–376. https://doi.org/10.1080/00288233.2016.1209529
59. Wang L, Chai B (2022) Fate of antibiotic resistance genes and changes in bacterial community with increasing breeding scale of layer manure. Front Microbiol 9;13:857046. https://doi.org/10.3389/fmicb.2022.857046. eCollection 2022. PMID: 35356511
60. Wang S, Feng X, Wang Y, Zheng Z, Li T, He S et al (2019) Characteristics of nitrogen loss in sloping farmland with purple soil in southwestern China during maize (Zea mays L.) growth stages. Catena 182:104169. https://doi.org/10.1016/j.catena.2019.104169
61. Wright GD (2010) The antibiotic resistome. Expert Opin Drug Discov 5(8):779–788. https://doi.org/10.1517/17460441.2010.497535
62. Xu X, Du X, Wang F, Sha J, Chen Q, Tian G, Zhu Z, Ge S, Jiang Y (2020) Effects of potassium levels on plant growth, accumulation and distribution of carbon, and nitrate metabolism in apple dwarf rootstock seedlings. Front Plant Sci 11. https://doi.org/10.3389/fpls.2020.00904
63. Yang G, Wagg C, Veresoglou SD, Rillig MC, Hempel S (2018) How soil biota drive ecosystem stability. Trends Plant Sci 23(12):1057–1067. https://doi.org/10.1016/j.tplants.2018.09.007
64. Yuan HZ, Ge TD, Wu XH, Liu SL, Tong CL, Qin HL et al (2012) Long-term field fertilisation alters the diversity of autotrophic bacteria based on the ribulose 1,5-biphosphate carboxylase/oxygenase (RubisCO) large subunit genes in paddy soil. Appl Microbiol Biot 95:1061–1071. https://doi.org/10.1007/s00253-011-3760-y

65. Zhang YG, Li DQ, Wang HM, Xiao QM, Liu XD (2006) Molecular diversity of nitrogen-fixing bacteria from the Tibetan Plateau, China. FEMS Microbiol Lett 260:134–142. https://doi.org/10.1111/j.1574-6968.2006.00317.x
66. Zhao ZB, He JZ, Geisen S et al (2019) Protist communities are more sensitive to nitrogen fertilisation than other microorganisms in diverse agricultural soils. Microbiome 7:33. https://doi.org/10.1186/s40168-019-0647-0

Antibiotics, Other Emerging Pollutants, and Pathogenic Microorganisms in Raw and Treated Sewage Sludge Reaching Soils

Avelino Núñez-Delgado, Ainoa Míguez-González, Raquel Cela-Dablanca, Ana Barreiro, María J. Fernández-Sanjurjo, Esperanza Álvarez-Rodríguez, Vanesa Santás-Miguel, Andrés Rodríguez-Seijo, and Manuel Arias-Estévez

Abstract There is huge and growing environmental and public health concern in relation to the potential spreading of emerging chemical pollutants, as well as pathogenic microorganism, through wastewater and subsequent application of sewage sludge on soils. In this chapter the authors review recent data on antibiotics and other emerging pollutants present in raw and treated sewage sludge. In addition, available data on pathogenic microorganisms, mainly bacteria, virus and fungus, is also reviewed for both raw and treated sludge. The reviewed data could be considered for assessing the eventual need of further complementary treatments, as well as of precautionary measures regarding the management of sewage sludge intended to agronomic recycling.

Keywords Antibiotics · Emerging pollutants · Bacteria · Virus · Fungus

1 Introduction

As indicated in [8], "emerging pollutants are defined as synthetic or naturally occurring chemicals that are not commonly monitored in the environment, but which have the potential to enter it and cause known or suspected adverse ecological and/or human health effects". These authors also indicate that "in some cases, release of emerging pollutants to the environment has likely occurred for a long time but may not have been recognized until new detection methods were developed, while in other cases, synthesis of new chemicals or changes in use and disposal of existing chemicals

A. Núñez-Delgado (✉) · A. Míguez-González · R. Cela-Dablanca · A. Barreiro · M. J. Fernández-Sanjurjo · E. Álvarez-Rodríguez
Department of Soil Science and Agricultural Chemistry, Engineering Polytechnic School, University of Santiago de Compostela, 27002 Lugo, Spain
e-mail: avelino.nunez@usc.es

V. Santás-Miguel · A. Rodríguez-Seijo · M. Arias-Estévez
Soil Science and Agricultural Chemistry, Fac. Sciences, University of Vigo, 32004 Ourense, Spain

A. Núñez-Delgado (ed.), *Frontier Studies in Soil Science*, Frontier Studies in Soil Science, https://doi.org/10.1007/978-3-031-50503-4_7

can create new sources of emerging pollutants". For [34], "emerging pollutants can be understood in a broad sense as any synthetic or naturally-occurring chemical or any microorganism that is not commonly monitored or regulated in the environment with potentially known or suspected adverse ecological and human health effects, mainly including chemicals found in pharmaceuticals, personal care products, pesticides, industrial and household products, metals, surfactants, industrial additives and solvents, with many of them used and released continuously into the environment even in very low quantities, and some of them potentially causing chronic toxicity, endocrine disruption in humans and aquatic wildlife and the development of bacterial pathogen resistance".

In fact, chemical emerging pollutants (and specifically antibiotics) are of increasing concern regarding environmental and public health issues, and it also happens with pathogenic microorganisms (which is clear when a pandemic disease takes place). This is true for different environmental compartments and materials, and sewage sludge (raw and/or treated, before and/or after being spread on soils) is of clear relevance as potential reservoir/vehicle for transmission in this regard [3, 17, 19].

In a previous work [23] we commented on antibiotics in sewage sludge, although the review did not cover other abiotic and biotic emerging contaminants.

In view of that, the authors of this chapter propose a review/update on the matter of antibiotics and other chemical emerging pollutants, as well as on pathogenic microorganisms, in raw and/or treated sewage sludge, focusing on recent data. The authors of the chapter think that this review could be a valuable aid for assessing the feasibility of eventual complementary treatments, as well as to make reflections on new alternatives dealing with the management of sewage sludge destined to be agronomically recycled.

2 Antibiotics and Other Emerging Pollutants in Raw and Treated Sewage Sludge

2.1 Antibiotics

As indicated above, the authors of this chapter have recently published a text dealing with effects of treatments of sewage sludge on antibiotics [23], including a table with data on concentrations for different antibiotics in sewage sludge that covered from 2003 to 2018. In view of the contents of that previous work, in the current chapter we will include very recent data, going from the start of 2019 to the end of June 2023.

Regarding number of publications, searching on Web of Science (WOS) for "antibiotics sewage-sludge" as topic, a total number of 645 results was reported for 2019–23 (while the number is 1,048 for 1900–2023). Among these references, we found as one of the most relevant that published by [35], dealing with the increased abundance of antibiotic resistance genes in agricultural soils receiving the application of sewage sludge, as well as the paper by [16], dealing with antibiotics removal and heavy metal immobilization during treatment of sewage sludge and pig manure. As comparison, Google Scholar (GS) provides 17,000 results for the same string ("antibiotics sewage-sludge") in the range 2019–23, and 43,200 results for the whole range of years covered by the tool.

Searching for "antibiotics-concentration sewage-sludge" on WOS, a total of 21 results is reported for 2019–2023. Among them, the paper by [29] shows the absence of effects on soil bacterial resistome when sewage sludge is spread on farmland, whereas the one by [13] deals with the fate of antibiotics present in a primary sludge during co-composting with palm wastes, with fluoroquinolones showing the highest risks, the paper by [38] reports that antibiotic residues may stimulate or suppress methane yield and microbial activity during high-solid anaerobic digestion (a method to dispose dewatered sludge),and the publication by [11] focuses on the fate of antibiotics and antibiotic resistance genes in a food waste treatment plant, showing that "human food might not be free of antibiotics and antibiotic resistance genes, and that food waste was an underestimated antibiotics resistance pool with various determinants, although derived hazards of food waste could be mitigated through biological treatment with well-planned daily operations". Just as comparison, GS provides 188 results for the same string ("antibiotics-concentration sewage-sludge") in the range 2019–23.

When the search topic is "antibiotics treated-sewage-sludge", 9 results are shown by WOS, including the one by [36], dealing with antibiotics and distribution of antibiotic resistance genes in microbial electrolysis cells (an emerging pollution control technology) treating sewage sludge, the paper by [32], showing good results for thermal hydrolysis pre-treatment and various anaerobic digestion conditions, regarding removal/decrease of antibiotics, antibiotic resistance genes, and mobile genetic elements in sewage sludge, and the one by [31], who studied retention-release of ciprofloxacin and azithromycin in biosolids and biosolids-amended soils, finding that "the target antibiotic bioaccessibilities from biosolids and finer-textured (typical agricultural) soils would be minimal and that biosolids (not soils) control desorption of the two biosolids-borne chemicals". GS provides 688 results for the same string ("antibiotics treated-sewage-sludge") in the range 2019–23.

However, just one result is found by WOS when searching for "antibiotics raw-sewage-sludge", while GS finds 240 results in the range 2019–23.

Slightly modifying the searching string to "antibiotics treated sewage-sludge", 104 results are reported by WOS (see Fig. 1a for number of publications by country), while GS provides 16,200 results for the period. And searching for "antibiotics raw sewage-sludge", the number of results found by WOS is 44 (see Fig. 1b for number of publications by country), while GS provides 14,400 results. In both cases, the highest number of results from WOS corresponds to China, followed by the USA, and Spain placed 3rd and 6th, respectively.

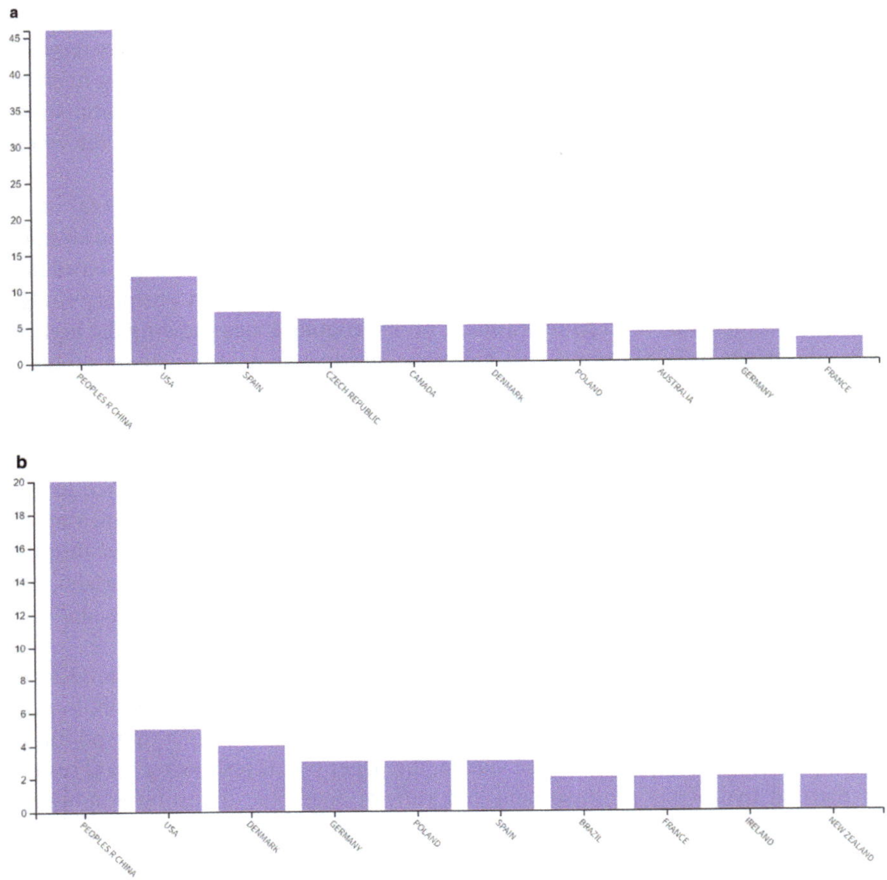

Fig. 1 Number of publications by country (period going from January 2019 to July 2023) provided by WOS for the searching strings "antibiotics treated sewage-sludge" (**a**), and "antibiotics raw sewage-sludge" (**b**)

2.2 Other Emerging Pollutants

To search for chemical emerging pollutants other than antibiotics, we should use searching strings for all chemical emerging pollutants, and then we should take into account previous results for antibiotics. However, in this case we will carry out the search for a more extended period than that performed above just for antibiotics, as in our previous work covering up to 2018 [23] we had not focused on chemical emerging pollutants other than these biocides.

Searching for "chemical emerging pollutants raw sewage-sludge" on WOS, the total number of results is 8 when covering from 1900 to 2023, coincident with that found when the search is restricted to 2010–2023. Among the papers, we consider relevant the one by [33], an editorial piece for a special issue focused on biowaste for energy recovery and environmental remediation, as well as that by [9], dealing with the removal of organic micropollutants and biological activities from sewage sludge by means of anaerobic digestion, and also, the paper by [15], dealing with sewage sludge ash. As comparison, GS provides 16,800 results for the period 2019–23.

Searching for "chemical emerging pollutants treated sewage-sludge" the total number of results provided by WOS is 27 from 1900 to 2023, the same number as that for the period 2010–2023. Among these publications, we consider especially relevant the one by [14], focused on spatial distribution of pharmaceuticals in a conventional wastewater treatment plant with sludge treatment reed beds technology, as well as the paper by [6], dealing with classic and emerging contaminants resulting from the application of biosolids to agricultural lands, and the publication by [27], focused on the assessment of industry-derived organic amendments for agricultural use. Just as comparison, GS provides 16,400 results for the range 2019–23.

Figure 2a shows, with data from WOS, the countries with higher results for the string "chemical emerging pollutants raw sewage-sludge", with the highest number of results in 2021 (Fig. 2b). While for the string "chemical emerging pollutants treated sewage-sludge" the results by countries are shown in Fig. 3a, and the years showing higher results were 2022 and 2021 (Fig. 3b).

Table 1 shows results provided by GS by range of years, for the searching strings "chemical emerging pollutants raw sewage-sludge" and "chemical emerging pollutants treated sewage-sludge". The number is clearly increasing for both strings, being higher for "treated".

3 Pathogenic Microorganisms in Raw and Treated Sewage Sludge

Biotic emerging pollutants are receiving a highly increased attention, mainly due to the COVID-19 pandemic caused by the SARS-CoV-2 coronavirus. In this regard, searching for "biotic emerging pollutants sludge" on WOS, 5 results are provided for the whole range of years covered by the tool, but in fact 4 of them deal with chemical emerging pollutants. As a comparison, GS provides 29,200 results.

Searching for "emerging pathogens sludge" on WOS, the number of results is 116 (58,800 results in GS).

To refine and differentiate, when the search on WOS is performed for "emerging pathogens raw sludge", the number of results is 12 (31,200 in GS). In WOS, some of the papers included deal with leachates from landfills (not really focused on sludge), or with wastewater, but not with sludge, or with drinking water, or on wastewater.

Changing the searching string in WOS to "emerging pathogens treated sludge", the number of results is 40 (46,400 results in GS). Sorting by relevance, one of the results in WOS is that by [30], dealing with recycling of sludge as fertilizer, and including some comments regarding pathogens, another one is that by [10],

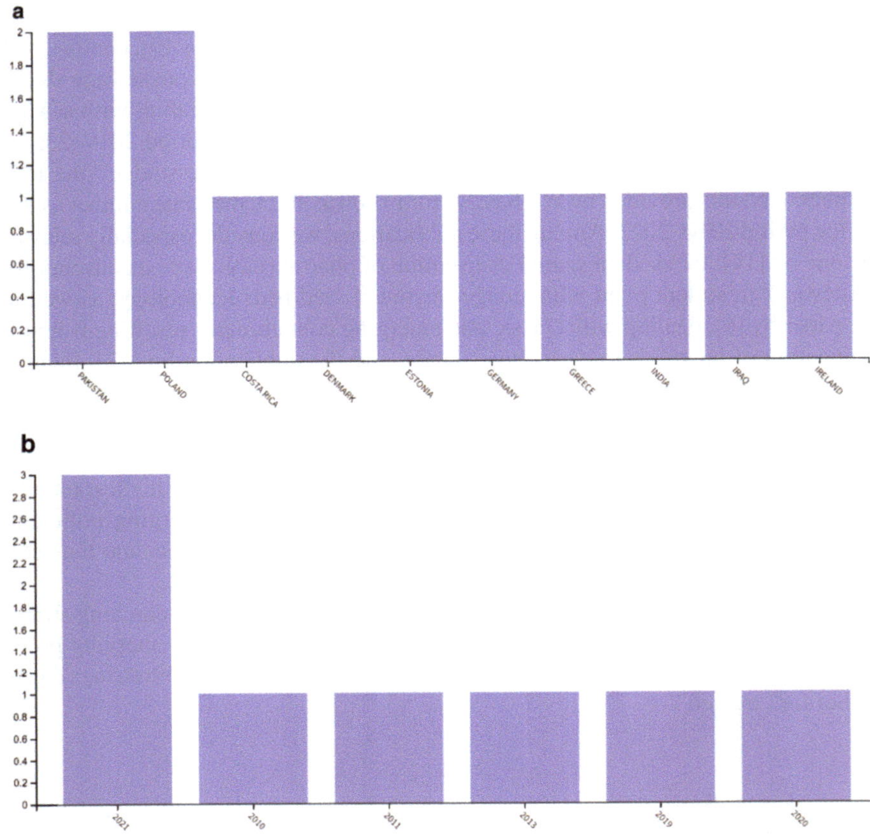

Fig. 2 Number of publications for the searching string "chemical emerging pollutants raw sewage-sludge" provided by WOS by country (**a**) and by year (**b**)

including some data on removal of pathogens, and also, a paper by [12], dealing with the pathogenic bacteria *Acinetobacter baumannii* and other pollutants in sludge.

Table 2 shows results provided by GS by range of years, for different searching strings. By range of years, it is clear that the number of results has drastically increased for all searching strings, reaching between 16,500 and 17,100 results in the period 2011–2023, thus more than doubling the results of the period 2001–10 for three of the four strings considered.

Regarding SARS-CoV-2 in sewage sludge, WOS shows 69 results for the whole period covered by the searching tool (number of results coincident with the period 2020–23). The author with the highest number of results corresponding to this searching string is one of the coauthors of this chapter (A. Núñez-Delgado). Some of the publications including data and comments in this regard are the ones by [4, 7, 21, 22].

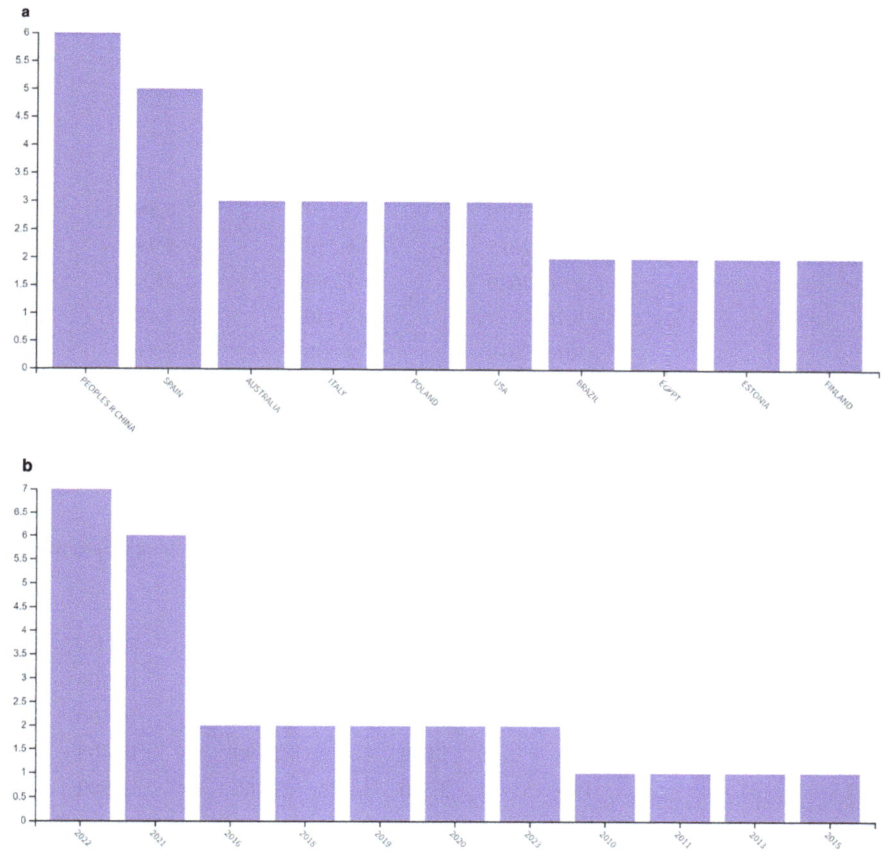

Fig. 3 Number of publications provided by WOS for the searching string "chemical emerging pollutants treated sewage-sludge", also by country (**a**) and year (**b**)

In addition, papers such as those by [20, 37] have evaluated survival on sewage sludge of bacteria and viruses causing different diseases, while there are various other interesting works regarding pathogens in sludge (for instance [18]). Also relevant, [2] studied some viruses in sewage sludge.

The authors of this chapter have also published more recent papers on the matter [1, 5, 24–26, 28].

Figure 4 shows images of sewage sludge sampled by the authors of the book for various research tasks performed in past years.

Table 1 Number of results provided by GS by range of years, for two different searching strings related to chemical emerging pollutants

Range of years	Searching string	
	CEPRSS	CEPTSS
1900–1960	18	24
1961–1980	165	269
1981–2000	1,230	3,040
2001–2010	4,250	6,560
2011–2015	5,550	8,980
2016–2020	10,600	15,800
2016–2017	3,440	5,480
2018–2019	4,530	7,210
2020–2021	17,000	19,100
2022–2023	17,600	19,600

CEPRSS: Chemical emerging pollutants raw sewage-sludge; CEPTSS: Chemical emerging pollutants treated sewage-sludge.

Table 2 Number of results provided by GS by range of years, for different searching strings related to biotic emerging pollutants

String	Range of years				
	1900–1960	1961–80	1981–2000	2001–10	2011–23
BEPS	58	767	4,280	10,500	16,700
EPS	38	201	3,610	7,490	16,500
EPRS	22	112	1,040	3,790	17,100
EPTS	35	196	3,540	7,160	16,900

BEPS: Biotic emerging pollutants sludge; EPS: Emerging pathogens sludge; EPRS: Emerging pathogens raw sludge; EPTS: Emerging pathogens treated sludge.

Fig. 4 Some raw (up) and treated (down) sewage sludge samples used by the authors of the book

4 Perspectives on Future Research in This Field

Recent data reviewed for the current chapter show that further research is still needed to achieve better results regarding removal of antibiotics from sludge, both for physic-chemical and biological processes. This is in line with that indicated in our previous chapter on antibiotics in sludge [23], where we concluded that "treatments with participation of microorganisms (related to biodegradation) are considered more efficient than other kinds of processes, when dealing with antibiotics removal from sewage sludge and bio-solids, although removal efficiency is not always as high as desired".

Regarding other chemical emerging pollutants in sewage sludge, a clearly increasing number of results is shown by searching tools, especially when using GS. Obviously, it is a growing field of research, even if the searching tools just show an approximate view of the works really focused on the matter. The number of results (and then probably the trend for coming years) is clearly higher for "chemical emerging pollutants treated sewage-sludge" than for the equivalent search focused on "raw sewage-sludge".

Biotic emerging pollutants in sewage sludge, and specifically emerging pathogens, are also of growing interest, mostly since the beginning of the COVID-19 pandemic, with clearly increasing results provided by searching tools. However, it should be noted that the numbers are drastically different when using the various searching tools available, and that the specificity (or quality) of the results provided is just partial, regarding the fitting to the theme. In any case, in recent years the number of results is higher for "emerging pathogens treated sludge" than for the equivalent search dealing with "raw sludge", and even higher than the more general searches "biotic emerging pollutants sludge", and "emerging pathogens sludge".

So, treated sewage sludge seems to concentrate increasingly higher efforts as regards research dealing with abiotic (chemical) and biotic (pathogens) emerging pollutants.

Finally, taking into account that in some cases various thousands of results can be found for a search covering just one year (or some months), and that the real "relevance" obtained when sorting by that item on the tools is sometimes very limited, those researchers looking into the theme here covered (or into other themes) should base their eventual selection of the most representative/interesting papers on their own selective views/readings (or on additional refined tools, when available), as we did for this chapter in addition to considering the results proposed by the tools.

Acknowledgements This work was supported by the Spanish "*Agencia Estatal de Investigación*" (State Investigation Agency) [grant number PID2021-122920OB-C21].

The sponsors had not involvement in study design; in the collection, analysis and interpretation of data; in the writing of the report, and in the decision to submit the article for publication.

References

1. Barreiro A, Fox A, Núñez-Delgado A (2023) New data on microorganisms in soil and other environmental compartments. Environ Res 216(2):114619. https://doi.org/10.1016/j.envres.2022.114619
2. Bibby K, Peccia J (2013) Identification of viral pathogen diversity in sewage sludge by metagenome analysis. Environ Sci Technol 47(4):1945–1951. https://doi.org/10.1021/es305181x
3. Boix C, Ibáñez M, Fabregat-Safont D, Morales E, Pastor L, Sancho JV, Sánchez-Ramírez JE, Hernández F (2016) Behaviour of emerging contaminants in sewage sludge after anaerobic digestion. Chemosphere 163:296–304. https://doi.org/10.1016/j.chemosphere.2016.07.098
4. Carraturo F, Del Giudice C, Morelli M, Cerullo V, Libralato G, Galdiero E, Guida M (2020) Persistence of SARS-CoV-2 in the environment and COVID-19 transmission risk from environmental matrices and surfaces. Environ Pollut 265(B):115010. https://doi.org/10.1016/j.envpol.2020.115010
5. Cela-Dablanca R, Santás-Miguel V, Fernández-Calviño D, Arias-Estévez M, Fernández-Sanjurjo MJ, Álvarez-Rodríguez E, Núñez-Delgado A (2021) SARS-CoV-2 and other main pathogenic microorganisms in the environment: situation in Galicia and Spain. Environ Res 197:111049. https://doi.org/10.1016/j.envres.2021.111049
6. Clarke RM, Cummins E (2015) Evaluation of "classic" and emerging contaminants resulting from the application of biosolids to agricultural lands: a review. Hum Ecol Risk Assess Int J 21(2):492–513. https://doi.org/10.1080/10807039.2014.930295
7. Farkas K, Hillary LS, Malham SK, McDonald JE, Jones DL (2020) Wastewater and public health: the potential of wastewater surveillance for monitoring COVID-19. Curr Opin Environ Sci Health 17:14–20. https://doi.org/10.1016/j.coesh.2020.06.001
8. Geissen V, Mol H, Klumpp E, Umlauf G, Nadal M, van der Ploeg M, van de Zee SE, Ritsema CJ (2015) Emerging pollutants in the environment: a challenge for water resource management. Int Soil Water Conserv Res 3(1):57–65. https://doi.org/10.1016/j.iswcr.2015.03.002
9. Gonzalez-Gil L, Papa M, Feretti D, Ceretti E, Mazzoleni G, Steimberg N, Pedrazzani R, Bertanza G, Lema JM, Carballa M (2016) Is anaerobic digestion effective for the removal of organic micropollutants and biological activities from sewage sludge? Water Res 102:211–220. https://doi.org/10.1016/j.watres.2016.06.025
10. Gurung K, Ncibi MC, Sillanpää M (2017) Assessing membrane fouling and the performance of pilot-scale membrane bioreactor (MBR) to treat real municipal wastewater during winter season in Nordic regions. Sci Total Environ 579:1289–1297. https://doi.org/10.1016/j.scitotenv.2016.11.122
11. He P, Yu Z, Shao L, Zhou Y, Lü F (2019) Fate of antibiotics and antibiotic resistance genes in a full-scale restaurant food waste treatment plant: Implications of the roles beyond heavy metals and mobile genetic elements. J Environ Sci 85:17–34. https://doi.org/10.1016/j.jes.2019.04.004
12. Higgins PG, Hrenovic J, Seifert H, Dekic S (2018) Characterization of *Acinetobacter baumannii* from water and sludge line of secondary wastewater treatment plant. Water Res 140:261–267. https://doi.org/10.1016/j.watres.2018.04.057
13. Khadra A, Ezzariai A, Merlina G, Capdeville MJ, Budzinski H, Hamdi H, Pinelli E, Hafidi M (2019) Fate of antibiotics present in a primary sludge of WWTP during their co-composting with palm wastes. Waste Manag 84:13–19. https://doi.org/10.1016/j.wasman.2018.11.009
14. Kołecka K, Gajewska M, Stepnowski P, Caban M (2019) Spatial distribution of pharmaceuticals in conventional wastewater treatment plant with sludge treatment reed beds technology. Sci Total Environ 647:149–157. https://doi.org/10.1016/j.scitotenv.2018.07.439
15. Krüger O, Grabner A, Adam C (2014) Complete survey of german sewage sludge ash. Environ Sci Technol 48(20):11811–11818. https://doi.org/10.1021/es502766x
16. Li C, Xie S, Wang Y, Pan X, Yu G, Zhang Y (2020) Simultaneous heavy metal immobilization and antibiotics removal during synergetic treatment of sewage sludge and pig manure. Environ Sci Pollut Res 27:30323–30332. https://doi.org/10.1007/s11356-020-09230-0

17. Mailler R, Gasperi J, Chebbo G, Rocher V (2014) Priority and emerging pollutants in sewage sludge and fate during sludge treatment. Waste Manag 34(7):1217–1226. https://doi.org/10.1016/j.wasman.2014.03.028
18. Martínez-Puchol S, Rusiñol M, Fernández-Cassi X, Timoneda N, Itarte M, Andrés C, Antón A, Abril JF, Girones R, Bofill-Mas S (2020) Characterisation of the sewage virome: comparison of NGS tools and occurrence of significant pathogens. Sci Total Environ 713:136604. https://doi.org/10.1016/j.scitotenv.2020.136604
19. Mohapatra DP, Cledón M, Brar SK, Surampalli RY (2016) Application of wastewater and biosolids in soil: occurrence and fate of emerging contaminants. Water Air Soil Pollut 227(77):1–14. https://doi.org/10.1007/s11270-016-2768-4
20. Nag R, Whyte P, Markey BK, O'Flaherty V, Bolton D, Fenton O, Richards KG, Cummins E (2020) Ranking hazards pertaining to human health concerns from land application of anaerobic digestate. Sci Total Environ 710:136297. https://doi.org/10.1016/j.scitotenv.2019.136297
21. Núñez-Delgado A (2020a) What do we know about the SARS-CoV-2 coronavirus in the environment? Sci Total Environ 727:138647. https://doi.org/10.1016/j.scitotenv.2020.138647
22. Núñez-Delgado A (2020b) SARS-CoV-2 in soils. Environ Res 190:110045. https://doi.org/10.1016/j.envres.2020.110045
23. Núñez-Delgado A, Pousada-Ferradás Y, Álvarez-Rodríguez E, Fernández-Sanjurjo MJ, Conde-Cid M, Nóvoa-Muñoz JC, Arias-Estévez M (2019) Chapter 1—effects of microbiological and non-microbiological treatments of sewage sludge on antibiotics as emerging pollutants present in wastewater: a review. In: Shah MP, Rodriguez-Couto S (eds) Microbial wastewater treatment. Elsevier, Amsterdam, Netherlands, pp 1–17. https://doi.org/10.1016/B978-0-12-816809-7.00001-4
24. Núñez-Delgado A, Bontempi E, Coccia M, Kumar M, Farkas K, Domingo JL (2021) SARS-CoV-2 and other pathogenic microorganisms in the environment. Environ Res 201:111606. https://doi.org/10.1016/j.envres.2021.111606
25. Núñez-Delgado A, Ahmed W, Bontempi E, Domingo JL (2022) The environment, epidemics, and human health. Environ Res 214(2):113931. https://doi.org/10.1016/j.envres.2022.113931
26. Núñez-Delgado A, Ahmed W, Romande JL, Dhama K, Domingo JL (2023) Scientific evidence on the origin of SARS-CoV-2. Environ Res 216(2):114631. https://doi.org/10.1016/j.envres.2022.114631
27. Pérez RA, Sánchez-Brunete C, Albero B, Miguel E, Tadeo JL, Alonso J, Lobo MC (2016) Quality assessment of three industry-derived organic amendments for agricultural use. Compost Sci Util 24(3):190–202. https://doi.org/10.1080/1065657X.2015.1124817
28. Race M, Ferraro A, Galdiero E, Guida M, Núñez-Delgado A, Pirozzi F, Siciliano A, Fabbricino M (2020) Current emerging SARS-CoV-2 pandemic: potential direct/indirect negative impacts of virus persistence and related therapeutic drugs on the aquatic compartments. Environ Res 188:109808. https://doi.org/10.1016/j.envres.2020.109808
29. Rutgersson C, Ebmeyer S, Lassen SB, Karkman A, Fick J, Kristiansson E, Brandt KK, Flach CF, Larsson DGJ (2020) Long-term application of Swedish sewage sludge on farmland does not cause clear changes in the soil bacterial resistome. Environ Int 137:105339. https://doi.org/10.1016/j.envint.2019.105339
30. Seleiman MF, Santanen A, Mäkelä PSA (2020) Recycling sludge on cropland as fertilizer—advantages and risks. Resour Conserv Recycl 155:104647. https://doi.org/10.1016/j.resconrec.2019.104647
31. Sidhu H, D'Angelo E, O'Connor G (2019) Retention-release of ciprofloxacin and azithromycin in biosolids and biosolids-amended soils. Sci Total Environ 650(1):173–183. https://doi.org/10.1016/j.scitotenv.2018.09.005
32. Sun C, Li W, Chen Z, Qin W, Wen X (2019) Responses of antibiotics, antibiotic resistance genes, and mobile genetic elements in sewage sludge to thermal hydrolysis pre-treatment and various anaerobic digestion conditions. Environ Int 133(A):105156. https://doi.org/10.1016/j.envint.2019.105156
33. Tsang YF, Kwon EE, Ok YS, Gao MT, Cao B (2018) Biowaste for energy recovery and environmental remediation. Process Saf Environ Prot 115. https://doi.org/10.1016/j.psep.2018.04.002

34. UNESCO (2020) Emerging pollutants in water and wastewater. https://en.unesco.org/emergingpollutantsinwaterandwastewater#:~:text=Emerging%20pollutants%20can%20be%20understood,ecological%20and%20human%20health%20effects
35. Urra J, Alkorta I, Mijangos I, Epelde L, Garbisu C (2019) Application of sewage sludge to agricultural soil increases the abundance of antibiotic resistance genes without altering the composition of prokaryotic communities. Sci Total Environ 647:1410–1420. https://doi.org/10.1016/j.scitotenv.2018.08.092
36. Zhang X, Li R (2020) Variation and distribution of antibiotic resistance genes and their potential hosts in microbial electrolysis cells treating sewage sludge. Biores Technol 315:123838. https://doi.org/10.1016/j.biortech.2020.123838
37. Zhao Q, Liu Y (2019) Is anaerobic digestion a reliable barrier for deactivation of pathogens in biosludge? Sci Total Environ 668:893–902. https://doi.org/10.1016/j.scitotenv.2019.03.063
38. Zhi S, Zhang K (2019) Antibiotic residues may stimulate or suppress methane yield and microbial activity during high-solid anaerobic digestion. Chem Eng J 359:1303–1315. https://doi.org/10.1016/j.cej.2018.11.050

Remediation of PTEs Contaminated Soils by Washing Treatment with Biodegradable Reagents

M. Race, A. Ferraro, D. Spasiano, A. Reyes, P. Papetti, G. Monteverde, A. Panico, and F. Pirozzi

Abstract This chapter discusses different aspects related to the soil washing techniques, in both ex-situ and in-situ configurations, and particularly focuses on the application and effectiveness of the remediation through biodegradable reagents. Nowadays, the use of these reagents is a key-factor to ensure a remediation process characterized by operating costs affordability and environmental sustainability. Low-molecular-weight organic acids (LMWOA), chelating agents, and biosurfactants can represent suitable extractant agents for the achievement of both high treatment efficiencies and minimized soil characteristics alteration. In this perspective, the characteristics of the main reagents involved in the washing process and the related mechanisms influencing the removal of Potentially Toxic Elements (PTEs) from soils are thoroughly analyzed. Indeed, the findings from previous studies highlight the need of preliminary investigations at lab scale to identify the soil properties and consequently select the best process operating conditions in order to perform feasible and effective treatments. Moreover, the literature review identifies the chelating agents as the most used extractants to date. Nonetheless, several researches aimed at further

M. Race (✉)
Department of Civil and Mechanical Engineering, University of Cassino and Southern Lazio, Via Gaetano di Biasio, 43, 03043 Cassino, Italy
e-mail: marco.race@unicas.it

A. Ferraro · D. Spasiano
Department of Civil, Environmental, Land, Building Engineering and Chemistry, Polytechnic University of Bari, Via E. Orabona 4, Bari 70125, Italy

A. Reyes
Mining Engineering Department, University of Antofagasta, 1240000 Antofagasta, Chile

P. Papetti · G. Monteverde
Department of Economics and Law, Territorial and Products Analysis Laboratory, University of Cassino and Southern Lazio, Via S. Angelo, Folcara, 03043 Cassino, Italy

A. Panico
Department of Engineering, University of Campania "L. Vanvitelli", Via Roma 29, 81031 Aversa, Italy

F. Pirozzi
Department of Civil, Architectural and Environmental Engineering, University of Naples "Federico II", Via Claudio 21, 80125 Naples, Italy

A. Núñez-Delgado (ed.), *Frontier Studies in Soil Science*, Frontier Studies in Soil Science, https://doi.org/10.1007/978-3-031-50503-4_8

deepening this topic also indicate the LMWOA and biosurfactants as promising and less impacting alternatives in the near future also due to their possible low-cost production through biological processes.

Keywords Soil washing · Soil flushing · Low-molecular-weight organic acids · Chelating agents · Biosurfactants

1 Introduction

Contamination of soils and sediments by Potentially Toxic Elements (PTEs) is an ascertained worldwide issue as also highlighted by several scientific studies over the last 30 years [38, 58]. This concern is not only due to the PTEs toxicity but also to their solubility in lipids. In fact, PTEs tend to concentrate either in the fatty tissue or in organisms from one trophic level to another along the food chain. Also, the very slow rate of PTEs excretion from fatty tissue is a potential cause for "biological magnification" occurrence [48].

Soils represent the main target of PTEs contamination from human activities and their high contaminant retention capacity leads to long-lasting pollution conditions in the environment [21, 63]. Besides this, PTEs contamination cannot be reduced by biological degradation as naturally occurs for organic compounds.

In the agricultural sector, soils contamination by PTEs prevents an optimal use of the land, thus causing serious economic damage [2] and altering the existing balance of natural components [13].

Also, influence by environmental factors can change both PTEs chemical form (speciation) and bioavailability with time thus possibly resulting in more hazardous form of contamination. The European Environment Agency (EEA) estimated 2.5 million of potentially contaminated sites in the EU. The 14% of these cases are characterized by an ascertained contamination then requiring the involvement of remediation processes (EEA, 2020).

For contaminated agricultural areas, optimal remediation approaches should be aimed at removing pollutants without affecting the original structure and composition of the soil. Generally, this result can be achieved by phytoremediation, but this technique might not be feasible for high contaminants concentration and very deep contamination in soils [23]. In a contaminated land, anthropogenic activities can be cause of very high PTEs concentrations occurrence in delimited areas. These are identified as hot spots [25] and their remediation requires more specific approaches than that based on treatments for low contamination levels (such as phytoremediation) [15].

According to this, the remediation of PTEs contaminated soils is an important challenge to deal with the involvement of suitable and efficient strategies. In this perspective, soil washing processes represent a promising technical solution if performed with extracting reagents which minimize the soil characteristics alteration and do not result in toxic by-products formation after treatment [54].

Soil Washing is a well-assessed chemical-physical technique for contaminants removal from the solid matrix through solubilization in an aqueous phase. It is an actual washing of the contaminated soils which are excavated and mixed with water solutions containing an extracting reagent (such as surfactants, biosurfactants, organic solvents, chelating reagents, acids or bases) in closed reactors [15].

This method has been successfully applied to different pollutant classes, such as semi-volatile organic compounds, fuels, PTEs as well as volatile organic compounds and pesticides. The removal efficiencies of inorganic compounds can be very high but vary as a function of the extracting reagent added to the washing solution. Also for PTEs removal, Soil Washing can be performed through different extracting reagents which differ for costs and efficiency. In particular, acid solutions (HCl, H_2SO_4, HNO_3) or complexing reagents such as citric acid, ammonium acetate, Nitrilotriacetic acid (NTA), Ethylenediaminetetraacetic acid (EDTA) and Ethylenediaminedisuccinic acid (EDDS) are most frequently used [18]. However, suitable reagent selection needs to be carried out by also considering the effect on soil characteristics after the treatment and the possible reuse of the remediated solid matrix. For instance, the soil fertility preservation and prevention of toxic by-products formation should be priority aspects in the treatment of soils for further agricultural reuse [56]. Accordingly, the involvement of biodegradable and non-toxic reagents indeed represents a key-solution for a feasible and environment friendly contaminated soil remediation processes. This chapter aims at deeply discussing the Soil Washing technique performed through biodegradable reagents and specifically provides detailed considerations about the chemical, biological, and engineering aspects related to the process.

2 Biodegradable Reagents

Several literature studies analyzed various reagents applicability in the remediation of contaminated sites. The main one involved in the remediation process are generally: (i) inorganic reagents, (ii) low-molecular-weight organic acids (LMWOA), (iii) chelating reagents. However, only LMWOAs and some synthetic chelating reagents are biodegradable and, therefore, suitable to perform environmentally sustainable treatments.

2.1 Low-Molecular-Weight Organic Acids

LMWOAs are very widespread since their microbial, plant, or animal origins. In the industrial field, their production is mainly achieved through chemical processes while only in modest quantities through biological ones [52]. LMWOAs are naturally found in soils due to production by lichens and algae. Moreover, LMWOAs are strongly linked to the carbon cycle, and usually, their concentration does not exceed 100 mM [3]. LMWOAs are characterized by the presence of at least one carboxylic acid group, which enables the formation of complexes with PTEs. For this reason, the number of scientific studies investigating the use of LMWOAs as soil-washing reagents has grown in the last decade [28, 56, 73]. The main advantage of LMWOAs application is their low toxicity and high biodegradability compared to inorganic acids and chemically produced organic acids. Unlike strong mineral acids, organic acids are usually weak and are not completely dissociated in water [34]. Furthermore, LMWOAs can be widely used in remediation processes, even at high concentrations, thanks to their high solubility (Table 1). The supply costs are strongly dependent on the production method and, obviously, on the degree of LMWOAs purity. For instance, biological methods are advantageous for the concurrent production of various LMWOAs facilitating their subsequent use as a mixture. In this case, the supply costs would be very low compared to other biodegradable reagents.

2.2 Aminopolycarboxylic Acids

Aminopolycarboxylic acids (AAs) are produced by both chemical and biological processes [9]. AAs are characterized by a high tendency to form complexes with PTEs. Therefore, these reagents are widely used in the remediation of PTEs contaminated soils. To date, the most commonly used AAs is the EDTA. However, EDTA is characterized by negligible biodegradability, and previous studies showed the potential toxicity of EDTA-Me complexes [27, 61]. Consequently, in recent decades, new AAs with high biodegradability and low toxicity have been investigated. These AAs are characterized by one or more N atoms bound to carboxyl groups by C atoms. Nonetheless, the production of biodegradable AAs is still not common at industrial scale. In fact, EDTA is still one of the AAs mainly used in soil remediation processes thus resulting in high costs for treatments carried out by involving other AAs. A further disadvantage of soil remediation through AAs is represented by the reagents reduced solubility. Then, treatments performed with high reagent concentrations are not feasible [65]. Accordingly, AAs application could be recommended for site contaminations not spatially extensive and characterized by modest pollutant concentrations.

2.3 Biosurfactants

Biosurfactants (BSs) are produced by the vital activities of bacteria, yeasts, and fungi (i.e., surfactin, sophorolipids etc.) but also by plants and animals (tannic acid) [53]. BSs enhance the desorption and solubilization rate of PTEs. Due to their different origins, they can be composed by different chemical compounds (i.e., carboxylic

Table 1 List of the most common biodegradable reagents used for remediation of contaminated sites

	pKa	Structure	Solubility [g/L]	Stability Constants (Ks)	Number of manuscripts 2013–2022	References
citric acid (CA)	2.79	$C_6H_8O_7$ or CH_2COOH-C(OH)COOH-CH_2COOH 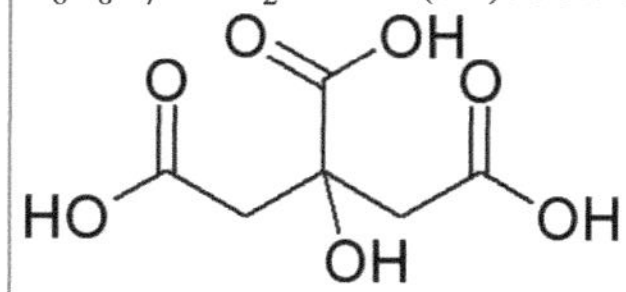	592	Cd: 3.5 Ni: 3.4 Cu: 5.9 Cr(III): 7.7 Pb: 4.1 Zn: 4.9 Mn: 2.6 Fe(III): 13.2 Ca: 4.9 Mg: 4.9	86	[45, 46, 60]
oxalic acid (OA)	pKa 1: 1.46; pKa 2: 4.40	$C_2H_2O_4$ or $(COOH)_2$ or HOOCCOOH 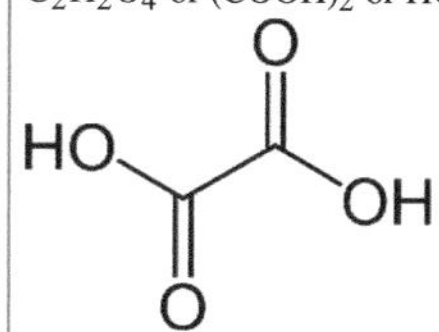	220	Cd: 3.0 Ni: 5.2 Cu: 6.3 Cr(III): 5.3 Pb: 4.0 Zn: 4.9 Mn: 3.9 Fe(III): 9.4 Mg: 2.6	46	[12, 51, 67]

(continued)

Table 1 (continued)

	pKa	Structure	Solubility [g/L]	Stability Constants (Ks)	Number of manuscripts 2013–2022	References
lactic acid (LA)	3.86	$C_3H_6O_3$ or $CH_3CHOHCOOH$ or $HC_3H_5O_3$ 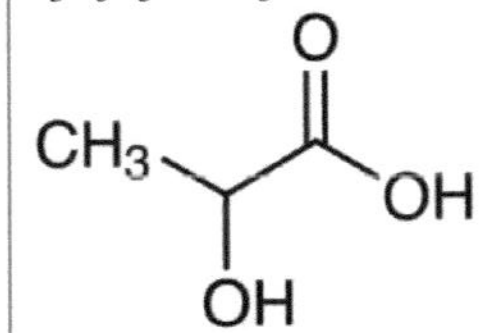	1000	Cd: 1.3 Ni: 1.6 Cu: 2.5 Cr(III): 3.3 Pb: 2.0 Zn: 1.6 Mn: 1.0 Ca: 0.9 Mg: 0.8	8	[29, 44, 55]
Acetic Acid (AA)	4.76	$C_2H_4O_2$ or CH_3COOH 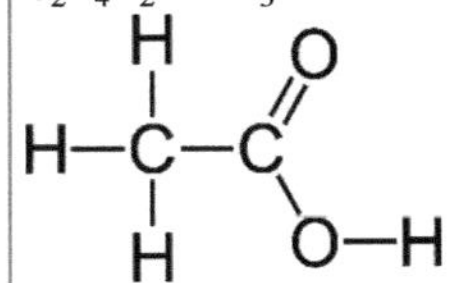	1000	Cd: 1.3 Ni: 0.7 Cu: 1.8 Cr(III): 3.7 Pb: 1.9 Zn: 0.9 Fe(III): Ca: 0.5 Mg: 0.5	29	[10, 60]

(continued)

Table 1 (continued)

	pKa	Structure	Solubility [g/L]	Stability Constants (Ks)	Number of manuscripts 2013–2022	References
Ethylenediamine-N,N'-disuccinic acid (EDDS)	pKa1 2.4, pKa2 3.9, pKa3 6.8, pKa4 9.8	$C_{10}H_{16}N_2O_8$ 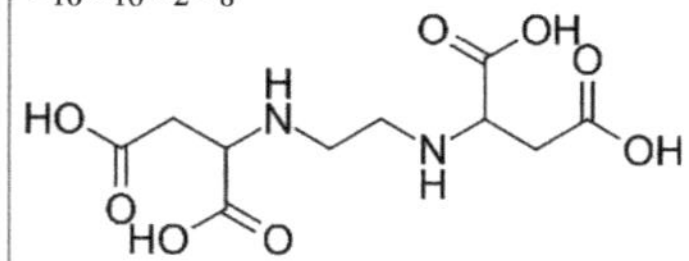	slightly soluble	Cd: 10.9 Ni: 16.7 Cu: 18.4 Cr(III): 10.8 Pb: 12.7 Zn: 13.4 Fe(III): 22 Ca: 4.6 Mg: 6.0	38	[50, 65]
Glutamic acid-N,N-diacetic acid (GLDA)	pKa1 2.6, pKa2 3.5, pKa3 5.0, pKa4 9.4	$C_9H_{13}NO_8$ 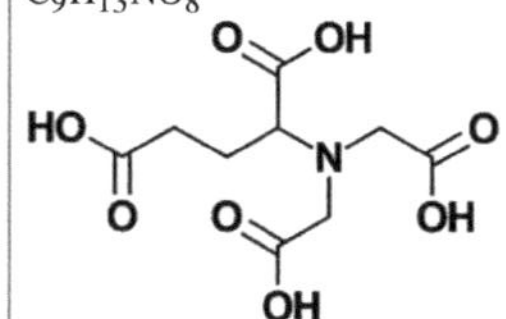	65	Cd: 10.3 Ni: 12.7 Cu: 13.0 Cr(III): 12.9 Pb: 11.6 Zn: 11.5 Mn: 7.6 Fe(III): 11.7 Ca: 5.2 Mg: 6.1	15	[1, 7 19, 36, 76]

acid with an aliphatic chain, glycolipid, etc.) as well as display different chemical structures.

BSs of natural origin are very suitable for the remediation of contaminated sites as they do not significantly alter the soil characteristics. However, the co-use of several biosurfactants during the chemical washing may be necessary, especially for highly contaminated soils. Their costs are currently relatively high. However, soil remediation with BSs may become economically feasible in the perspective of an increase of their production through biological processes [75].

3 Washing Treatments

Chemical washing on soils can be carried out both in situ and ex-situ [15]. In the first case, the soil is not excavated by site of origin and process is defined as Soil Flushing. In the ex-situ remediation, the soil is instead excavated prior the treatment which is as Soil Washing. The reagents involved can be the same for both approaches and dependent on the type of target contaminant. PTEs, radioactive elements and some organic compounds (i.e., PAHs, PCBs etc.) are generally removed by Soil Washing.

3.1 Soil Washing

Soil Washing is a remediation treatment based on physical and/or chemical mechanisms [40]. Physical mechanisms consist of dry (or wet) grading methods to separate the fine fraction (most contaminated) from the coarse material which can be recovered. Instead, the chemical mechanisms allow the pollutant removal through solubilization from the solid matrix to the liquid phase represented by the washing solution [26]. In this case, the chemical reagents in the washing solution enhance the pollutant removal. The treatment cycle (Fig. 1) includes sequential operating units, each of them affected by different factors depending on characteristics of both soils and contaminants. The main steps are as follow:

(1) Removal of soil from the origin site;
(2) Granulometric separation: the separation technique can vary according to the soil particle sizes to collect for further treatment; the most common are the drum screening, hydrocyclones, etc.;
(3) Treatment of the contaminated fraction: the reagent is chosen according to: (i) the selectivity towards the target contaminants, (ii) economy, and (iii) low toxicity; moreover, studies are required to evaluate the best operating conditions (i.e., contact time, pH, soil/solution ratio, reagent dosage);
(4) Treatment of the exhausted solution: processes aimed at treating industrial process waters are used; however, whenever possible, attempts are conducted to reuse both the aqueous solution and the reagent.

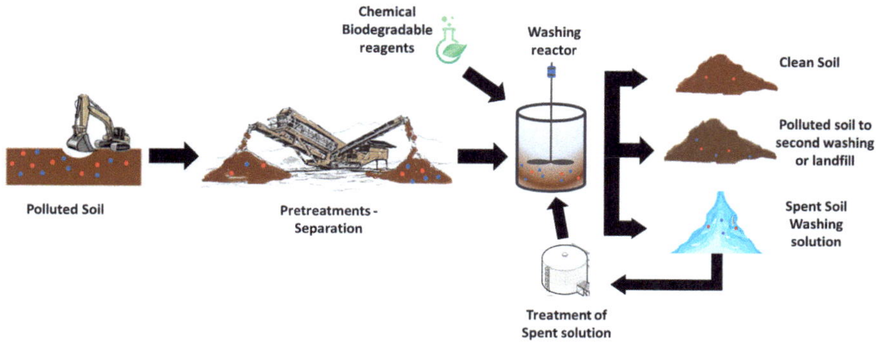

Fig. 1 Scheme of the soil washing process

3.2 *Soil Flushing*

In the Soil Flushing process, being an in-situ approach, the previously discussed steps 1 and 2 are not carried out. Furthermore, in this case, proper selection of the involved reagent is essential to avoid the pollution of the deep layers of soil as well as the aquifer [43]. Soil Flushing requires the construction of wells (vertical or sloped) for the washing solutions injection within the soil. Moreover, the wells allow the extraction of the solutions containing the dissolved contaminants after the treatment. For the Soil Flushing, soil permeability plays an important role as it affects the flows of the injected washing solution [5]. Then, a geological study of the contaminated area before treatment is essential (Fig. 2).

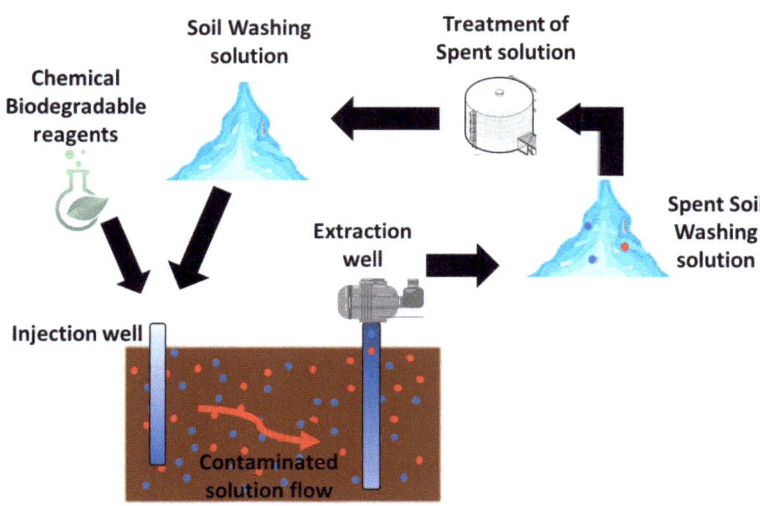

Fig. 2 Scheme of the soil flushing process

4 PTEs-Reagents Interactions

The interactions between PTEs and reagents can differ according to their mutual characteristics. However, the pivotal principle of all successful mechanisms is the solubilization of PTEs. This section describes the individual mechanisms that can occur between reagents and PTEs.

4.1 PTEs Interaction Mechanisms with Low-Molecular-Weight Organic Acids (LMWOAs)

The LMWOAs mainly studied in previous literature are reported in Table 1. LMWOAs can dissolve, solubilize and desorb PTEs from the soil particles [4]. The interaction between PTEs and LMWOAs results in the formation of soluble complexes. LMOWAs have different stability and complexation constants for each single metal (Table 1). Then, the involvement of LMWOAs for soil remediation should be decided according to the specific contamination detected in the solid matrix.

The use of LMWOAs at high concentrations can change the soil pH resulting in the non-selective dissolution of several elements. The pH variation can affect the balance of the formed complexes by varying the speciation of the single LMWOA and, therefore, altering its capacity to form complexes [35].

However, the extraction of the single PTE can also be influenced by its bioavailability, which is generally determined through sequential extraction procedure but not always discussed in scientific studies. More in detail, PTEs can be bound to different fractions of the soil: exchangeable, oxidizable, reducible and residual. The quantity of PTEs contained in the residual fraction cannot be leached in any environmental condition. However, some study also reported the decrease of PTEs in the residual fraction due to the dissolution of inorganic complexes. For instance, [77] observed vanadium reduction by more than 50%. On the contrary, PTEs dissolution occurs more easily from the mobile fractions represented by the exchangeable, reducible, and oxidizable ones. In general, LMWOAs solubilize elements from these fractions [74]. PTEs can be released from the first fraction (exchangeable) in slightly acidic conditions. Concerning the reducible fraction, it is basically due to the reduction of the iron/manganese oxides is the main phenomenon which leads to the PTEs release. Finally, for the oxidizable fraction, LMWOAs can dissolve the soil's organic matter thus allowing the PTEs solubilization [16]. The solubilization effect is not observed with High-molecular-weight organic acids (HMWOAs) which instead are generally effective in the stabilization of the contaminants within a solid matrix [74]. Based on the above discussion, it is clear that selection of proper analyses aimed at characterizing the PTEs bioavailability in soils could play a fundamental role for the efficiency of the treatment. Then, single sequential extraction analysis could be unsuitable especially considering the variable chemical speciation possibly displayed by metals.

Remediation process on soils contaminated by different PTEs can be carried out at various operating conditions. Table 2 shows the main results from scientific literature. The characteristics of both soils and contaminants as well as the washing methods can significantly affect the removal efficiency. High concentration of LMWOA results in high PTEs removal efficiencies. However, concurrent removal of nutrients (i.e., K, P, Ca, Mg, etc.) can also be observed from soil since LMWOA are non-selective extractants. According to this, a previous ecotoxicological study suggested the addition of nutrient-rich salts after the treatment process in order to restore the soil initial characteristics [56]. Due to the strong selectivity of LMWOAs toward specific pollutants (Table 1), the use of a low reagent dosage could be very effective for the removal of target contaminants. In fact, for citric acid, the Cu removal efficiency is usually more significant than that of the other PTEs such as Cd and Zn. Among the LMOWAs, oxalic acid (OA) showed effective applicability especially at high concentrations. OA forms insoluble complexes with PTEs (i.e., Pb) and other elements in the soil. Nonetheless the formed OA complex can easily resolubilize at slightly acidic pH thus increasing the fraction of potentially releasable metals. [62], observed a concentration increase of Cu and Pb in the exchangeable fraction due to the formation of oxalates. Accounting for Tartaric Acid and Lactic Acid, their use in the washing process is less frequent despite the high removal efficiencies achievable. Currently, scientific studies on washing process performed through LMOWA mixtures are lacking in literature. Indeed, this aspect could be worth of future investigations by also studying the effects on the soil remediation process through washing solution characterized by combination of LMOWAs from biological treatments carried out different application fields (such as wastewater treatment).

The solid/liquid (S/L) ratio (v/v) indirectly affects the PTEs-LMOWA molar ratio. High values of the S/L ratio lead to enhanced diffusion phenomena [20]. As reported in Table 2, treatment efficiency increase can be observed at increasing S/L ratio values. However, in full-scale applications, too high S/L ratio could hinder the process feasibility due to the consequent requirement of large washing solution volumes and their subsequent treatment. According to the results observed in Table 2, S/L ratio of 1/10 seems in most of the case an optimal operating condition. LMOWAs are readily biodegradable but after their complexation with PTEs, the biodegradability rate slows down [49]. [22] observed that the citric acid is not biodegradable after complex formation with Cu at CA/Cu molar ratio equal to 1/1. On the contrary, CA is still totally biodegradable if the CA/Cu molar ratio increases to 2/1. The biodegradation of LMOWAs can also occur due to microorganisms present in the soil. Nonetheless, monitoring of potential changes occurring in the distribution of microorganisms community within soil after treatment with LMOWAs would be strongly recommended [57]. Therefore, based on the above reported discussion, the selection of a proper LMWOA (and its concentration) for the soil treatment should be done by also evaluating the biodegradability and toxicity of the formed complexes besides the PTEs removal efficiencies potentially achievable.

Table 2 Removal efficiencies for PTEs by varying the operating conditions and the reagents

Soil	PTE	Concentration (mg kg^{-1})	Parameter	Removal efficiences [%]	References
pH 7.4; CEC:3.6 meq/100 g; Sand(%): 97; Silt(%):3; Clay(%) 0; OM(%): 0.2	Cd	500 (Artificially contaminated)	0.001 – 0.1 mol l-1 S/L = 1/10 T = 3 h	AA: 5.2–31.1	[28]
	Pb	500 (artificially contaminated)		AA: 42.2–100	
	Ni	500 (artificially contaminated)		AA: < 1	
pH 6.1; CEC:1.6%; Sand(%): 78; Silt(%):13; Clay(%) 9; OM(%): 0.2	Ni	438 (Artificially contaminated)	0.05 mol l-1 S/L = 1/25 T = 6 h	CA: 46 TA: 28	[72]
	Cu	498 (Artificially contaminated)		CA: 50 TA: 30	
	Zn	376 (Artificially contaminated)		CA: 44 TA: 27	
	Cd	340 (Artificially contaminated)		CA: 38 TA: 19	
	Pb	293 (Artificially contaminated)		CA: 31 TA: 16	
pH 5.1; CEC:1.6%; Sand(%): 49; Silt(%):30; Clay(%) 21; OM(%): 5.4	Pb	2797	0.03–0.16 mol l-1 S/L = 1/25 T = 6-9 h	CA: 83 TA: 88	[70]
	Cd	1821		CA: 96 TA: 97	
pH 6.3; CEC:12.2%; Sand(%): 53; Silt(%):24; Clay(%) 23; OM(%): 7.1	Pb	1036	0.03–0.40 mol l-1 S/L = 1/25 T = 12-24 h	CA: 80 TA: 87 CA + OA: 70	
	Cd	794		CA: 100 TA: 99.9 CA + OA:82	
	Cu	926		CA: 88 TA: 97 CA + OA: 99.9	

(continued)

Table 2 (continued)

Soil	PTE	Concentration (mg kg^{-1})	Parameter	Removal efficiences [%]	References
	Zn	1029		CA: 95 TA: 98 CA + OA: 99.9	
pH 7.5–8.4; CEC: 1.2–4.0 meq/100 g; Sand(%): 60; Silt(%):30; Clay(%) 10; OM(%): 7.1	Cd	7	0.05 mol l-1 S/L = 1/10 T = 3 h	CA: 40 OA: 9	[54]
	Cr	238		CA: 2 OA: 2	
	Cu	1241		CA: 24 OA: 15	
	Hg	2		CA: 13 OA: 14	
	Pb	21,560		CA: 5 OA: < 1	
	Zn	3729		CA: 22 OA: 5	
	As	18		CA: 5 OA: 5	
pH 8; CEC: -; Sand(%): 100; Silt(%):0; Clay(%) 0; OM(%): 0	U	0.006 (Artificially contaminated)	0.01NaCl + 0.001 mol l-1 S/L = 1/5 T = 1 h	NaCl + CA: 95	[33]
pH 6.25; CEC: 12.40 cmol kg – 1; Sand(%): 52; Silt(%):8; Clay(%) 40; OM(%): 1.9	Cd	17	0.070 mol l-1 S/L = 1/10 T = 2 h	CA: 24 EDDS: 24 GLDA: 61	[69]
	Pb	1640		CA: 35 EDDS: 44 GLDA: 89	
	Zn	2589		CA: 40 EDDS: 83 GLDA: 38	
pH 7.0; CEC: 18 cmol kg – 1; Sand(%): 65; Silt(%):17; Clay(%) 18; OM(%): 2.4	Cd	33	0.070 mol l-1 S/L = 1/10 T = 2 h	CA: 46 EDDS: 43 GLDA: 67	
	Pb	681		CA: 32 EDDS: 72 GLDA: 85	
	Zn	368		CA: 43 EDDS: 34 GLDA: 40	

(continued)

Table 2 (continued)

Soil	PTE	Concentration (mg kg^{-1})	Parameter	Removal efficiences [%]	References
pH: 8.5 CEC: 6 (meq/100 g); Sand (%): 19 Silt (%): 47 Clay (%): 34	Cr	769	0.2 $Na_3C_6H_5O_7$ + 0.2 mol l-1 S/L = 1/2 T = 20 h	CA: 49	[39]
pH: 8.0 CEC: - Sand (%): 99 Silt (%): < 1 Clay (%): < 1; TOC 23: g kg-1	As	49	25%v/v S/L = 1/10 T = 1 h	LA: 79	[59]
	Cu	80		LA: 74	
	Cr	35		LA: 71	
pH: 7.4 - CEC: - Sand (%): 60; Silt (%): 28; Clay (%): 12; OM (%): 27	Cr	3051	0.8 - 1 mol l-1 S/L = 1/10 T = 48 h	LA: 85 MA: 67 OA: 74	[56]
			0.1 mol l-1 S/L = 1/10 T = 48 h	LA: 2 MA: 5 OA: 3 CA: 6	
pH: 5.2 - CEC: 15.82 cmol kg-1; OM (%): 2.9	Cd	27	0.1 mol l-1 S/L = 1/25 T = 48 h	MA: 53 TA: 33 CA: 69 NTA: 58 EDDS: 44 CTAB: 1	[73]
			10%v/v S/L = 1/25 T = 48 h	LA:62 GLDA: 14	
	Pb	619	0.1 mol l-1 S/L = 1/25 T = 48 h	MA: 35 TA: 3 CA: 49 NTA: 31 EDDS: 21 CTAB: 1	
			10%v/v S/L = 1/25 T = 48 h	LA: 60 GLDA: 4	
	As	290	0.1 mol l-1 S/L = 1/25 T = 48 h	MA: 7 TA: 7 CA: 10 NTA: 0 EDDS: 10 CTAB: 0	
			10%v/v S/L = 1/25 T = 48 h	LA: 17 GLDA: 6	

(continued)

Table 2 (continued)

Soil	PTE	Concentration (mg kg^{-1})	Parameter	Removal efficiences [%]	References
pH: 6.3 - CEC: 19 cmol kg-1; OC (%): 19	Zn	1037	10 g l-1 S/L = 1/10 T = 4 h	OA: 59 TA: 57 CA: 54	[11]
pH: 7.8 - CEC: 2.8 cmol kg-1; Sand (%): 78; Silt (%): 10; Clay (%): 12; OM (%): 10	Cu	6511	0.10 g l-1 S/L = 1/25 T = 24-72 h	Surfactin: 17	[42]
	Pb	4955		Surfactin: < 1	
	Zn	15,090		Surfactin: 24	
	Cu	6511	0.015 g l-1 S/L = 1/25 T = 24-72 h	Saponin: 36	
	Pb	4955		Saponin: 47	
	Zn	15,090		Saponin: 18	
pH: 6.3; Sand (%): 40; Silt (%): 8; Clay (%): 52; OC (%): 2	Cd	18	0.05 mol l-1 S/L = 1/10 T = 1 h	GLDA: 39 ISA: 12	[68]
	Pb	1222		GLDA: 58 ISA: 34	
	Zn	2648		GLDA: 22 ISA: 13	
pH: 7.1; Sand (%): 18; Silt (%): 17; Clay (%): 65; OC (%): 2.4	Cd	43	0.05 mol l-1 S/L = 1/10 T = 1 h	GLDA: 81 ISA: 40	
	Pb	232		GLDA: 79 ISA: 45	
	Zn	1286		GLDA: 64 ISA: 61	
pH: 4.8 - CEC: 13 cmol kg-1; Sand (%): 40; Silt (%): 40; Clay (%): 20; OM (%): 4	Cd	5	0.06 mol l-1 S/L = 1/2 T = 18 h	GLDA: 76 EDDS: 20 IDS: 31	[24]
	Pb	809		GLDA: 58 EDDS: 7 IDS: 25	
	Zn	448		GLDA: 29 EDDS: 23 IDS: 22	
Organic-rich soil reference material BCR 700 from EC-JRCIRMM	Cd	132	0.05 mol l-1 S/L = 1/10 T = 1 h	GLDA: 62 MGDA: 49 EDDS: 36 IDSA: 35	[8]
	Cu	169		GLDA: 44 MGDA: 35 EDDS: 37 IDSA: 36	
	Ni	373		GLDA: 18 MGDA: 13 EDDS: 14 IDSA: 14	

(continued)

Table 2 (continued)

Soil	PTE	Concentration (mg kg^{-1})	Parameter	Removal efficiences [%]	References
	Pb	288		GLDA: 17 MGDA: 10 EDDS: 9 IDSA: 11	
	Zn	1584		GLDA: 32 MGDA: 25 EDDS: 23 IDSA: 23	
smelter site, no other information is available	Cr	1060	0.1 mol l-1 S/L = 1/10 T = 24 h	AA: < 1 TA: 4 CA: 5 MA: 5 OA: 6	[77]
	V	2015		AA: 27 TA: 48 CA: 48 MA: 48 OA: 52	
pH: 6.6—CEC: 11.7 cmol kg-1; Sand (%): 86; Silt (%): 6; Clay (%): 8; OM (%): 0.5	Pb	842	0.5 mol l-1 S/L = 1/10 T = 1 h	TA: 63	[47]
	Cu	438		TA: 42	
	Zn	375		TA: 49	
pH: 7.8—CEC: 2.4 cmol dm-3; Silt-loam; C(%): 1.2	As	114 (Artificially contaminated)	2% S/L = 1/20 T = 48 h	Rhamnolipid: 53	[14]
	Cd	122 (Artificially contaminated)		Rhamnolipid: 90	
	Zn	3339 (Artificially contaminated)		Rhamnolipid: 80	
pH: 6.4—CEC: 14.6 meq per 100 g; Sand (%): 15; Silt(%): 65; Clay(%): 20; OM(%): 20	Cu	140	Soil flushing 10 g 0.05 ml g-1 min-1	Rhamnolipid: 17	[41]
	Zn	4854		Rhamnolipid: 6	
	Ni	76		Rhamnolipid: 16	
pH: 7.7—CEC: 73 meq per 100 g; Sand (%): 27; Silt(%): 26; Clay(%): 47; C(%): 0.45	Cd	16	Soil flushing 50 g 0.2 ml min-1 T: 36	Rhamnolipid: 19	[32]
	Pb	19		Rhamnolipid: 8	

(continued)

Table 2 (continued)

Soil	PTE	Concentration (mg kg^{-1})	Parameter	Removal efficiences [%]	References
pH: 6.3—CEC: 66 meq per 100 g; Sand (%): 30; Silt(%): 28; Clay(%): 42; C(%): 0.38	Cd	435 (Artificially contaminated)		Rhamnolipid: 91	
	Pb	905 (Artificially contaminated)		Rhamnolipid: 87	

4.2 *Aminopolycarboxylic Acids*

Aminopolycarboxylic Acids (AAs) are synthetic chelating substances. AAs can form complexes with very stable and water-soluble PTEs, thus avoiding PTEs precipitation [37]. To date, the most used AAs at full scale are scarcely biodegradable (i.e., EDTA) but many scientific studies investigated the possible application of biodegradable AAs in the remediation of contaminated sites. The removal mechanisms of PTEs from soils are similar to those discussed for LMWOAs. Compared to a LMWOA washing, the removal efficiencies achievable in the AAs washing at same reagent concentration are higher due to the high stability constants values of the formed complexes (Table 1) [73]. However, unlike LMWOAs, the pH variation plays a minor role with AAs since no soil pH alteration are significantly observed. In general, the efficiency of AAs washing is affected by: (i) physicochemical characteristics of the soil, (ii) process parameters, and (iii) characteristics of PTEs contamination.

Despite the very high removal efficiencies which can be generally achieved with AAs, selection of the most suitable reagent is done by evaluating the stability constants of the complexes that they could form from the interaction with the target PTE [6, 64].

After the AAs washing, possible consequences observable in the treated soil are: (i) removal of some nutritional elements (i.e., Ca and Mg) in addition to the PTEs and (ii) partial adsorption of the formed PTEs-AAs on the soil particles. The AAs biodegradability after the complex formation with the PTE as well as the toxicity of biodegradable PTEs-AAs complexes are not thoroughly reported in scientific literature. Nonetheless, an exemplifying study showed the reduction of both overall biodegradability and biodegradation rate of EDDS after complexation with Cu [66].

4.3 *Biosurfactant*

According to their definition, surfactants are a mixture of surface-active agents created by Antara products in 1950. Surfactants are usually amphiphilic organic compounds (they contain both hydrophobic and hydrophilic groups). Therefore, surfactants are soluble in both organic solvents and water [30]. Surfactants reduce the surface tension of water by adsorbing at the liquid–gas interface and reduce the interfacial tension between oil and water by adsorbing at the liquid–liquid interface. Surfactants can be either anionic or cationic and both are successfully used to remediate sites contaminated by PTEs [31]. Usually, anionic surfactants are preferred to the cationic ones because they are more soluble in water and form complexes with positively charged PTEs (i.e., Cd, Pb, Cu, etc.). Instead, the remediation of soils with PTEs in oxyanions form (i.e., Cr and As) requires cationic surfactants. Surfactants are generally produced through chemical processes. However, in recent years, research has increasingly moved toward the BSs production. These are generated by microorganisms, animals, or plants (i.e., rhamnolipids can come from Pseudomonas aeruginosa) [76].

The most common BSs used to remediate contaminant sites are rhamnolipids (glycolipids) and saponin. Rhamnolipids reduces the solution's surface tension and are characterized by the presence of carboxyl groups that complex the PTEs with stability constants greater than those between PTEs and soil particles. Table 2 shows the results obtained by using rhamnolipids highlighting the achievement of very high removal efficiencies (about 90%). Moreover, with the use of rhamnolipids, no alterations of the treated soils characteristics (i.e., pH variations) are observed. On the contrary, an increase in the microbial population was reported [32]. Accounting for the PTEs removal based on their bioavailability in soil, [41] observed As, Cd, and Zn solubilization from both exchangeable and carbonate-bound fractions.

Another BS widely used in the soil remediation treatment is saponin, which can display PTEs removal efficiencies comparable to rhamnolipids. In general, BSs have a high biodegradability. However, when BSs are complexed to PTEs are more slowly biodegradable than CA [71].

5 Conclusions and Future Perspectives

The remediation of contaminated sites is a relevant topic and its significance is constantly increasing: the fertile soil resource is becoming rare and proper agricultural soils preservation/reclamation strategies are nowadays required. Remediation treatments can be useful for PTEs removal, but at the same time, can reduce the soil fertility up to the occurrence of very adverse environmental conditions. The use of LMWOAs, AAs, and BSs could represent a feasible approach because of the high extraction rates achievable, compared to inorganic acids, and the lack of soil characteristics alteration if properly used.

However, further studies are needed to evaluate their applicability at full-scale and to evaluate any long-term ecotoxicity effects as well. In this perspective, it also worth noticing that laboratory scale tests can still represent a fundamental source for preliminary identification of optimal washing reagent and operating conditions to perform an efficient and environmentally sustainable soil washing process.

References

1. Abdelgawad KZ, Mahmoud M, Elkatatny S, Abdulraheem A, Patil S (2019) Reaction kinetics and coreflooding study of high-temperature carbonate reservoir stimulation using GLDA in seawater. Energies 12:3407
2. Abrahams PW, Thornton I (1994) The contamination of agricultural land in the metalliferous province of southwest England: implications to livestock. Agric Ecosyst Environ 48:125–137
3. Adeleke R, Nwangburuka C, Oboirien B (2017) Origins, roles and fate of organic acids in soils: A review. South African J. Bot. 108:393–406
4. Ash C, Tejnecký V, Borůvka L, Drábek O (2016) Different low-molecular-mass organic acids specifically control leaching of arsenic and lead from contaminated soil. J Contam Hydrol 187:18–30
5. Atteia O, Del Campo Estrada E, Bertin H (2013) Soil flushing: a review of the origin of efficiency variability. Rev. Environ. Sci. Bio/Technology 12:379–389
6. Bari ASMF, Lamb D, MacFarlane GR, Rahman MM (2022) Soil washing of arsenic from mixed contaminated abandoned mine soils and fate of arsenic after washing. Chemosphere 296:134053
7. Begum ZA, Rahman IMM, Sawai H, Tate Y, Maki T, Hasegawa H (2012) Stability constants of Fe (III) and Cr (III) complexes with dl-2-(2-carboxymethyl) nitrilotriacetic acid (GLDA) and 3-hydroxy-2, 2′-iminodisuccinic acid (HIDS) in aqueous solution. J Chem Eng Data 57:2723–2732
8. Begum ZA, Rahman IMM, Tate Y, Sawai H, Maki T, Hasegawa H (2012) Remediation of toxic metal contaminated soil by washing with biodegradable aminopolycarboxylate chelants. Chemosphere 87:1161–1170
9. Bucheli-Witschel M, Egli T (2001) Environmental fate and microbial degradation of aminopolycarboxylic acids. FEMS Microbiol Rev 25:69–106
10. Bunting JW, Thong KM (1970) Stability constants for some 1: 1 metal–carboxylate complexes. Can J Chem 48:1654–1656
11. Chen Y, Zhang S, Xu X, Yao P, Li T, Wang G, Gong G, Li Y, Deng O (2016) Effects of surfactants on low-molecular-weight organic acids to wash soil zinc. Environ Sci Pollut Res 23:4629–4638
12. Clayton GD, Clayton FE (1981) Patty's industrial hygiene and toxicology. Vol. 2A. Toxicology. John Wiley & Sons, Inc., Baffins Lane, Chichester, Sussex PO19 1DU
13. Dahlin S, Witter E, Mårtensson A, Turner A, Bååth E (1997) Where's the limit? Changes in the microbiological properties of agricultural soils at low levels of metal contamination. Soil Biol Biochem 29:1405–1415
14. Dahrazma B, Mulligan CN (2007) Investigation of the removal of heavy metals from sediments using rhamnolipid in a continuous flow configuration. Chemosphere 69:705–711
15. Dermont G, Bergeron M, Mercier G, Richer-Laflèche M (2008) Soil washing for metal removal: a review of physical/chemical technologies and field applications. J Hazard Mater 152:1–31. https://doi.org/10.1016/j.jhazmat.2007.10.043
16. Ding Q, Wu HL, Xu Y, Guo LJ, Liu K, Gao HM, Yang H (2011) Impact of low molecular weight organic acids and dissolved organic matter on sorption and mobility of isoproturon in two soils. J Hazard Mater 190:823–832
17. EEA, E.E.A., (2020) Soil contamination widespread in Europe
18. Fabbricino M, Ferraro A, Del Giudice G, D'Antonio L (2013) Current views on EDDS use for ex situ washing of potentially toxic metal contaminated soils. Rev. Environ. Sci. Bio/Technology 12:391–398
19. FEEDAP, E.P. on A. and P. or S. used in A.F., Rychen, G., Aquilina, G., Azimonti, G., Bampidis, V., Bastos, M. de L., Bories, G., Chesson, A., Cocconcelli, P.S., Flachowsky, G., (2018) Safety and efficacy of Kelforce®(l-glutamic acid, N, N-diacetic acid, tetrasodium salt (GLDA-Na4)) as a feed additive for chickens for fattening. EFSA J 16:e05279

20. Ferraro A, Fabbricino M, van Hullebusch ED, Esposito G, Pirozzi F (2016) Effect of soil/ contamination characteristics and process operational conditions on aminopolycarboxylates enhanced soil washing for heavy metals removal: a review. Rev. Environ. Sci. Bio/technology 15:111–145
21. Ferraro A, Parisi A, Barbone E, Race M, Mali M, Spasiano D, Fratino U (2023) Characterising contaminants distribution in marine-coastal sediments through multivariate and nonparametric statistical analyses: a complementary strategy supporting environmental monitoring and control. Environ Monit Assess 195:1–19
22. Francis AJ, Dodge CJ, Gillow JB (1992) Biodegradation of metal citrate complexes and implications for toxic-metal mobility. Nature 356:140–142
23. Gabriele I, Race M, Papirio S, Papetti P, Esposito G (2022) Phytoremediation of a pyrene-contaminated soil by Cannabis sativa L. at different initial pyrene concentrations. Chemosphere 300, 134578
24. Gluhar S, Kaurin A, Lestan D (2020) Soil washing with biodegradable chelating agents and EDTA: Technological feasibility, remediation efficiency and environmental sustainability. Chemosphere 257:127226
25. Goovaerts P (1997) Geostatistics for natural resources evaluation. Oxford University Press
26. Griffiths RA (1995) Soil-washing technology and practice. J Hazard Mater 40:175–189
27. Gusiatin ZM, Kulikowska D, Klik B (2020) New-generation washing agents in remediation of metal-polluted soils and methods for washing effluent treatment: A review. Int J Environ Res Public Health 17:6220
28. Gzar HA, Abdul-Hameed AS, Al-Taie A (2014) Extraction of lead, cadmium and nickel from contaminated soil using acetic acid. Open J soil Sci 4:207–214
29. Hamada YZ, Carlson B, Dangberg J (2005) Interaction of malate and lactate with chromium (III) and iron (III) in aqueous solutions. Synth React Inorg Met Chem 35:515–522
30. Jene Q, Pearson JC, Lowe CR (1997) Surfactant modified enzymes: solubility and activity of surfactant-modified catalase in organic solvents. Enzyme Microb Technol 20:69–74
31. Jiménez-Castañeda ME, Medina DI (2017) Use of surfactant-modified zeolites and clays for the removal of heavy metals from water. Water 9:235
32. Juwarkar AA, Nair A, Dubey KV, Singh SK, Devotta S (2007) Biosurfactant technology for remediation of cadmium and lead contaminated soils. Chemosphere 68:1996–2002
33. Kantar C, Honeyman BD (2006) Citric acid enhanced remediation of soils contaminated with uranium by soil flushing and soil washing. J Environ Eng 132:247–255
34. Khan SH, Iqbal J (2016) Recent advances in the role of organic acids in poultry nutrition. J Appl Anim Res 44:359–369
35. Kim J-O, Lee Y-W, Chung J (2013) The role of organic acids in the mobilization of heavy metals from soil. KSCE J Civ Eng 17:1596–1602
36. Kołodyńska D (2011) Chelating agents of a new generation as an alternative to conventional chelators for heavy metal ions removal from different waste waters. Expand. issues Desalin. 339–370.
37. Lestan D, Luo C, Li X (2008) The use of chelating agents in the remediation of metal-contaminated soils: a review. Environ Pollut 153:3–13. https://doi.org/10.1016/j.envpol.2007.11.015
38. Li C, Wang H, Liao X, Xiao R, Liu K, Bai J, Li B, He Q (2022) Heavy metal pollution in coastal wetlands: A systematic review of studies globally over the past three decades. J Hazard Mater 424:127312
39. Li G, Yang X, Liang L, Guo S (2017) Evaluation of the potential redistribution of chromium fractionation in contaminated soil by citric acid/sodium citrate washing. Arab J Chem 10:S539–S545
40. Liu J, Zhao L, Liu Q, Li J, Qiao Z, Sun P, Yang Y (2021) A critical review on soil washing during soil remediation for heavy metals and organic pollutants. Int J Environ Sci Technol 1–24
41. Lopes CSC, Teixeira DB, Braz BF, Santelli RE, de Castilho LVA, Gomez JGC, Castro PV, R., Seldin, L., Freire, D.M.G., (2021) Application of rhamnolipid surfactant for remediation of toxic metals of long-and short-term contamination sites. Int J Environ Sci Technol 18:575–588

42. Maity JP, Huang YM, Hsu C-M, Wu C-I, Chen C-C, Li C-Y, Jean J-S, Chang Y-F, Chen C-Y (2013) Removal of Cu, Pb and Zn by foam fractionation and a soil washing process from contaminated industrial soils using soapberry-derived saponin: a comparative effectiveness assessment. Chemosphere 92:1286–1293
43. Martel R, Gélinas PJ, Saumure L (1998) Aquifer washing by micellar solutions: 3 Field test at the Thouin Sand Pit (L'Assomption, Quebec, Canada). J Contam Hydrol 30:33–48
44. Merck Index, M., 1996. 7943 and 9877, Merck & Co.
45. Mertens M, Höss S, Neumann G, Afzal J, Reichenbecher W (2018) Glyphosate, a chelating agent—relevant for ecological risk assessment? Environ Sci Pollut Res 25:5298–5317
46. Mo A, Wang Y, Liu Y, Zhou W, Zhou Q, Yasin A, Yang H (2018) Enhanced viscosity of poly (acrylamide) solution in the presence of chromium citrate triggered by release of CO2. Chinese J Chem Phys 31:117–122
47. Moon DH, Park J-W, Koutsospyros A, Cheong KH, Chang Y-Y, Baek K, Jo R, Park J-H (2016) Assessment of soil washing for simultaneous removal of heavy metals and low-level petroleum hydrocarbons using various washing solutions. Environ Earth Sci 75:1–8
48. Murtaza G, Shehzad MT, Kanwal S, Farooqi ZUR, Owens G (2022) Biomagnification of potentially toxic elements in animals consuming fodder irrigated with sewage water. Environ Geochem Health 44:4523–4538
49. Nowak G, Nowak J (1990) Turnover of 14 C-labelled oat residues and small molecular organic compounds in two soils under different levels of mineral nutrition. Plant Soil 122:67–77
50. Orama M, Hyvönen H, Saarinen H, Aksela R (2002) Complexation of [S, S] and mixed stereoisomers of N, N′-ethylenediaminedisuccinic acid (EDDS) with Fe (III), Cu (II), Zn (II) and Mn (II) ions in aqueous solution. J. Chem. Soc. Dalt. Trans. 4644–4648
51. Ostroushko AA, Russkikh OV (2017) Oxide material synthesis by combustion of organicinorganic compositions. Наносистемы: физика, химия, математика 8, 476–502
52. Papagianni M (2011) 1.09—Organic Acids, in: Moo-Young, M.B.T.-C.B. (Second E. (Ed.), Academic Press, Burlington, pp 109–120. https://doi.org/10.1016/B978-0-08-088504-9.00011-8
53. Paria S (2008) Surfactant-enhanced remediation of organic contaminated soil and water. Adv Colloid Interface Sci 138:24–58
54. Peters RW (1999) Chelant extraction of heavy metals from contaminated soils. J Hazard Mater 66:151–210
55. Portanova R, Lajunen LHJ, Tolazzi M, Piispanen J (2003) Critical evaluation of stability constants for alpha-hydroxycarboxylic acid complexes with protons and metal ions and the accompanying enthalpy changes. Part II. Aliphatic 2-hydroxycarboxylic acids (IUPAC Technical Report). Pure Appl Chem 75:495–540
56. Race M, Marotta R, Fabbricino M, Pirozzi F, Andreozzi R, Guida M, Siciliano A, (2019) Assessment of optimal conditions for the restoration and recovery of agricultural soil. J. Hazard. Mater. 373, 801–809. https://doi.org/10.1016/j.jhazmat.2019.04.007
57. Renella G, Landi L, Nannipieri P (2004) Degradation of low molecular weight organic acids complexed with heavy metals in soil. Geoderma 122:311–315
58. Reyes A, Cuevas J, Fuentes B, Fernández E, Arce W, Guerrero M, Letelier MV (2021) Distribution of potentially toxic elements in soils surrounding abandoned mining waste located in Taltal. Northern Chile. J. Geochemical Explor. 220:106653
59. Reynier N, Blais J-F, Mercier G, Besner S (2013) Treatment of arsenic-, chromium-, copper-and pentachlorophenol-polluted soil using flotation. Water, Air, Soil Pollut 224:1514
60. Serjeant EP, Dempsey B (1979) Ionisation constants of organic acids in aqueous solution. IUPAC Chem. data Ser. 23:160–190
61. Sillanpää M, Oikari A (1996) Assessing the impact of complexation by EDTA and DTPA on heavy metal toxicity using microtox bioassay. Chemosphere 32:1485–1497
62. Sun B, Zhao FJ, Lombi E, McGrath SP (2001) Leaching of heavy metals from contaminated soils using EDTA. Environ Pollut 113:111–120
63. Tarvainen T, Reyes A, Sapon S (2020) Acceptable soil baseline levels in Taltal, Chile, and in Tampere. Finland. Appl. Geochemistry 123:104813

64. Tsang DCW, Hartley NR (2014) Metal distribution and spectroscopic analysis after soil washing with chelating agents and humic substances. Environ Sci Pollut Res 21:3987–3995
65. USDA (2010) Technical Evaluation Report. Tech. Eval. Rep. (S, S)-Ethylenediaminedisuccinic Acid (free acid) Crop Prod.
66. Vandevivere PC, Saveyn H, Verstraete W, Feijtel TCJ, Schowanek DR (2001) Biodegradation of metal−[S, S]-EDDS complexes. Environ Sci Technol 35:1765–1770
67. Vinokurov EG, Bondar', V. V, (2003) Prediction of stability constants for Cr (III) and Cr (II) complexes. Russ J Coord Chem 29:66–72
68. Wang G, Pan X, Zhang S, Zhong Q, Zhou W, Zhang X, Wu J, Vijver MG, Peijnenburg WJGM (2020) Remediation of heavy metal contaminated soil by biodegradable chelator–induced washing: Efficiencies and mechanisms. Environ Res 186:109554
69. Wang G, Zhang S, Xu X, Zhong Q, Zhang C, Jia Y, Li T, Deng O, Li Y (2016) Heavy metal removal by GLDA washing: Optimization, redistribution, recycling, and changes in soil fertility. Sci Total Environ 569–570:557–568. https://doi.org/10.1016/J.SCITOTENV.2016.06.155
70. Wasay SA, Barrington SF, Tokunaga S (1998) Remediation of soils polluted by heavy metals using salts of organic acids and chelating agents. Environ Technol 19:369–379
71. Wen J, Stacey SP, McLaughlin MJ, Kirby JK (2009) Biodegradation of rhamnolipid, EDTA and citric acid in cadmium and zinc contaminated soils. Soil Biol Biochem 41:2214–2221
72. Wuana RA, Okieimen FE, Imborvungu JA (2010) Removal of heavy metals from a contaminated soil using organic chelating acids. Int J Environ Sci Technol 7:485–496
73. Xu L, Dai H, Skuza L, Xu J, Shi J, Wei S (2022) Co-high-efficiency washing agents for simultaneous removal of Cd, Pb and As from smelting soil with risk assessment. Chemosphere 300:134581
74. Yao W, Huang L, Yang Z, Zhao F (2022) Effects of organic acids on heavy metal release or immobilization in contaminated soil. Trans Nonferrous Met Soc China 32:1277–1289
75. Zhang H, Xu Y, Kanyerere T, Wang Y, Sun M (2022) Washing reagents for remediating Heavy-Metal-Contaminated Soil: A Review. Front. Earth Sci
76. Zheng X-J, Li Q, Peng H, Zhang J-X, Chen W-J, Zhou B-C, Chen M (2022) Remediation of heavy metal-contaminated soils with soil washing: a review. Sustainability 14:13058
77. Zou Q, Gao Y, Yi S, Jiang J, Aihemaiti A, Li D, Yang M (2019) Multi-step column leaching using low-molecular-weight organic acids for remediating vanadium-and chromium-contaminated soil. Environ Sci Pollut Res 26:15406–15413

Natural Barriers to Antibiotic Entry into the Trophic Chain: Some Examples of the Role of Soils and By-Products

Ventura Castillo-Ramos, Raquel Cela-Dablanca, Ainoa Míguez-González, Ana Barreiro, Avelino Núñez-Delgado, María J. Fernández-Sanjurjo, and Esperanza Álvarez-Rodríguez

Abstract This chapter explores the vital intersection of environmental science and pharmaceutical pollution mitigation. Withing the overall problem, specific examples are provided, investigating the adsorption behaviors of three prominent antibiotics—amoxicillin, trimethoprim, and ciprofloxacin—across various bio-adsorbents. In that way, this study delves into the mechanisms governing the interactions between these emerging contaminants and environmentally sustainable remediation agents. Also, as examples, some by-products (specifically, pine bark and oak ash) are revealed as promising bio-adsorbents for the effective retention of antibiotics, holding potential as green barriers against the mobilization of pharmaceutical pollutants into soils and essential water resources. This chapter take into account global issues related to environmental pollution due to emerging pollutants, and specifically antibiotics (thus including a variety of groups, such as anticoccidials and many others), but is mainly conceived to go ahead considering previous studies of the research group, contributing to pave the way for future exploration of binary and tertiary adsorption systems, offering valuable insights into the evolving landscape of pharmaceutical pollution remediation of soils, both at a local scale and globally.

Keywords Antibiotics pollution · Bio-adsorbents · Emerging pollutants · Environmental pollution remediation · Soil pollution · Water pollution

---

V. Castillo-Ramos (✉)
Department of Inorganic Chemistry, Faculty of Science, University of Granada, 18071 Granada, Spain
e-mail: vcastillo@ugr.es

R. Cela-Dablanca · A. Míguez-González · A. Barreiro · A. Núñez-Delgado · M. J. Fernández-Sanjurjo · E. Álvarez-Rodríguez
Department of Soil Science and Agricultural Chemistry, Engineering Polytechnic School, University of Santiago de Compostela, 27002 Lugo, Spain

A. Núñez-Delgado (ed.), *Frontier Studies in Soil Science*, Frontier Studies in Soil Science, https://doi.org/10.1007/978-3-031-50503-4_9

1 Introduction

1.1 Background

Anti-infective drugs, particularly antibiotics, represent a pivotal pharmacological category extensively employed in therapeutic applications. Their utility extends beyond antimicrobial treatment, encompassing prophylactic measures against transmissible diseases in humans, animals, and plants. In certain regions, antibiotics have even served as growth promoters in animal husbandry [8]. Nevertheless, due to the associated health hazards, the European Union took a decisive step in 2006 by prohibiting the use of antibiotics as growth promoters in livestock [7].

The widespread utilization of antibiotics in recent years has resulted in their pervasive presence and subsequent detection in water bodies and cultivated soils, posing significant threats to both human well-being and ecosystems health [59]. Given their incomplete absorption in the gastrointestinal tract, a substantial portion of these biocides, as much as 90%, is excreted from the body via feces and urine [23, 51]. Consequently, antibiotics find their way into solid excreta, veterinary effluents, wastewater, and the sludge produced by wastewater treatment plants (WWTPs) after their use in human medicine [49]. Pertinently, most WWTPs exhibit inefficiencies in the inactivation or removal of antibiotics, with removal efficiencies ranging from 30 to 90%. Consequently, a significant proportion of these compounds is either released into effluents or accumulates in sludges [27, 35]. The use of WWTP sludge and irrigation with wastewater effluents to enhance soil fertility may introduce these pollutants into crop soils [38, 58]. This process raises concerns related to the emergence of antibiotic-resistant bacteria, ecotoxicity, and their ingress into the food chain through water and agricultural products. Even at low concentrations, these compounds are considered emerging pollutants due to their potential adverse impacts on human and animal health [44]. The surge in bacterial resistance has led to increased medical expenses, extended hospital stays, and elevated mortality rates [6, 26, 34].

Numerous research works have focused on the development of systems to mitigate the presence of antibiotics in soils [21]. Additionally, advanced oxidation, hydrolysis, photodegradation, biodegradation, and adsorption represent prominent techniques for antibiotic removal from water. Nevertheless, advanced oxidation processes may generate toxic by-products [22], and membrane technologies often lack the requisite reliability for antibiotic removal [15]. Conversely, adsorption stands out as a straightforward, cost-effective, highly efficient, environmentally friendly, and sustainable method [1, 3, 32, 55]. The adsorption of antibiotics onto soils holds promise in mitigating the entry of these pollutants into the food chain [11]. In this context, the use of (bioadsorbent materials to sequester antibiotics and prevent their release into the environment emerges as a feasible alternative that could enhance their retention in soils [12, 47]. Among sorbent materials, activated carbon is commonly employed, albeit with considerable cost and regeneration challenges [19]. On the other hand, bio-adsorbents such as mussel shells and waste/by-products from the forestry industry, including cork, pine needles, and biomass ash, offer potential as candidates for this purpose [18, 30, 41]. Incorporating these residues and by-products into soil or utilizing them in specially designed modules within WWTPs could curtail the risk of antibiotic dispersion into the environment. Furthermore, such an approach would contribute to the valorization of these by-products, mitigating their environmental impact [4, 28, 31]. A study conducted by Conde-Cid et al. [17] demonstrated that pine bark enhances the adsorption and reduces the desorption of three types of sulfonamides. These materials have previously exhibited promising results as adsorbents for heavy metals, inorganic anions, and antibiotics, including tetracyclines and sulfonamides [37, 42, 43, 45, 50, 52, 53, 54].

1.2 Objectives and Significance

The study of adsorption and desorption behaviors of pharmaceuticals in the environment has gained significant importance due to the potential ecological and human health risks associated with their presence in natural systems. This chapter goes ahead previous studies of the research team, focusing on elucidating the intricate interactions between three commonly used antibiotics in human medicine, namely amoxicillin (AMX), trimethoprim (TMP), and ciprofloxacin (CIP), and their interactions with three distinct bio-adsorbents: pine bark, mussel shell, and oak ash.

Rationale for Antibiotic Selection: The selection of AMX, TMP, and CIP for this study is based on their widespread use in clinical settings, leading to their increased prevalence in environmental compartments. These antibiotics belong to different classes, allowing for a comprehensive examination of adsorption behavior.

Adsorption Mechanisms: Understanding the mechanisms governing the adsorption process is essential to assess the efficiency of these bio-adsorbents. These mechanisms may involve physical adsorption, chemical reactions, electrostatic interactions, and complexation with functional groups present on the adsorbent surfaces.

Building on Previous Research: This study builds upon previous research works (such as [11, 12, 13, 48]) which have investigated the adsorption capabilities of these bio-adsorbents for other contaminants. By extending the scope to antibiotics, this chapter aims to contribute to a broader understanding of their applicability and effectiveness in mitigating emerging pharmaceutical pollutants.

Implications for Environmental Risk Reduction: The significance of this research lies on its potential to inform strategies for mitigating the dispersion of pharmaceuticals in the environment. Given the ecological and human health concerns associated with the presence of antibiotics in natural systems, the findings presented herein hold promise for developing sustainable and effective solutions to reduce these risks.

Through a systematic exploration of the adsorption and desorption behaviors of antibiotics on by-products used as bio-adsorbents, this chapter seeks to provide valuable insights into the fate of pharmaceuticals in the environment and the role of bio-adsorbents in their retention in Galician soils, which could be extended globally in certain aspects. Ultimately, these insights contribute to the broader goal of safeguarding environmental and public health by reducing the environmental presence of pharmaceutical contaminants.

2 Methods

2.1 Study Design

Experimental Setup: Batch tests were conducted under controlled conditions at a temperature of 25 °C to investigate the adsorption and desorption behaviors of three distinct antibiotics: amoxicillin (AMX), ciprofloxacin (CIP), and trimethoprim (TMP). Each antibiotic was evaluated separately in three replicate adsorption tests.

Adsorption Experiments: In each adsorption test, the antibiotic of interest was subjected to seven different concentrations, ranging from 0 μmol L^{-1} to 600 μmol L^{-1}. A total of 0.5 g of the chosen bio-adsorbent material was mixed with 10 mL of a 0.005 M $CaCl_2$ solution, which served as a background electrolyte to maintain constant ionic strength. These mixtures, one for each antibiotic concentration, were stirred continuously for 48 h. This duration was determined based on prior kinetic experiments, ensuring adequate time to reach equilibrium [11, 48]. Following the adsorption step, the samples were subjected to centrifugation at 4000 rpm for 15 min. Subsequently, the resulting supernatants were meticulously filtered through 0.45 μm nylon-type syringe filters to remove any particulate matter.

Antibiotic Concentration Measurement: Antibiotic concentrations in the filtered supernatants were quantified using High-Performance Liquid Chromatography (HPLC). The analytical system consisted of a Thermo-Fisher Model LPG 3400 SD equipment (USA) equipped with a HPG-3400 quaternary pump (USA), a WPS3000 autosampler (USA), a thermostated compartment for the TM-120 column, and an ultraviolet–visible detector from the UltiMate 3000 series (USA). Data processing was performed using the Chromeleon software on a connected computer. For HPLC analysis, a Luna C18 column (150 mm long, 4.6 mm internal diameter, 5 μm particle size) provided by Phenomenenex (Madrid, Spain) was employed. Additionally, a pre-column (4 mm long, 2 mm internal diameter, 5 μm particle size) packed with the same material as the column was used. The injection volume was set at 50 μL, and a flow rate of 1.5 mL min^{-1} was maintained.

Mobile Phases and Gradient Conditions: The composition of the mobile phases and gradient conditions varied depending on the specific antibiotic being analyzed. For AMX, the mobile phase consisted of acetonitrile (phase A) and 0.01 M phosphoric acid with a pH of 2 (phase B). The linear gradient transitioned from 5 to 15% of phase A and from 95 to 85% of phase B over 4 min. For TMP and CIP, the same mobile phases as for AMX were used, but with a different linear gradient. This gradient ranged from 5 to 32% of phase A and from 95 to 68% of phase B over 10.5 min.

Detection Wavelengths: To quantify the concentrations of TMP and CIP, a wavelength of 212 nm was employed, while a wavelength of 230 nm was utilized for AMX.

Desorption Experiments: After the completion of the adsorption step, desorption tests were conducted under the same temperature conditions (25 °C). Similar to the adsorption phase, three replicate desorption tests were carried out for each antibiotic. In the desorption experiments, the remaining bio-adsorbent material from the adsorption process was accurately weighed. Subsequently, 10 mL of a 0.005 M $CaCl_2$ solution was added to the material, and the samples were allowed to stir for 48 h. After this equilibration period, the samples underwent centrifugation and filtration using the same procedures as those applied during the adsorption phase. The resulting equilibrium solution was analyzed to determine the concentration of the corresponding antibiotic, following procedures analogous to those employed in the adsorption tests.

2.2 *Selection of By-Products Used for the Antibiotics Retention in Galician Soils and Characterization*

The selection of appropriate biosorbents is a critical aspect of this study, as it directly impacts the efficiency of the adsorption and desorption processes. The biosorbents employed in this research were sourced from the forestry and agro-food industry, chosen for their potential to effectively adsorb pharmaceutical pollutants. The following biosorbents were selected for their unique properties and origins:

Oak Ash (Lugo, Spain)

- Source: Oak ash was obtained from the combustion of oak logs in a combustion boiler located in Lugo, Spain.
- Rationale: Oak ash was chosen for its inherent characteristics, which may include a high surface area and the presence of adsorption-active sites resulting from the combustion process.

Mussel Shell (Illa De Arousa, Pontevedra, Spain)

- Source: Crushed Mussel shell was provided by Abonomar S.L., a company situated in Illa de Arousa, Pontevedra, Spain.
- Particle Size: The mussel shell material used in this study exhibited a particle size of less than 1 mm.
- Rationale: Mussel shells are rich in calcium carbonate, a compound known for its adsorption capabilities. In addition a variety of other potentially active compounds has been detected and reported in previous papers by the research team. The small particle size of the mussel shell material used would enhance the accessibility of adsorption sites.

Pine Bark (Madrid, Spain)

- Source: Pine bark was supplied by Geolia, a company based in Madrid, Spain.
- Particle size: Prior to experimentation, the pine bark material was processed in the laboratory. It was crushed and subsequently sieved through a 0.63 mm mesh to ensure uniformity and consistency.
- Rationale: Pine bark has been recognized for its adsorption potential in previous investigations performed by the research team, and the crushing and sieving process aimed to optimize its particle size for effective adsorption studies.

The selection of these biosorbents was guided by their availability, potential adsorption capacity, and previous research [9, 10, 12, 16, 36] indicating their suitability for adsorbing various contaminants. These distinct biosorbents were chosen to provide a comprehensive evaluation of their performance in the adsorption and desorption of pharmaceutical pollutants, contributing valuable insights to the study's objectives. It should be noted that, among the works indicated above, the main starting point and background reference for the current chapter is a specific paper recently published [36], which is complemented with various other to achieve a broader view of the theme, as usually expected for a book chapter.

Characterizations

The pH was assessed in both water and a KCl solution employing a soil-to-liquid ratio of 1:2.5, with contact durations of ten minutes and two hours, respectively, utilizing a CRISON model 2001 pH-meter (Crison, Barcelona, Spain) [29]. Elemental analysis was conducted to measure carbon, and subsequently, organic matter (OM) and nitrogen content, utilizing a TRUSPEC CHNS equipment, LECO model (USA). Exchange cations were displaced from the bio-adsorbents through the application of a 1 M NH_4Cl extractant, maintaining a 1:10 adsorbent-to-solution ratio for a duration of 12 h [39]. The quantification of aluminum (Al), calcium (Ca), magnesium (Mg), sodium (Na), and potassium (K) was achieved using atomic absorption/emission spectrophotometry (Analyst 200, PerkinElmer, USA) with the addition of 1% $LaCl_3$ to mitigate potential interferences. The summation of these exchange cations provided the effective cation exchange capacity (eCEC) [25]. Extraction of non-crystalline iron (Fe_{ox}) and aluminum (Al_{ox}) fractions was executed utilizing an ammonium oxalate buffered solution adjusted to pH = 3 [5]. These extracted fractions were subsequently measured employing the same atomic absorption spectrophotometer. Refer to Table 1 within the Results section for a comprehensive compilation of these physicochemical parameters, as consistent with previous publications related to the respective sorbent materials [10].

Table 1 Physicochemical characterization of the three bio-adsorbents

Parameter	Unit	Bio-adsorbents		
		Oak ash	Mussel shell	Pine bark
OM	%	22.76	19.66	83.76
N	%	0.22	0.21	0.08
C	%	13.23	11.43	48.70
C/N		60.14	55.65	608.75
pH $_{(H2O)}$		11.31	9.39	3.99
pH $_{(KCl)}$		13.48	9.04	3.42
eCEC	cmolc kg^{-1}	361.15	30.25	14.92
Ca_e	cmolc kg^{-1}	95.00	24.75	5.38
Mg_e	cmolc kg^{-1}	3.26	0.72	2.70
Na_e	cmolc kg^{-1}	12.17	4.37	0.46
K_e	cmolc kg^{-1}	250.65	0.38	4.60
Al_e	cmolc kg^{-1}	0.07	0.03	1.78
Fe_{ox}	mg kg^{-1}	4233	171	74
Al_{ox}	mg kg^{-1}	8328	178	315

OM: Organic Matter; N: total nitrogen; C: total carbon; C/N: carbon to nitrogen ratio; eCEC: effective cation exchange capacity; Ca_e, Mg_e, Na_e, K_e, Al_e: Exchangeable cations; Al_{ox}, Fe_{ox}: Al and Fe extracted with ammonium oxalate

2.3 *Data Analysis*

Adsorption Models

To characterize the adsorption behavior of the antibiotics onto the selected bio-adsorbents, three well-established models were employed: the Freundlich model (Eq. 1), the Langmuir model (Eq. 2), and the Linear model (Eq. 3) [56].

Freundlich Model

$$q_e = K_f C_e^n \tag{1}$$

- q_e (μmol kg^{-1}): Adsorbed amount of antibiotic per unit mass of the adsorbent (the difference between the added amount and what remains in the equilibrium solution).
- C_e (μmol L^{-1}): Concentration of the antibiotic in the equilibrium solution.
- K_f (L^n $μmol^{1-n}$ kg^{-1}): Freundlich constant related to the adsorption capacity.
- n (dimensionless): Parameter related to the intensity of adsorption.

Langmuir Model

$$q_e = \frac{q_{max} K_L C_e}{1 + K_L C_e} \tag{2}$$

- q_e (μmol kg^{-1}): Adsorbed amount of antibiotic per unit mass of the adsorbent.
- C_e (μmol L^{-1}): Concentration of the antibiotic in the equilibrium solution.
- q_{max} (μmol kg^{-1}): Langmuir's maximum adsorption capacity.
- K_L (L μmol^{-1}): Langmuir constant related to the adsorption energy.

Linear Model

$$K_d = \frac{q_e}{C_e} \tag{3}$$

- K_d (L kg^{-1}): Partition coefficient in the Linear model.

2.3.1 Model Fitting and Statistical Analysis

The fitting of these adsorption models was accomplished using the SPSS software (version 18). This allowed for the assessment of how well these models described the experimental data and provided insights into the adsorption mechanisms.

Furthermore, bivariate statistical Pearson correlations were conducted to explore potential relations between adsorption and desorption parameters and the physicochemical characteristics of the bio-adsorbents. This statistical analysis aimed to identify key factors influencing the adsorption and desorption processes and to provide a comprehensive understanding of the interactions between the antibiotics and the selected bio-adsorbents.

The application of these models and statistical analyses facilitates a deeper insight into the adsorption and desorption mechanisms and their dependence on various parameters, contributing to the comprehensive evaluation of the study's outcomes.

3 Results

In this section, we present the findings of the study, which encompass the adsorption of various antibiotics on the selected bio-adsorbents. The results are depicted in Figs. 1, 2, and 3, and they are represented in two forms: as adsorption values in μmol kg^{-1} and as percentages. Furthermore, equilibrium adsorption curves and modeling data are included in Table 2 to provide a comprehensive overview of the experimental outcomes. Furthermore, Fig. 4 shows the desorption behavior of the set of antibiotics from the bio-adsorbents of study. Table 1 shows the physicochemical results derived from the characterization of the three bio-adsorbents.

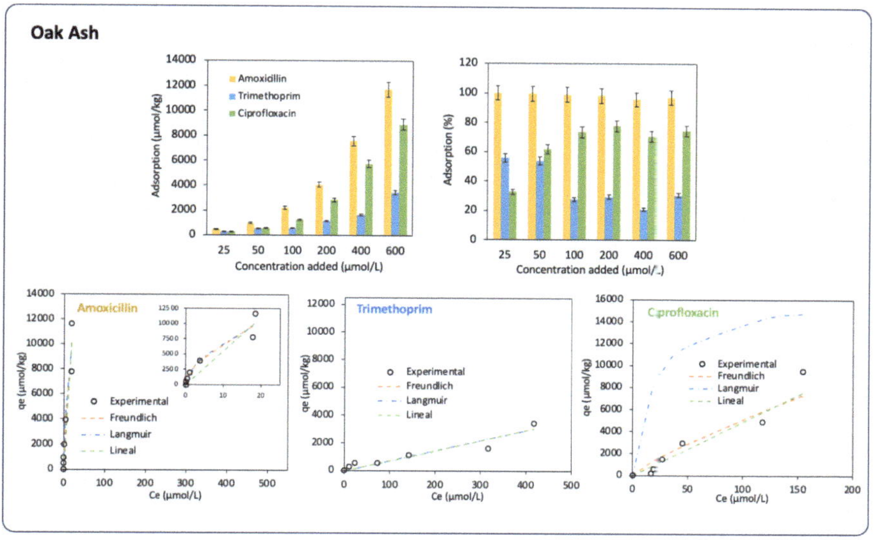

Fig. 1 Adsorption of AMX, TMP and CIP onto oak ash in µ mol kg^{-1} and percentage (up) and modelling of equilibrium adsorption curves of the three antibiotics onto the oak ash bio-adsorbent (bottom)

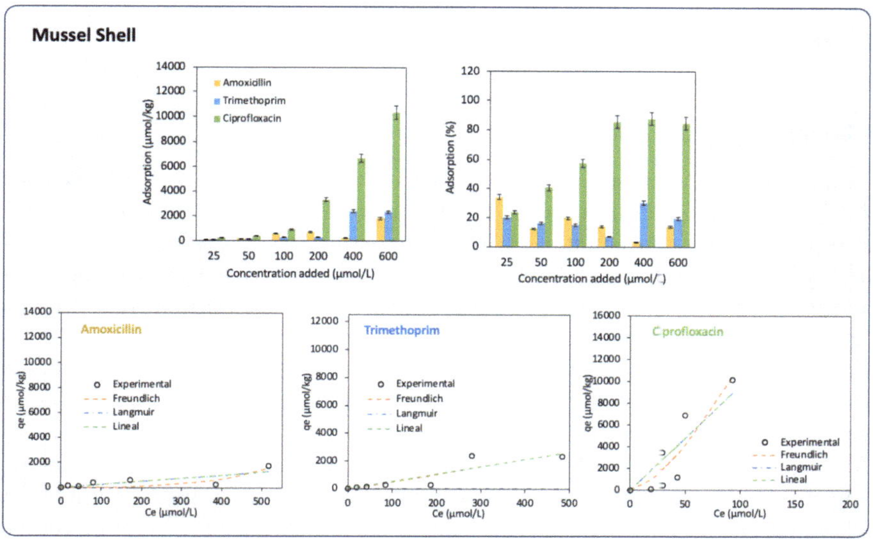

Fig. 2 Adsorption of AMX, TMP and CIP onto mussel shell in µ mol kg^{-1} and percentage (up) and modelling of equilibrium adsorption curves of the three antibiotics onto the mussel shell bio-adsorbent (bottom)

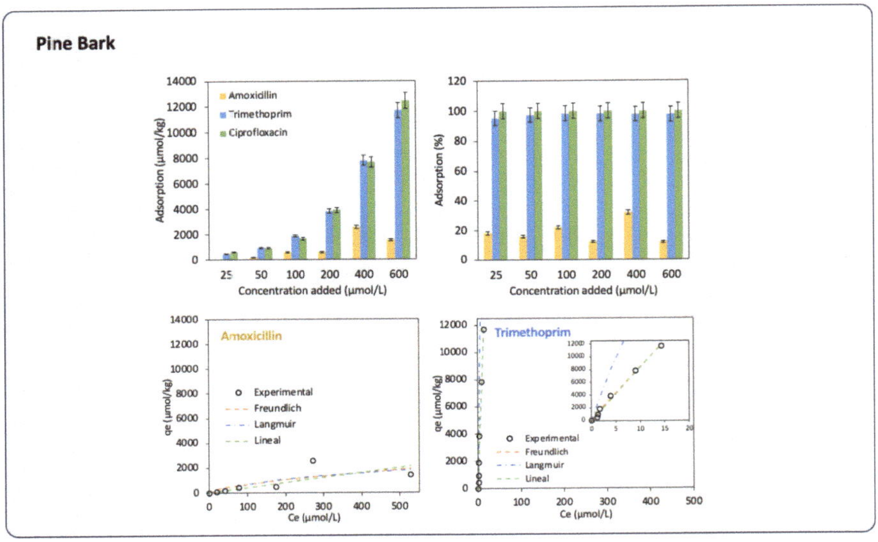

Fig. 3 Adsorption of AMX, TMP and CIP onto pine bark in µ mol kg^{-1} and percentage (up) and modelling of equilibrium adsorption curves of the three antibiotics onto the pine bark bio-adsorbent (bottom)

4 Discussion

Adsorption (Figs. 1, 2 and 3)

For amoxicillin, the highest adsorption capacity was exhibited by oak ash. Remarkably, oak ash achieved near-complete adsorption of AMX, reaching nearly 100% adsorption for all concentrations added. In contrast, pine bark and mussel shell displayed significantly lower AMX adsorption capacities, typically remaining below 20%. Conversely, for ciprofloxacin and trimethoprim, pine bark proved to be the superior adsorbent, consistently achieving adsorption percentages close to 100% for all concentrations added (Fig. 3). Oak ash and mussel shell also demonstrated notable adsorption potential for CIP, particularly at higher antibiotic concentrations, with adsorption percentages reaching approximately 70% and 80%, respectively. However, in the case of TMP, oak ash exhibited limited adsorption capacity, retaining only 50% of the antibiotic at lower concentrations and not exceeding 30% at higher concentrations. Mussel shell generally displayed less than 20% adsorption for any of the TMP concentrations added.

The adsorption mechanisms of AMX are complex, influenced by its three pK_a values, causing it to behave as a cation, anion, or zwitterion under different pH conditions. The amino group, with a pK_{a1} of 2.67, influences the molecule's charge at pH levels below this value. Within the pH range between pK_{a1} and pK_{a2} (7.11), the zwitterionic form predominates, while at pH levels between pK_{a2} and pK_{a3} (9.63),

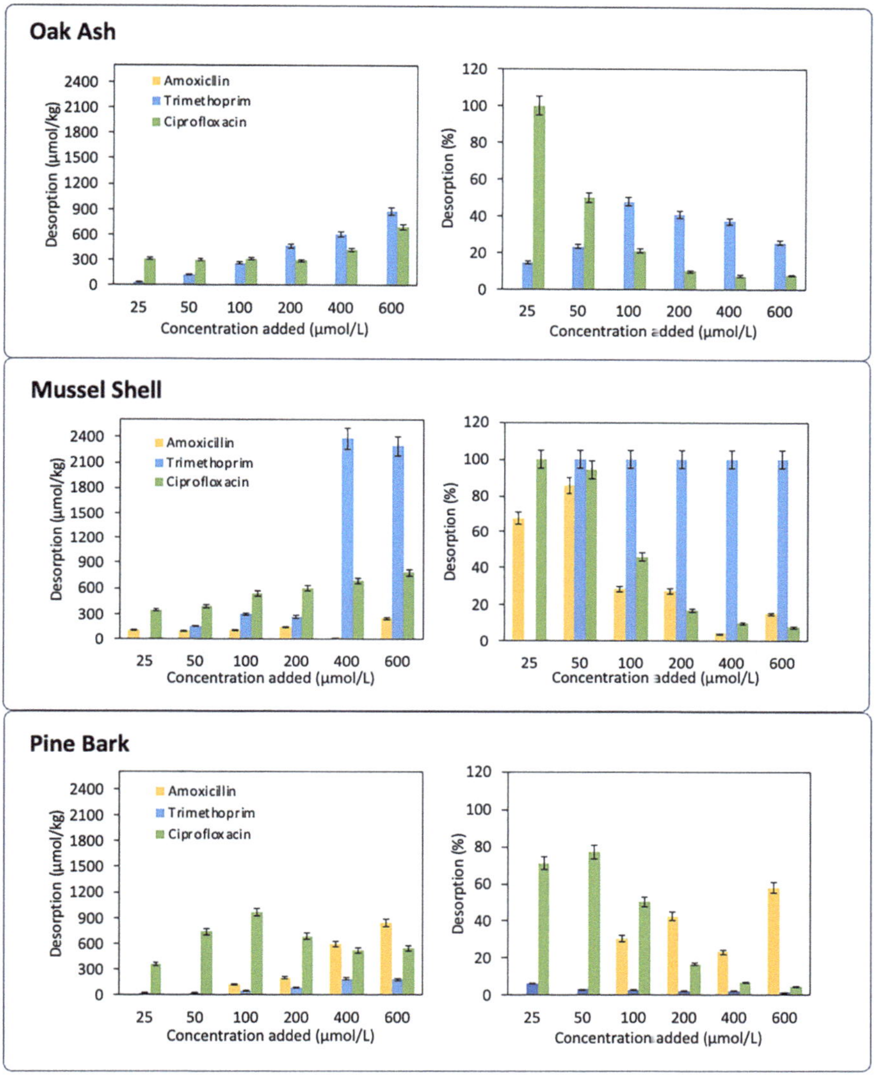

Fig. 4 Desorption of AMX, TMP and CIP from oak ash, mussel shell and pine bark in µ mol kg^{-1} (left) and percentage (right)

the deprotonated form of the carboxyl and amine groups becomes dominant [20]. Beyond the pK_{a3}, AMX carries a double negative charge due to the deprotonation of the phenolic hydroxyl group [40]. The diverse characteristics of the bio-adsorbents result in varying retention mechanisms, including electrostatic interactions, cationic bridges, and hydrogen bonds [2].

Table 2 Parameters of the Freundlich model (K_f, expressed in L^n kg^{-1} μ mol^{1-n}, and n, dimensionless), Langmuir model (K_L, expressed in L μ mol^{-1}, and q_{max}, expressed in μ mol kg^{-1}) and Linear model (K_d, expressed in L kg^{-1}) for amoxicillin (AMX), ciprofloxacin (CIP) and trimethoprim (TMP) adsorption onto the three bio-adsorbents of study; –: error value too high for fitting

Bio-adsorbent	Antibiotic	Freundlich					Langmuir					Linear model		
		K_f	Error	n	Error	R^2	K_L	Error	q_{max}	Error	R^2	K_d	Error	R^2
Oak ash	AMX	1824.03	685.92	0.58	0.14	0.94	0.104	0.076	14,909	4359	0.93	555.28	55.84	0.89
	CIP	–	–	0.81	0.36	0.73	0.050	0.009	–	–	0.75	48.81	8.82	0.71
	TMP	–	–	1.02	0.28	0.91	–	–	–	–	–	7.26	0.66	0.91
Mussel shell	AMX	2.767 10^{-6}	0	3.22	2.00	0.69	–	–	–	–	–	2.50	0.52	0.64
	CIP	–	–	1.47	0.42	0.81	–	–	–	–	–	96.56	16.25	0.76
	TMP	–	–	1.03	0.40	0.80	–	–	–	–	–	5.30	0.81	0.80
Pine bark	AMX	–	–	0.62	0.38	0.58	–	–	–	–	–	4.05	0.49	0.49
	CIP	–	–	–	–	–	–	–	–	–	–	–	–	–
	TMP	981.88	137.41	0.93	0.06	0.99	0.030	0.009	76,429	4.6 10^4	0.99	833.48	23.10	0.99

Oak ash, characterized by its alkaline pH (11.31) and high concentrations of Fe and Al non-crystalline minerals, possesses a negative charge. In an environment with a pH of 11.31, the amino, carboxyl, and phenol groups of AMX become deprotonated, acquiring a substantial negative charge. In this context, calcium present in oak ash acts as a cationic bridge between the adsorbent and the adsorbate. A similar mechanism can be extrapolated to mussel shell, although its adsorption capacity is notably lower, possibly due to its lower content of non-crystalline Fe and Al, as well as exchangeable calcium, and a less alkaline pH (9.39) compared to oak ash. This results in fewer reactive surfaces and negative charges. Besides, the reduced content of variable charge components in mussel shell, such as organic matter or non-crystalline minerals, might affect the adsorption process. However, other mechanisms, such as electrostatic attraction between the protonated groups of AMX and the negative charges of the adsorbent, may also contribute.

Conversely, the low adsorption of AMX onto pine bark may be attributed to its highly acidic pH (3.99). At this pH, the organic components and non-crystalline minerals in pine bark carry positive charges. AMX, acting as a zwitterion at this pH, exhibits protonated amino groups and deprotonated carboxyl groups, with the former likely prevailing due to the medium's acidity. Electrostatic repulsions may account for this low adsorption. In a previous study, Cela-Dablanca et al. [12] observed higher adsorption of AMX with oak ash (89.71%) than mussel shell (48.15%), and much lower adsorption with pine bark (29.38%).

For CIP, it has pK_a values of 5.90 ± 0.15 for the carboxylic acid group and 8.89 ± 0.11 for the N-basic group, enabling it to exist as a cation, zwitterion, or anion depending on soil and water pH conditions. Pine bark emerged as the bio-adsorbent with the highest adsorption capacity for CIP, primarily due to its lower pH compared to the pK_{a1} of CIP, resulting in positive charges on both the adsorbent and CIP, leading to electrostatic repulsion. Pine bark's high organic matter content, rich in carboxyl and hydroxyl groups, forms hydrogen bridges with CIP's carbonyl and hydroxyl groups. The presence of benzene rings in CIP facilitates electron acceptance due to the high electronegativity of fluorine in its structure, receiving electrons from the hydroxyl groups of pine bark's organic components. Additionally, under the acidic conditions in this study, free H^+ ions in the medium contribute to the formation of hydrogen bonds between the adsorbent surface and CIP molecules. While some authors have highlighted the maximum adsorption efficiency for CIP between pH 4 and 7 [14], oak ash and mussel shell also demonstrated considerable adsorption capacity for CIP, particularly at higher antibiotic concentrations (Figs. 1 and 2). In the case of oak ash, cationic bridges facilitated by exchangeable calcium likely play a key role, as CIP acts as an anion due to oak ash's alkaline pH (11.31). In contrast, mussel shell's adsorption mechanisms are likely similar to those observed for AMX. Additionally, at the pH of mussel shell (9.39), CIP might still retain some positive charge, enabling electrostatic attractions. The study by Chandrasekaran et al. [14] also indicated a higher adsorption of CIP in simple systems compared to AMX, similar to the observations for pine bark and mussel shell in this study but differing for oak ash.

In the present research, pine bark was identified as the only effective bio-adsorbent for TMP retention, while oak ash and mussel shell displayed limited adsorption capacity, especially at higher antibiotic concentrations (Figs. 1, 2 and 3). TMP possesses a single pK_a of 7.3, primarily existing as a cation (80%) and, to a lesser extent, as a zwitterion (20%). The high adsorption of TMP on pine bark can be attributed to hydrogen bonding, facilitated by pine bark's numerous carboxyl and hydroxyl groups interacting with the NH_2 groups in TMP's heterocyclic aromatic ring. Furthermore, under acidic conditions, as in this study, free H^+ ions in the medium contribute to hydrogen bond formation between the adsorbent surface and TMP molecules, similar to the mechanism observed for CIP. A positive and significant correlation ($p < 0.05$) between carbon content and TMP adsorption was identified across all bio-adsorbents used in this study. As pine bark boasts the highest organic compound content, it exhibited the highest TMP adsorption percentage compared to other bio-adsorbents.

Equilibrium Modelling (Figs. 1, 2 and 3)

In the analysis of equilibrium adsorption processes, it was observed that the determination of appropriate modeling parameters is crucial. However, it should be noted that in cases where error values associated with the fitting are not excessively high, the R^2 values demonstrate notable insights into the suitability of different models for characterizing adsorption phenomena.

For trimethoprim, R^2 values for all models were found to be consistently similar across various bio-adsorbents, ranging from 0.80 to 0.99. Nevertheless, the Freundlich model was observed to provide a more favorable description of AMX adsorption in oak ash ($R^2 = 0.94$) and mussel shell ($R^2 = 0.69$). Conversely, the Freundlich and Linear models exhibited comparable fits for AMX adsorption onto pine bark, with R^2 values of 0.69 and 0.64, respectively. As for Ciprofloxacin (CIP), the Freundlich and Linear models demonstrated the best fits for oak ash, yielding R^2 values of 0.73 and 0.71, and for mussel shell, resulting in R^2 values of 0.81 and 0.76, respectively. However, it is worth noting that CIP adsorption onto pine bark achieved a maximum adsorption percentage of 100% across all antibiotic concentrations, rendering modeling of this process unfeasible.

Of significance is the observation that the Langmuir model was associated with significantly high errors when predicting most parameters and for most of the sorbents. This phenomenon may be attributed to the failure to achieve the thermodynamic saturation condition of the bio-adsorbents due to the utilization of insufficiently high antibiotic concentrations to ensure the saturation state [57].

When focusing on TMP adsorption, especially concerning the Linear model, partition coefficient values (K_d) exhibited a range from 5.3 to 833 L kg^{-1}. The highest K_d values were observed for TMP adsorption onto pine bark, accompanied by adsorption percentages near 100% for all concentrations (Figs. 1, 2 and 3). A comparative analysis of these results with prior studies indicated K_d values for TMP adsorption on agricultural soils within the range of 10–48 L kg^{-1} [48]. Similarly, some Australian soils exhibited scores ranging 9–311 L kg^{-1} [33], consistent with findings in the current investigation. In addition, K_f and K_L values corresponding to TMP adsorption on pine bark were reported, signifying the superior affinity of pine bark for this antibiotic. However, it should be noted that fitting the model was not feasible for oak ash and mussel shell due to exceedingly high error values associated with the fitting process. In prior research, K_f levels for TMP adsorption on agricultural soils ranged from 29–125 $L^n\ kg^{-1}\ \mu mol^{1-n}$ [48]. Interestingly, this is comparable but significantly lower than the values obtained in the present investigation for pine bark, which reached 981 $L^n\ kg^{-1}\ \mu mol^{1-n}$. Additionally, the Freundlich n parameter for TMP adsorption on pine bark was found to be less than 1 (specifically 0.93), indicating the existence of heterogeneous adsorption sites and a concave adsorption curve [11]. In contrast, the n parameter values for the other two adsorbents were slightly higher than 1, suggesting alternative characteristics.

Turning attention to AMX, it was noted that oak ash exhibited the highest K_f values at 1824 $L^n\ kg^{-1}\ \mu mol^{1-n}$. Conversely, mussel shell demonstrated a notably lower score of 2.767×10^{-6} $L^n\ kg^{-1}\ \mu mol^{1-n}$. However, it is important to acknowledge that fitting the model to pine bark was not possible due to excessively high error values. In previous studies, which focused on AMX retention in corn and vineyard soils amended with bio-adsorbents, K_f values fell within the range of 1–139 $L^n\ kg^{-1}\ \mu mol^{1-n}$ [12]. The present work revealed considerably higher K_f values, especially for oak ash, indicating the enhanced affinity of these bio-adsorbents for AMX. Notably, the incorporation of oak ash into corn soil was found to double the K_f level [12]. Additionally, the Freundlich model's n parameter was less than 1 for oak ash and pine bark, resulting in concavity in the AMX adsorption curve.

The Linear model analysis highlighted that the K_d value for AMX adsorption onto oak ash was two orders of magnitude higher compared to mussel shell and pine bark. Specifically, oak ash exhibited a K_d value of 555 $L\ kg^{-1}$, while mussel shell and pine bark displayed values of 2.5 $L\ kg^{-1}$ and 4.0 $L\ kg^{-1}$, respectively, in accordance with expectations. These values aligned with results from previous studies on vineyard and corn soils. In addition, the K_L Langmuir parameters were reported, resulting in a value of 0.104 $L\ \mu mol^{-1}$ for oak ash. Unfortunately, fitting the model was not possible for the other adsorbents due to high errors associated with estimation. Nevertheless, this value was consistent with results obtained from prior research on AMX adsorption in agricultural soils.

Desorption (Fig. 4)

In general, AMX and CIP display relatively low desorption from oak ash and mussel shell, particularly when high concentrations of antibiotics were added. Notably, TMP exhibited the least desorption from pine bark. AMX demonstrated the highest levels of desorption from pine bark, reaching 839 μ mol kg^{-1} when an antibiotic concentration of 600 $\mu mol\ L^{-1}$ was introduced. For CIP, the highest desorption occurred from pine bark, with a value of 966.78 μ mol kg^{-1} observed at an added concentration of 100 $\mu mol\ L^{-1}$. TMP exhibited notable desorption levels, especially when interacting with mussel shell, resulting in desorption values of 2380.186 and 2294.927 μ mol kg^{-1} for added concentrations of 400 and 600 $\mu mol\ L^{-1}$, respectively. In the case of oak ash, desorption values were intermediate between those of mussel shell and pine bark, with the maximum TMP desorption reaching 883.45 μ mol kg^{-1} for an added concentration of 600 μmol L^{-1}, whereas pine bark exhibited minimal desorption, releasing only 189.60 μ mol kg^{-1} for an added concentration of 400 μmol L^{-1}.

When the values are expressed in terms of percentage, AMX exhibited the highest desorption from mussel shell at a concentration of 50 μmol L^{-1} (80%), while pine bark displayed the highest desorption for an added concentration of 600 μmol L^{-1} (60%). Conversely, oak ash demonstrated the most substantial irreversible adsorption of this antibiotic, with desorption close to 0% (Fig. 4). For CIP, high desorption percentages were observed at low concentrations for both oak ash and mussel shell, with medium values for desorption percentages from pine bark at those concentrations. However, at the two highest concentrations added, CIP desorption remained consistently less than 10% for all three adsorbent materials.

It is important to note that for CIP, a higher percentage of adsorption and lower desorption were observed when higher concentrations were added, implying that as the concentration of this antibiotic increases, adsorption is favored, and the bonds with the adsorbent grow stronger. This observation aligns with the cooperative model of adsorption suggested by several authors [24, 46], where it is posited that a solute, when retained by a site on a homogeneous adsorbent surface, can influence consecutive active sites, facilitating new adsorptions.

With respect to TMP, desorption was nearly 100% for mussel shell, while it remained minimal for pine bark, peaking at 6.29% desorption for an added concentration of 25 μmol L^{-1}. Mussel shell and pine bark exhibited low AMX adsorption capacity (always less than 30% of what was added) (Figs. 2 and 3) and desorption that could reach up to 80% for mussel shell and up to 60% for pine bark (Fig. 4). Conversely, oak ash displayed the highest AMX adsorption while releasing negligible amounts. This can be attributed to oak ash's higher content of Fe and Al non-crystalline minerals, as well as the greater amount of calcium, resulting in stronger cationic bridges during adsorption, particularly compared to mussel shell. A prior study conducted by Cela-Dablanca et al. [12] also concluded that the desorption of AMX from oak ash was minimal, highlighting its efficacy for retaining this antibiotic.

CIP exhibited high desorption percentages in general across all bio-adsorbents, particularly at lower concentrations, while desorption percentages decreased to below 10% at higher concentrations. Thus, at the two highest concentrations of this antibiotic, all three bio-adsorbents displayed a high adsorption capacity (between 75 and 100%) (Figs. 1, 2 and 3) and low desorption (<10%), with pine bark exhibiting the lowest desorption (Fig. 4). Nevertheless, in terms of the amount of CIP desorbed (in μmol kg^{-1}), the values remained similar across the entire concentration range, suggesting that a constant quantity of CIP is desorbed from the bio-adsorbents. This may indicate the involvement of various adsorption mechanisms in the adsorption process, as discussed above.

Regarding TMP, Fig. 4 demonstrates that the levels of μ mol kg^{-1} desorbed increase with the added concentration, particularly evident with mussel shell, which desorbs the most significant quantity of this antibiotic. This trend corresponds with the findings in a previous study examining TMP adsorption [33]. Pine bark emerges as an effective retention material for TMP, as it absorbs nearly all of the added antibiotic and exhibits minimal desorption (<5%), making the retention process highly irreversible.

In previous studies, these same three bio-adsorbents (oak ash, pine bark, and mussel shell) demonstrated high retention capacities for tetracyclines (oxytetracycline, chlortetracycline, and tetracycline), with pine bark and oak ash delivering superior performances [18]. Additionally, a study of sulfonamides (sulfadiazine, sulfamethazine, and sulfachloropyridazine) highlighted pine bark's exceptional retention capacity (almost 100%), while oak ash and mussel shell did not perform as favorably in retaining these antibiotics [16]. Wood ash and mussel shell emerged as the bio-adsorbents with the highest sorption capacity and retention for the antibiotic cefuroxime [10].

5 Limitations and Future Directions

While the discussions and insights provided in this chapter offer valuable contributions to the understanding of adsorption behaviors of antibiotics on various bio-adsorbents, several limitations in the current research should be acknowledged.

- First, the experimental conditions should consider exploring the behavior of antibiotics on bio-adsorbents in more realistic scenarios, such as undisturbed soil and water systems, to provide a comprehensive understanding of the practical applications and implications of these findings.
- Second, expanding the investigation to include a broader spectrum of antibiotics and a wider range of bio-adsorbent materials would provide a more comprehensive database for the assessment of antibiotic adsorption in diverse contexts. In this regard, the research team is currently investigating anticoccidials.
- Furthermore, investigating the rate at which adsorption occurs and the thermodynamic factors influencing these processes could offer additional insights into the interactions between antibiotics and bio-adsorbents.

The ecological implications of antibiotics leaching into ecosystems and the potential effects on microbial communities remain largely unexplored. Investigating the broader ecological consequences of antibiotic adsorption on bio-adsorbents is essential for understanding the long-term impact on ecosystems. Also, future research can address the use of mixed bio-adsorbents and employing continuous-flow column experiments, which may better mimic real-world adsorption scenarios. At higher scales, lysimetric installations and in situ measurements performed in undisturbed environments, carried out in a continuous mode by using portable lab high-tech equipment, would provide valuable data, both as actual information and as input or verification for high-level modeling, in order to go steps ahead in this methodologically rather classical (up to now) field of study.

6 Conclusion

In this chapter, considering as starting point the overall problem of pollution due to emerging pollutants, and specifically that caused by antibiotics (including many of those previously studied by the research team, and others that are being currently investigated, such as anticoccidials), as specific examples we have delved into the adsorption behaviors of three distinct antibiotics, namely AMX, CIP, and TMP, across various bio-adsorbents. The observed differences in the adsorption patterns of these antibiotics can be primarily attributed to their unique properties, such as polarity, hydrophobicity, molecular structure, and speciation with pH. These distinctive attributes govern the interactions between the antibiotics and the reactive surfaces of the chosen bio-adsorbents.

Notably, our findings highlight the efficacy of specific bio-adsorbents in the retention of these pharmaceuticals. Pine bark emerged as a highly suitable bio-adsorbent for the adsorption of CIP and TMP, exhibiting an impressive adsorption rate of nearly 100% for CIP across all tested concentrations. Furthermore, the desorption of these two antibiotics from pine bark was notably lower when compared to the other bio-adsorbents. This can be attributed to the formation of robust hydrogen bonds between the functional groups of the drugs and the bio-adsorbent, a phenomenon facilitated by the acidic conditions found in TMP and the presence of benzene rings that serve as electron acceptor groups in CIP.

Conversely, oak ash proved to be the optimal material for the retention of AMX. It demonstrated the lowest level of desorption for AMX, primarily due to the establishment of strong cationic bridges between the bio-adsorbent and the antibiotic. This was further facilitated by the alkaline pH and the elevated content of calcium and non-crystalline substances within the bio-adsorbent.

Considering the insights gleaned from this research, the utilization of pine bark and oak ash as environmentally friendly remediation agents, to be incorporated into agricultural soils, or in appropriate modules linked to waste or wastewater treatment systems, appears to be a promising approach in addressing the entry of these emerging pollutants—AMX, CIP, and TMP—into the food chain.

In conclusion, this chapter underscores the potential of nature-based solutions as a means to mitigate the environmental impact of pharmaceutical contaminants, emphasizing the importance of understanding the adsorption dynamics of these antibiotics on bio-adsorbents for developing effective strategies to safeguard both ecosystems and human health. Although the main focus of the chapter was on the three specific antibiotics indicated above, the strategy of the research would be appropriate for other emerging pollutants, such as anticoccidials, which are currently the objective of the research group. Using certain waste and by-products as low-cost sorbents can be seen as an easily and potentially effective affordable strategy to face environmental pollution. After obtaining results to lead the application of this basic strategy, further steps could be directed to modify the raw materials or substitute them by high-tech materials/procedures, which would be justified to provide better outcomes when needed, and when the ratio cost/effectiveness is clearly favorable.

Acknowledgement This work was supported by the Spanish "*Agencia Estatal de Investigación*" (State Investigation Agency) [grant number PID2021-122920OB-C21].

The sponsors had not involvement in study design; in the collection, analysis and interpretation of data; in the writing of the report, and in the decision to submit the article for publication.

References

1. Adewuyi A (2020) Chemically modified biosorbents and their role in the removal of emerging pharmaceutical waste in the water system. Water 12(6):1551
2. Anastopoulos I, Pashalidis I, Orfanos AG, Manariotis ID, Tatarchuk T, Sellaoui L, Bonilla-Petriciolet A, Mittal A, Núñez-Delgado A (2020) Removal of caffeine, nicotine and amoxicillin from (waste) waters by various adsorbents. A review. J Environ Manag 261:110236
3. Azhar SHM, Abdulla R, Jambo SA, Marbawi H, Gansau JA, Faik AAM, Rodrigues KF (2017) Yeasts in sustainable bioethanol production: a review. Biochem Biophys Rep 10:52–61
4. Bădescu IS, Bulgariu D, Ahmad I, Bulgariu L (2018) Valorisation possibilities of exhausted biosorbents loaded with metal ions–a review. J Environ Manag 224:288–297
5. Blackmore AV (1978) Interpretation of electrical conductivity in a clay soil containing salts. Soil Res 16(3):311–318
6. Brown ED, Wright GD (2016) Antibacterial drug discovery in the resistance era. Nature 529(7586):336–343
7. Casewell M, Friis C, Marco E, McMullin P, Phillips I (2003) The European ban on growth-promoting antibiotics and emerging consequences for human and animal health. J Antimicrob Chemother 52(2):159–161
8. Castanon JIR (2007) History of the use of antibiotic as growth promoters in European poultry feeds. Poult Sci 86(11):2466–2471
9. Cela-Dablanca R, Conde-Cid M, Ferreira-Coelho G, Arias-Estévez M, Fernández-Calviño D, Núñez-Delgado A, Fernández-Sanjurjo MJ, Álvarez-Rodríguez E (2020) Adsorption of tetracycline and sulfadiazine onto three different bioadsorbents in binary competitive systems. Processes 9(1):28
10. Cela-Dablanca R, Nebot C, Rodríguez López L, Fernández-Calviño D, Arias-Estévez M, Núñez-Delgado A, Fernández-Sanjurjo MJ, Álvarez-Rodríguez E (2021) Efficacy of different waste and by-products from forest and food industries in the removal/retention of the antibiotic cefuroxime. Processes 9(7):1151
11. Cela-Dablanca R, Barreiro A, López LR, Santás-Miguel V, Arias-Estévez M, Núñez-Delgado A, Álvarez-Rodríguez E, Fernández-Sanjurjo MJ (2022a) Relevance of sorption in bio-reduction of amoxicillin taking place in forest and crop soils. Environ Res 208:112753
12. Cela-Dablanca R, Barreiro A, Rodríguez-López L, Santás-Miguel V, Arias-Estévez M, Fernández-Sanjurjo MJ, Álvarez-Rodríguez E, Núñez-Delgado A (2022b) Amoxicillin retention/release in agricultural soils amended with different bio-adsorbent materials. Materials 15(9):3200
13. Cela-Dablanca R, Barreiro-Buján A, Ferreira-Coelho G, López LR Santás-Miguel V, Arias-Estévez M, Núñez-Delgado A, Fernández-Sanjurjo MJ, Álvarez-Rodríguez E (2022c) Competitive adsorption and desorption of tetracycline and sulfadiazine in crop soils. Environ Res 214:113726
14. Chandrasekaran A, Patra C, Narayanasamy S, Subbiah S (2020) Adsorptive removal of Ciprofloxacin and Amoxicillin from single and binary aqueous systems using acid-activated carbon from Prosopis juliflora. Environ Res 188:109825
15. Cheng D, Ngo HH, Guo W, Liu Y, Chang SW, Nguyen DD, Nghiem LD, Zhou J, Ni B (2018) Anaerobic membrane bioreactors for antibiotic wastewater treatment: performance and membrane fouling issues. Biores Technol 267:714–724

16. Conde-Cid M, Cela-Dablanca R, Ferreira-Coelho G, Fernández-Calviño D, Núñez-Delgado A, Fernández-Sanjurjo MJ, Arias-Estévez M, Álvarez-Rodríguez E (2021) Sulfadiazine, sulfamethazine and sulfachloropyridazine removal using three different porous materials: Pine bark, "oak ash" and mussel shell. Environ Res 195:110814
17. Conde-Cid M, Fernández-Calviño D, Núñez-Delgado A, Fernández-Sanjurjo MJ, Arias-Estévez M, Álvarez-Rodríguez E (2020) Influence of mussel shell, oak ash and pine bark on the adsorption and desorption of sulfonamides in agricultural soils. J Environ Manage 261:110221
18. Conde-Cid M, Ferreira-Coelho G, Arias-Estevez M, Alvarez-Esmoris C, Nóvoa-Muñoz JC, Nunez-Delgado A, Fernandez-Sanjurjo MJ, Alvarez-Rodriguez E (2019) Competitive adsorption/desorption of tetracycline, oxytetracycline and chlortetracycline on pine bark, oak ash and mussel shell. J Environ Manag 250:109509
19. Crisafully R, Milhome MAL, Cavalcante RM, Silveira ER, de Keukeleire D, Nascimento RF (2008) Removal of some polycyclic aromatic hydrocarbons from petrochemical wastewater using low-cost adsorbents of natural origin. Biores Technol 99(10):4515–4519
20. de Oliveira T, Fernandez E, Fougère L, Destandau E, Boussafir M, Sohmiya M, Sugahara Y, Guégan R (2018) Competitive association of antibiotics with a clay mineral and organoclay derivatives as a control of their lifetimes in the environment. ACS Omega 3(11):15332–15342
21. Ding H, Wu Y, Zou B, Lou Q, Zhang W, Zhong J, Lu L, Dai G (2016) Simultaneous removal and degradation characteristics of sulfonamide, tetracycline, and quinolone antibiotics by laccase-mediated oxidation coupled with soil adsorption. J Hazard Mater 307:350–358
22. Du L, Liu W (2012) Occurrence, fate, and ecotoxicity of antibiotics in agro-ecosystems. a review. Agron Sustain Dev 32:309–327
23. Duan H, Li X, Mei A, Li P, Liu Y, Li X, Li W, Wang C, Xie S (2021) The diagnostic value of metagenomic next-generation sequencing in infectious diseases. BMC Infect Dis 21(1):1–13
24. Foo KY, Hameed BH (2010) Insights into the modeling of adsorption isotherm systems. Chem Eng J 156(1):2–10
25. Fox RL, Kamprath EJ (1970) Phosphate sorption isotherms for evaluating the phosphate requirements of soils. Soil Sci Soc Am J 34(6):902–907
26. García RC (2022) Medioambiente y resistencias: la prespectiva One Health en el Plan Nacional frente a la Resistencia a los Antibióticos. Revista de Salud Ambiental 22(Especial Congreso):96–97
27. Gogoi A, Mazumder P, Tyagi VK, Chaminda GGT, An AK, Kumar M (2018) Occurrence and fate of emerging contaminants in water environment: a review. Groundw Sustain Dev 6:169–180
28. González Fernández LA, Navarro Frómeta AE, Carranza Álvarez C, Flores Ramírez R, Díaz Flores PE, Castillo Ramos V, Sánchez Polo M, Carrasco Marín F, Medellín Castillo NA (2023) Valorization of sargassum biomass as potential material for the remediation of heavy-metals-contaminated waters. Int J Environ Res Public Health 20(3):2559
29. Guitian Ojea F, Carballas T (1976) Técnicas de análisis de suelos. Pico Sacro
30. Ighalo JO, Zhou Y, Zhou Y, Igwegbe CA, Anastopoulos I, Raji MA, Iwuozor KO (2022) A review of pine-based adsorbents for the adsorption of dyes. Biomass-Deriv Mater Environ Appl 319–332
31. Karić N, Maia AS, Teodorović A, Atanasova N, Langergraber G, Crini G, Ribeiro ARL, Đolić M (2022) Bio-waste valorisation: agricultural wastes as biosorbents for removal of (in) organic pollutants in wastewater treatment. Chem Eng J Adv 9:100239
32. Karoui S, Arfi RB, Mougin K, Ghorbal A, Assadi AA, Amrane A (2020) Synthesis of novel biocomposite powder for simultaneous removal of hazardous ciprofloxacin and methylene blue: Central composite design, kinetic and isotherm studies using Brouers–Sotolongo family models. J Hazard Mater 387:121675
33. Li J, Zhang H (2017) Factors influencing adsorption and desorption of trimethoprim on marine sediments: mechanisms and kinetics. Environ Sci Pollut Res 24:21929–21937
34. Liu X, Steele JC, Meng X-Z (2017) Usage, residue, and human health risk of antibiotics in Chinese aquaculture: a review. Environ Pollut 223:161–169

35. Michael I, Rizzo L, McArdell CS, Manaia CM, Merlin C, Schwartz T, Dagot C, Fatta-Kassinos D (2013) Urban wastewater treatment plants as hotspots for the release of antibiotics in the environment: a review. Water Res 47(3):957–995
36. Míguez-González A, Cela-Dablanca R, Barreiro A, Rodríguez-López L, Rodríguez-Seijo A, Arias-Estévez M, Núñez-Delgado A, Fernández-Sanjurjo MJ, Castillo-Ramos V, Álvarez-Rodríguez E (2023) Adsorption of antibiotics on bio-adsorbents derived from the forestry and agro-food industries. Environ Res 116360
37. Otero M, Cutillas-Barreiro L, Nóvoa-Muñoz JC, Arias-Estévez M, Fernández-Sanjurjo MJ, Álvarez-Rodríguez E, Núñez-Delgado A (2015) Cr (VI) sorption/desorption on untreated and mussel-shell-treated soil materials: fractionation and effects of pH and chromium concentration. Solid Earth 6(2):373–382
38. Pan M, Chu LM (2016) Adsorption and degradation of five selected antibiotics in agricultural soil. Sci Total Environ 545:48–56
39. Peech M (1947) Methods of soil analysis for soil-fertility investigations (Issue 757). US Department of Agriculture
40. Pezoti O, Cazetta AL, Bedin KC, Souza LS, Martins AC, Silva TL, Júnior OOS, Visentainer JV, Almeida VC (2016) NaOH-activated carbon of high surface area produced from guava seeds as a high-efficiency adsorbent for amoxicillin removal: kinetic, isotherm and thermodynamic studies. Chem Eng J 288:778–788
41. Philippou K, Anastopoulos I, Pashalidis I, Hosseini-Bandegharaei A, Usman M, Kornaros M, Omirou M, Kalderis D, Milojković JV, Lopičić ZR (2021) The application of pine-based adsorbents to remove potentially toxic elements from aqueous solutions. Sorbents Mater Control Environ Pollut 113–133
42. Quintáns-Fondo A, Ferreira-Coelho G, Arias-Estévez M, Nóvoa-Muñoz JC, Fernández-Calviño D, Álvarez-Rodríguez E, Fernández-Sanjurjo MJ, Núñez-Delgado A (2019) Chromium VI and fluoride competitive adsorption on different soils and by-products. Processes 7(10):748
43. Quintáns-Fondo A, Ferreira-Coelho G, Paradelo-Núñez R, Nóvoa-Muñoz JC, Arias-Estévez M, Fernández-Sanjurjo MJ, Álvarez-Rodríguez E, Núñez-Delgado A (2016) Promoting sustainability in the mussel industry: mussel shell recycling to fight fluoride pollution. J Clean Prod 131:485–490
44. Rahbar Shahrouzi J, Molaee S, Ebadi A, Towfighi F, Bakhti F (2019) Investigation of effective parameters on adsorption of amoxicillin from aqueous medium onto activated carbon. Adv Environ Technol 5(2):107–114
45. Ramírez-Pérez AM, Paradelo M, Nóvoa-Muñoz JC, Arias-Estévez M, Fernández-Sanjurjo MJ, Álvarez-Rodríguez E, Núñez-Delgado A (2013) Heavy metal retention in copper mine soil treated with mussel shells: batch and column experiments. J Hazard Mater 248:122–130
46. Ringot D, Lerzy B, Chaplain K, Bonhoure J-P, Auclair E, Larondelle Y (2007) In vitro biosorption of ochratoxin a on the yeast industry by-products: comparison of isotherm models. Biores Technol 98(9):1812–1821
47. Rodríguez-González L, Núñez-Delgado A, Álvarez-Rodríguez E, Díaz-Raviña M, Arias-Estévez M, Fernández-Calviño D, Santás-Miguel V (2023) Direct toxicity of six antibiotics on soil bacterial communities affected by the addition of bio-adsorbents. Environ Pollut 121161
48. Rodríguez-López L, Santás-Miguel V, Cela-Dablanca R, Núñez-Delgado A, Álvarez-Rodríguez E, Pérez-Rodríguez P, Arias-Estévez M (2022) Ciprofloxacin and trimethoprim adsorption/desorption in agricultural soils. Int J Environ Res Public Health 19(14):8426
49. Rodriguez-Mozaz S, Vaz-Moreira I, Della Giustina SV, Llorca M, Barceló D, Schubert S, Berendonk TU, Michael-Kordatou I, Fatta-Kassinos D, Martinez JL (2020) Antibiotic residues in final effluents of European wastewater treatment plants and their impact on the aquatic environment. Environ Int 140:105733
50. Romar-Gasalla A, Santás-Miguel V, Nóvoa-Muñoz JC, Arias-Estévez M, Álvarez-Rodríguez E, Núñez-Delgado A, Fernández-Sanjurjo MJ (2018) Chromium and fluoride sorption/desorption on un-amended and waste-amended forest and vineyard soils and pyritic material. J Environ Manag 222:3–11

51. Russell JN, Yost CK (2021) Alternative, environmentally conscious approaches for removing antibiotics from wastewater treatment systems. Chemosphere 263:128177
52. Seco-Reigosa N, Bermúdez-Couso A, Garrido-Rodríguez B, Arias-Estévez M, Fernández-Sanjurjo MJ, Álvarez-Rodríguez E, Núñez-Delgado A (2013) As (V) retention on soils and forest by-products and other waste materials. Environ Sci Pollut Res 20:6574–6583
53. Seco-Reigosa N, Cutillas-Barreiro L, Nóvoa-Muñoz JC, Arias-Estévez M, Fernández-Sanjurjo MJ, Álvarez-Rodríguez E, Núñez-Delgado A (2014) Mixtures including wastes from the mussel shell processing industry: retention of arsenic, chromium and mercury. J Clean Prod 84:680–690
54. Seco-Reigosa N, Peña-Rodríguez S, Nóvoa-Muñoz JC, Arias-Estévez M, Fernández-Sanjurjo MJ, Álvarez-Rodríguez E, Núñez-Delgado A (2013) Arsenic, chromium and mercury removal using mussel shell ash or a sludge/ashes waste mixture. Environ Sci Pollut Res 20:2670–2678
55. Singh R, Singh TS, Odiyo JO, Smith JA, Edokpayi JN (2020) Evaluation of methylene blue sorption onto low-cost biosorbents: equilibrium, kinetics, and thermodynamics. J Chem 2020:1–11
56. Sukul P, Lamshöft M, Zühlke S, Spiteller M (2008) Sorption and desorption of sulfadiazine in soil and soil-manure systems. Chemosphere 73(8):1344–1350
57. Tran HN, You S-J, Chao H-P (2016) Thermodynamic parameters of cadmium adsorption onto orange peel calculated from various methods: a comparison study. J Environ Chem Eng 4(3):2671–2682
58. Yang C-W, Hsiao W-C, Chang B-V (2016) Biodegradation of sulfonamide antibiotics in sludge. Chemosphere 150:559–565
59. Zhang Y-J, Hu H-W, Gou M, Wang J-T, Chen D, He J-Z (2017) Temporal succession of soil antibiotic resistance genes following application of swine, cattle and poultry manures spiked with or without antibiotics. Environ Pollut 231:1621–1632

Salt-Affected Soils

Muhammad Shaaban

Abstract In this Chapter, the author makes a review on soils affected by degradation processes due to excess of salt, focusing on causes, types, and consequences of salt-affected soils. Finally, some alternatives to help solving these issues are presented.

Keywords Land degradation · Saline soils · Saline-sodic soils · Sodic soils · Soil degradation

1 Introduction

Land degradation by salts is a major threat to sustainable crop production in many arid and semi-arid regions of the world [3, 31]. Salt-affected soils are a significant challenge faced by farmers and land managers worldwide. They are characterized by high levels of soluble salts, principally sodium, which can have harmful effects on plant development and growth [31]. It is estimated that global arable lands over 1100 Mha are affected by salinity and sodicity, of which approximately 60% are saline, 26% sodic and 14% saline-sodic [7]. Salt-affected soils are widely located on all continents, and the most influenced regions are the Middle East, North Africa, Australia, and Eurasia. Poorly managed irrigation is a main reason for the development pf salt-affected soils, particularly in arid climates. It is estimated that 20 to 50% of irrigated arable lands are salt-affected at global level [26]. This chapter aims to provide an outline of salt-affected soils, including their sources, types, impacts, and reclamations measures.

M. Shaaban (✉)
College of Agriculture, Henan University of Science and Technology, Luoyang, China
e-mail: shabanbzu@hotmail.com

A. Núñez-Delgado (ed.), *Frontier Studies in Soil Science*, Frontier Studies in Soil Science, https://doi.org/10.1007/978-3-031-50503-4_10

2 Causes of Salt-Affected Soils

There are numerous factors that involve in the development of salt-affected soils.

2.1 Natural Factors

Natural processes such as soil formation, water movement, and weathering of parent material can result in salts accumulation in the soil profile. Arid and semi-arid climates with low rainfall and high evaporation rates are mainly vulnerable to salt accretion in soils.

(i) ***Soil parent material***

The geological composition of the parent material can affect the occurrence of salts in soils. Regions with naturally high salt contents in their sediments and rocks are more susceptible to to salt-affected soils [22, 24].

(ii) ***Amount and distribution of rainfall and climate***

The precipitation patterns in any region highly influence impact soil salinity. In arid and semi-arid climates, where evaporation exceeds precipitation, salts accumulate as a result of water evaporation, causing higher levels of soil salinity [2]. A simple illustration of salinity builds up and its impact on soil properties is presented in Fig. 1.

(iii) ***Groundwater***

High groundwater table is also an important natural factor, which regulates soil salinity. High groundwater levels, particularly in areas with shallow water tables, can bring the salts to the upper soil layers. Consequently, water evaporation leaves the salts, leading to development of salt-affected soils [19].

(iv) ***Topography***

The drainage patterns, slope, and shape of the lands can impact the movement and accumulation of salts in soils. Areas with depressions, poor drainage, or inadequate access to natural water sources are at higher jeopardy for salt-affected soils [13].

Fig. 1 A simple illustration of salinity builds up and its impact on soil properties

2.2 Human Interventions

Anthropogenic activities such as inappropriate irrigation and water management, extreme use of fertilizers, derisory drainage system, and scarcity land management practices can aggravate salt problems. Improper water management, over-irrigation, and inadequate leaching can cause the soil salinity.

(i) ***Irrigation techniques***

Improper irrigation practices, such as over-irrigation or the use of poor-quality water can substantially hasten the development of salt-affected soils. Over-irrigation without adequate drainage may cause salts accumulation and reduce soil fertility [8].

(ii) ***Derisory drainage systems***

Unsatisfactory or poorly designed drainage system can restrict the water movement in soil profile, resulting in waterlogging and subsequently build-up of soil salinity. If the drainage or water outlets are not proper, salts can accumulate in soil surface, leading to soil salinity [20].

(iii) ***Use of synthetic fertilizers***

Indiscriminate and excessive usage of chemical fertilizers can ascribe to soil salinity. Farmers use fertilizers for high crop yields, which contain high contents of salts, and their use over the time, can contribute to soil salinity. Moreover, runoff from fields treated with agro-chemicals in neighboring areas can also upsurge soil salinity [18].

(iv) ***Industrialization and urbanization***

Urban development and industrialization can introduce salts into the sewage water and environment. Disposal of industrial water, improper management of municipal waste, and the use of salt-based de-icing materials on roads can highly ascribe to development of soil salinity [33].

3 Types of Salt-Affected Soils

Salt-affected soils can be categorized into three broad categories based on their dominant salts:

(i) ***Saline soils***

Saline soils are characterized by high contents of soluble salts, primarily sulfate and chloride ions, which hamper plant growth due to osmotic stress. These soils generally develop in arid and semi-arid regions, where low rainfall and high evaporation rates lead to accumulation of salts. Saline soils may also arise from poor irrigation practices or excessive fertilizer use [31]. Normally, the water infiltration and soil permeability are poor in saline soils. Saline soils are characterized by an electrical conductivity (EC) of 4 deci-Siemens per meter or greater, as well as an exchangeable sodium percentage of less than 15. The pH of these soils typically falls below 8.5, and the sodium adsorption ratio remains below 15. A distinctive feature of these soils is the presence of white color encrustation on the surface, resulting in a salt crust that lacks vegetation [15, 28].

(ii) ***Sodic soils***

Sodic soils are characterized by a high sodium content. Sodium ions replace calcium and magnesium, causing the dispersion of soil particles and leading to the formation of aggregates that are unstable when wet. Consequently, sodic soils exhibit poor structural stability and low water infiltration rates [3]. They often have a high pH and are prone to erosion and waterlogging. Sodic soils are typically found in regions with high sodium levels in irrigation water or in areas where sodium-rich parent materials predominate. The exchangeable sodium percentage (ESP) of sodic soils is typically 15 or higher, while the electrical conductivity (EC) tends to be less than 4. Additionally, the sodium adsorption ratio (SAR) of these soils is typically 15 or higher. Furthermore, the pH levels in these soils usually fall within the range of 8.5 to 10 [29].

(iii) ***Saline-sodic soils***

These soils exhibit characteristics of both saline and sodic soils, making them particularly challenging for agriculture. They have high levels of both soluble salts and exchangeable sodium, posing dual challenges to plant growth. The combination of salt accumulation and dispersion of soil particles reduces water availability and limits nutrient uptake by plants [4]. Saline-sodic soils occur in areas with both salinity and sodicity problems, such as coastal regions or areas irrigated with saline water. Soluble salts have a tendency to cause soil particles to flocculate, whereas exchangeable sodium tends to disperse the soil, thereby reducing its permeability to water and air. Soils with an exchangeable sodium percentage (ESP) of 15 or higher, an electrical conductivity (EC) of 4 or higher, exhibit pH levels that depend on whether soluble salts or exchangeable sodium dominate. If exchangeable sodium dominates, the pH is typically above 5.8. Conversely, if soluble salts are dominant, the pH usually falls below 8.5 [24].

Table 1 presents a general criterion for the categorization of salt-affected soils. This criterion is used to classify different types of soils based on their salt content and level of salinity. The purpose of this classification is to provide a systematic way to analyze and understand the characteristics of these soils. By categorizing salt-affected soils, it becomes easier to develop appropriate strategies for their management and mitigation. This criterion takes into account various factors such as electrical conductivity (EC), exchangeable sodium percentage (ESP), soil pH, and salt concentration. Each factor is assigned a specific range or value, which helps determine the severity and classification of the salt-affected soil. It is important to note that this is a general criterion and may be subject to modification or adaptation based on specific regional or local conditions.

Table 1 A general criterion for the categorization of salt-affected soils

Soil type	pH	Electrical conductivity (dS/m)	Sodium adsorption ratio	Exchangeable sodium percentage
Saline	<8.5	≥4	<15	<15
Sodic	8.5–10	<4	≥15	≥15
Saline-Sodic	Depends on dominant salt	≥4	≥15	≥15
Normal	<8.5	<4	<15	<15

4 Consequences of Salt-Affected Soils

The presence of salt-affected soils can have numerous detrimental effects.

4.1 Physical Properties of Soils

(i) ***Deterioration of soil structure***

One of the significant consequences of salt-affected soils is the dilapidation of soil structure. Extreme salt accumulation can disturb soil aggregates, leading to poor soil porosity, abridged water infiltration, and augmented surface crusting. This degradation impairs soil aeration, root penetration, and nutrient uptake, therefore destructively affecting plant growth and overall soil health [9].

(ii) ***Decreased water availability***

Salt-affected soils display high electrical conductivity (EC) due to the occurrence of salts, which restricts with water availability to plants. High salt concentrations in the soil solution produce an osmotic potential that decreases the capacity of roots to extract water. This water stress can result in reduced crop yields and even plant death in exciting cases [15].

(iii) ***Soil texture***

Salinity alters soil particle arrangement, resulting in changes in soil texture. Clay particles tend to flocculate, while sand particles become dispersed, affecting water holding capacity and permeability [6, 34].

(iv) ***Soil erosion***

Saline soils are prone to increased erosion due to reduced stability and binding capacity, leading to soil degradation and loss of organic matter [6].

4.2 Chemical Properties

(i) ***Soil pH***

Salinity influences soil pH, causing soil alkalinization. High salt concentrations contribute to the accumulation of basic cations, affecting nutrient availability and microbial activity [31].

(ii) ***Nutrient uptake***

Salts in the soil can also interrupt the balance of essential nutrients in plants. Excessive salts can hinder nutrient uptake or induce nutrient imbalances, leading to nutrient deficiencies. Elevated sodium levels, for example, can lead to sodium toxicity and interfere with the uptake of potassium, calcium, and magnesium, which are vivacious for proper metabolism and plant growth. This interruption in nutrient balance can result in nutrient deficiencies and abridged crop productivity [10].

(iii) ***Ion toxicity***

High salt levels result in the accumulation of toxic ions (e.g., sodium, chloride, boron) in the soil, inhibiting plant growth and development [27].

(iv) ***Organic matter***

Salinity impacts the decomposition of organic matter, reducing microbial activity and nutrient cycling. This leads to a decline in soil fertility and organic carbon contents [29].

4.3 Biological Properties

Salt-affected soils have a harmful impact on soil microorganisms, including fungi, bacteria, and beneficial soil fauna. Elevated levels of salts can dislocate the microbial balance and decrease microbial diversity, which are vital for nutrient recycling, organic matter decomposition, and soil fertility. This disturbance in the soil microbiome can further exacerbate the negative consequences of salt-affected soils on plant growth.

(i) ***Microbial diversity***

Salinity affects the abundance and diversity of soil microorganisms. High salt levels can suppress beneficial microbial populations, impacting nutrient cycling, disease suppression, and soil health [20]. Saline soils often experience reduced biodiversity due to their inhospitable conditions for many organisms. This can lead to a loss of ecological resilience and ecosystem functions. High salt concentrations can limit the growth and survival of many microbial species. This restricts the diversity of microbes capable of thriving under this harsh condition. Consequently, the overall microbial community composition in salt-affected soils becomes dominated by salt-tolerant or halophilic microorganisms. These microorganisms have developed unique adaptive mechanisms to survive and thrive in high saline environments [31, 36].

(ii) ***Microbial abundance and activity***

Soil salinity affects microbial richness, as the high salt content constrains the growth and reproduction of certain microorganisms. The occurrence of salts upsurges the osmotic stress on microbes, plummeting their metabolic activity. This repressive consequence can lead to abridged microbial biomass and a slower nutrient cycling rate. However, salt-tolerant microbial populations, like halotolerant bacteria and fungi, have evolved stratagems to sustain their activity and perform indispensable functions under these challenging conditions. These salt-tolerant microbes attribute meaningfully to the nutrient cycling and overall constancy of salt-affected soils [21].

(iii) ***Functional variations***

Soil salinity impacts the functional potential of microbial communities. Certain microbial activities, such as nitrogen fixation, organic matter decomposition, and phosphorus solubilization, can be reduced in salt-affected soils. The diminished functional diversity of microbial communities under high salinity may negatively influence soil fertility and plant production. Nevertheless, it is imperative to note that not all functional groups are equally affected by salt stress. Some groups of microbes, such as halophilic bacteria and archaea, have specialized adaptations that enable them to perform specific functions, even under high saline conditions [35].

(iv) ***Microbial communications***

Microbes in salt-affected soils experience changed communications due to the variations in community composition and physiological rejoinders. Competition amid microorganisms for restricted resources rises as the overall microbial diversity declines. In some cases, salt-tolerant microbes form supportive connotations to advance their chances of existence. Besides, some microbes release extracellular substances, like exopolysaccharides, that help alleviate salt stress by acting as osmo-protectants. These complex microbial interactions play a key protagonist in determining the community structure and functionality of salt-affected soils [1].

4.4 Impact on Crop Yield

The consequences of salt-affected soils on crop production are profound. High salt contents deter seed germination, diminish seedling vigor, and limit overall crop growth [25]. Saline soils are specifically challenging for salt-sensitive crops, leading to reduced yields and economic losses for farmers. Even moderately salt-affected soils can still have a substantial impact on crop productivity. Salinity inhibits plant growth and productivity by affecting water and nutrient uptake, causing osmotic stress, and disrupting hormonal balance. This results in reduced crop yields and poor plant quality [23].

(i) ***Oxidative stress***

Salt-affected soils induce oxidative stress in plants. The accumulation of salt ions triggers the production of reactive oxygen species (ROS), including hydrogen peroxide, superoxide radicals, and singlet oxygen. The excess ROS can harm cellular components like membrane lipids, proteins, and DNA, leading to cellular disfunction and even cell death. Plants respond by activating antioxidant defense mechanisms to scavenge these ROS [14].

(ii) ***Ion toxicity***

The excessive salt content in soil causes an accumulation of toxic ions, particularly sodium (Na^+) and chloride (Cl^-), which negatively disturb plant physiology. High levels of Na^+ interrupt essential ion balance and restrict nutrient's uptake, especially magnesium (Mg^{2+}), calcium (Ca^{2+}), and potassium (K^+). This imbalance impacts several physiological processes, such as hormonal regulation photosynthesis, enzyme activity, and nutrient transport, leading to abridged growth and crop yield [14].

(iii) ***Osmotic stress***

Osmotic stress on plants is a devastating impact of salt-affected soils due to excess of salts. High soil salinity levels create a water potential gradient that limits water uptake by plants. This imbalance disrupts cellular osmotic regulation, leading to water scarcity within plant tissues. Subsequently, plants experience dehydration, resulting in wilting, reduced leaf area, and stunted growth [17].

(iv) ***Disruption of plant metabolism***

Salt stress often affects the metabolic pathways of plants and crops. It can impact important processes such as photosynthesis, respiration, and protein synthesis, leading to a decline in overall plant productivity [12]. Reduced photosynthetic rates, lower chlorophyll content, and altered stomatal conductance are common effects of salt stress on plant metabolism. These changes result in reduced energy production, carbohydrate accumulation, and altered nutrient metabolism [32].

(v) ***Growth limitations and yield reduction***

The cumulative effects of osmotic stress, ion toxicity, oxidative stress, and disrupted plant metabolism contribute to significant growth limitations and yield reductions in salt-affected soils. Reduced plant vigor, increased susceptibility to pests and diseases, premature leaf senescence, and smaller root systems are common manifestations of plants struggling to withstand salt stress. These conditions ultimately lead to decreased crop yields and economic losses in agricultural production [5].

4.5 Environmental Challenges

In addition to the instantaneous effects on soil and crops, salt-affected soils also pose environmental challenges. When salts accumulate, they can leach into groundwater supplies or be carried away by runoff, leading to salinization of freshwater resources. Saline groundwater can degrade water quality and pose risks to human and animal health. Additionally, salt-affected soils contribute to soil erosion and can have negative impacts on nearby ecosystems [11].

5 Amelioration of Salt-Affected Soils

There are various amelioration techniques available to alleviate salinity issues and reinstate the productivity and fertility of salt-affected soils.

(i) ***Soil drainage***

Improving soil drainage is one of the prime methods to ameliorate salt-affected soils. By removing excess water, salts can be flushed out of the root zone and leached away. This technique is mainly effective for soils with high water tables or poor drainage. Techniques such as installing tile drains, constructing subsurface drainage systems, and contouring the land to redirect surface runoff can markedly improve soil drainage [29].

(ii) ***Irrigation management***

Appropriate irrigation management practices play a central role in remedying salt-affected soils. Suitable irrigation scheduling helps avert waterlogging and excessive salt accumulation in the soil. Techniques such as precision irrigation (drip or sprinkler irrigation), use of saline-tolerant crops, and adopting deficit irrigation strategies can curtail water and salt-related problems, indorsing sustainable soil improvement [16].

(iii) ***Addition of soil organic matter***

Adding the organic matter in salt-affected soils can contribute to their amelioration. Organic matter acts as a natural amendment to recover soil structure, upsurge water-holding capacity, and improve nutrient availability. Incorporating organic residues, such as crop residues and compost, into the soil augments microbial activity and nutrient cycling while reducing soil salinity [16, 29].

(iv) ***Soil amendments***

The application of soil amendments can help rectify salt-affected soils by altering their chemical and physical properties. Gypsum, a calcium sulfate compound, is commonly used to improve soil structure and aid in leaching excess salts. Additionally, amendments like organic matter, elemental sulfur, and lime can be effective in adjusting soil pH, promoting nutrient availability, and improving soil health [3, 30].

(v) ***Crop rotation and saline tolerance***

Employing a well-designed crop rotation approach can assuage salt-affected soil problems. Some crops are more tolerant to salinity than others, and rotating salt-tolerant crops can help manage soil salt stress. These crops can extract salts from the soil and accumulate them in their aerial parts, which can then be removed with harvesting. This method, combined with suitable soil management practices, helps decrease salt build-up progressively [6, 12].

(vi) ***Phyto-remediation***

Phyto-remediation refers to the use of specific plant species to eliminate or counteract contaminants in soil. Some salt-tolerant plants, known as halophytes, have the capability to accumulate salts in their tissues by extracting them from the soil. By cultivating these plant species in salt-affected soils, farmers can reduce soil salinity levels over time. Phyto-remediation is a long-term approach to correcting salt-affected soils and requires vigilant plant selection and management [1, 14].

6 Conclusions and Summary

The summary of the chapter is presented in Fig. 2. The chapter on salt-affected soils serves as a detailed exploration into the complex issue of soil salinity and its influence on agricultural productivity. Throughout the chapter, numerous aspects such as the causes of soil salinity, the effects on plant growth, and potential management strategies have been discussed. In conclusion, it is evident that salt-affected soils pose significant challenges to sustainable agriculture and food production. Salinity, whether caused by natural processes or human activities, leads to the accumulation of soluble salts, negatively distressing soil structure, nutrient availability, and water uptake by plants. Consequently, crop yields, as well as the overall health and vigor of vegetation, are severely compromised. Effective management of salt-affected soils requires a holistic approach that integrates both preventive and remedial measures. Prevention strategies involve proper irrigation practices, the use of salt-tolerant crop varieties, and implementing effective drainage systems to minimize salt buildup. Remediation techniques include leaching excess salts through drainage, the use of amendments like gypsum to displace sodium ions, and the adoption of precision agriculture technologies to optimize water and nutrient application. Additionally, it is vital to concede that salt-affected soils are not a one-size-fits-all problem, as the severity and nature of salinity can vary greatly depending on regional conditions and soil characteristics. Therefore, site-specific approaches and tailored management plans are necessary for long-term success. Overall, the chapter highlights the importance of understanding and addressing the challenges posed by salt-affected soils to ensure sustainable agricultural practices. By implementing appropriate strategies and leveraging scientific knowledge, it is probable to alleviate the harmful effects of soil salinity and safeguard a more robust, productive, and food-secure future.

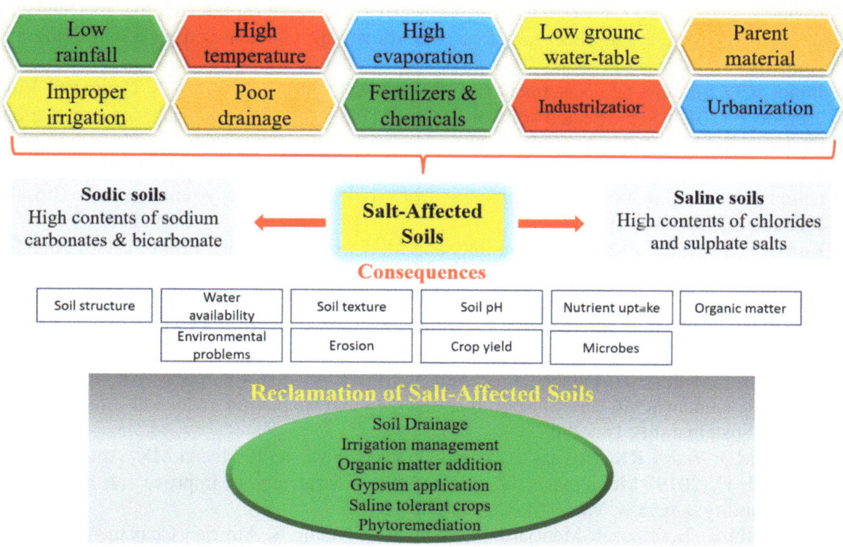

Fig. 2 Summary of salt-affected soils. Causes, types, consequences, and reclamation approaches

References

1. Bhardwaj S, Kumar P (2020) Salinity stress, its physiological response and mitigating effects of microbial bio inoculants and organic compounds. J Pharmacogn Phytochem 9:1297–1303
2. Corwin DL (2021) Climate change impacts on soil salinity in agricultural areas. Eur J Soil Sci 72:842–862
3. Devkota KP, Devkota M, Rezaei M, Oosterbaan R (2022) Managing salinity for sustainable agricultural production in salt-affected soils of irrigated drylands. Agric Syst 198:103390
4. Dong X, Li M, Lin Q, Li G, Zhao X (2019) Soil Na+ concentration controls salt-affected soil organic matter components in Hetao region China. J Soils Sediments 19:1120–1129
5. Egamberdieva D, Wirth S, Bellingrath-Kimura SD, Mishra J, Arora NK (2019) Salt-tolerant plant growth promoting rhizobacteria for enhancing crop productivity of saline soils. Front Microbiol 10:2791
6. El-Ramady H, Faizy S, Amer MM, Elsakhawy TA, Omara AE-D, Eid Y, Brevik E (2022) Management of salt-affected soils: a photographic mini-review. Environ, Biodivers Soil Secur 6:61–79
7. FAO I, ISRIC I (2012) Jrc: Harmonized world soil database (version 1.2). FAO, Rome, Italy and IIASA, Laxenburg, Austria
8. Garcia-Franco N, Wiesmeier M, Hurtarte LCC, Fella F, Martínez-Mena M, Almagro M, Martínez EG, Kögel-Knabner I (2021) Pruning residues incorporation and reduced tillage improve soil organic matter stabilization and structure of salt-affected soils in a semi-arid Citrus tree orchard. Soil Tillage Res 213:105129
9. Gondek M, Weindorf DC, Thiel C, Kleinheinz G (2020) Soluble salts in compost and their effects on soil and plants: a review. Compos Sci Util 28:59–75
10. Hailu B, Mehari H (2021) Impacts of soil salinity/sodicity on soil-water relations and plant growth in dry land areas: a review. J Nat Sci Res 12:1–10
11. Huang J, Hartemink AE (2020) Soil and environmental issues in sandy soils. Earth Sci Rev 208:103295

12. Huang L, Li Z, Liu Q, Pu G, Zhang Y, Li J (2019) Research on the adaptive mechanism of photosynthetic apparatus under salt stress: new directions to increase crop yield in saline soils. Ann Appl Biol 175:1–17
13. Junior JCO, Furquim SA, Nascimento AF, Beirigo RM, Barbiero L, Valles V, Couto EG, Vidal-Torrado P (2019) Salt-affected soils on elevated landforms of an alluvial megafan, northern Pantanal, Brazil. CATENA 172:819–830
14. Karakas S, Dikilitas M, Tıpırdamaz R (2020) Phytoremediation of salt-affected soils using halophytes. In: Handbook of Halophytes: from molecules to ecosystems towards Biosaline agriculture, 1–18
15. Kumar P, Sharma PK (2020) Soil salinity and food security in India. Front Sustain Food Syst 4:533781
16. Kumar R, Singh A, Bhardwaj AK, Kumar A, Yadav RK, Sharma PC (2022) Reclamation of salt-affected soils in India: progress, emerging challenges, and future strategies. Land Degrad Dev 33:2169–2180
17. Kumawat C, Kumar A, Parshad J, Sharma SS, Patra A, Dogra P, Yadav GK, Dadhich SK, Verma R, Kumawat GL (2022) Microbial diversity and adaptation under salt-affected soils: a review. Sustainability 14:9280
18. Meena MD, Yadav RK, Narjary B, Yadav G, Jat H, Sheoran P, Meena MK, Antil R, Meena B, Singh H (2019) Municipal solid waste (MSW): Strategies to improve salt affected soil sustainability: a review. Waste Manage 84:38–53
19. Nabiollahi K, Taghizadeh-Mehrjardi R, Shahabi A, Heung B, Amirian-Chakan A, Davari M, Scholten T (2021) Assessing agricultural salt-affected land using digital soil mapping and hybridized random forests. Geoderma 385:114858
20. Negacz K, Malek Ž, de Vos A, Vellinga P (2022) Saline soils worldwide: Identifying the most promising areas for saline agriculture. J Arid Environ 203:104775
21. Nguyen BT, Trinh NN, Bach Q-V (2020) Methane emissions and associated microbial activities from paddy salt-affected soil as influenced by biochar and cow manure addition. Appl Soil Ecol 152:103531
22. Omuto C, Vargas R, El Mobarak A, Mohamed N, Viatkin K, Yigini Y (2020) Mapping of salt-affected soils–technical manual
23. Paul PLC, Bell RW, Barrett-Lennard EG, Kabir E (2020) Variation in the yield of sunflower (Helianthus annuus L.) due to differing tillage systems is associated with variation in solute potential of the soil solution in a salt-affected coastal region of the Ganges Delta. Soil Tillage Res 197:104489
24. Paz AM, Amezketa E, Canfora L, Castanheira N, Falsone G, Gonçalves MC, Gould I, Hristov B, Mastrorilli M, Ramos T (2023) Salt-affected soils: field-scale strategies for prevention, mitigation, and adaptation to salt accumulation. Ital J Agron 18
25. Phuong NTK, Khoi CM, Ritz K, Linh TB, Minh DD, Duc TA, Sinh NV, Linh TT, Toyota K (2020) Influence of rice husk biochar and compost amendments on salt contents and hydraulic properties of soil and rice yield in salt-affected fields. Agronomy 10:1101
26. Pitman MG, Läuchli A (2002) Global impact of salinity and agricultural ecosystems. Salin: Environ Plants Mol 3:20
27. Roy S, Chowdhury N (2020) Salt stress in plants and amelioration strategies: a critical review. Abiotic stress in plants
28. Shaaban M (2019) Crop tolerance against salt stress. J Plant Biol Crop Res 2:1011
29. Shaaban M, Abid M, Abou-Shanab R (2013) Amelioration of salt affected soils in rice paddy system by application of organic and inorganic amendments. Plant Soil Environ 59:227–233
30. Shaaban M, Abid M, Qi-An P (2013b) Short term influence of gypsum, farm manure and commercial humic acid on physical properties of salt affected soil in rice paddy system. J Chem Soc Pak 34
31. Shaaban M, Wu Y, Núñez-Delgado A, Kuzyakov Y, Peng Q-A, Lin S, Hu R (2023) Enzyme activities and organic matter mineralization in response to application of gypsum, manure and rice straw in saline and sodic soils. Environ Res 224:115393

32. Souana K, Taïbi K, Abderrahim LA, Amirat M, Achir M, Boussaid M, Mulet JM (2020) Salt-tolerance in Vicia faba L. is mitigated by the capacity of salicylic acid to improve photosynthesis and antioxidant response. Sci Hortic 273:109641
33. Velmurugan A, Swarnam P, Subramani T, Meena B, Kaledhonkar M (2020) Water demand and salinity. Desalination-Chall Oppor, 1–11
34. Wang X, Li Y, Wang H, Wang Y, Biswas A, Chau HW, Liang J, Zhang F, Bai Y, Wu S (2022) Targeted biochar application alters physical, chemical, hydrological and thermal properties of salt-affected soils under cotton-sugarbeet intercropping. CATENA 216:106414
35. Zhang G, Bai J, Tebbe CC, Zhao Q, Jia J, Wang W, Wang X, Yu L (2021) Salinity controls soil microbial community structure and function in coastal estuarine wetlands. Environ Microbiol 23:1020–1037
36. Zhang G, Bai J, Zhai Y, Jia J, Zhao Q, Wang W, Hu X (2023) Microbial diversity and functions in saline soils: A review from a biogeochemical perspective. J Adv Res

Beyond the Obvious: Exploring Peat Vibrational Spectroscopy (FTIR-ATR) Data Using Principal Components Analysis on Transposed Data Matrix (tPCA), Store Mosse Bog (Sweden)

Antonio Martínez Cortizas, Mohamed Traoré, Olalla López-Costas, Jenny K. Sjöström, and Malin E. Kylander

Abstract Peatlands are a major soil carbon reservoir, despite only covering about 3% of the land surface. Waterlogged peatlands act as sinks of atmospheric carbon, through accumulation of partially decomposed plant remains. But there are concerns that boreal peatlands may shift from sinks to sources due to enhanced peat mineralization with climate change. Characterizing the molecular composition of the peat is key to understanding the responses of peatlands to climate change, as it is expected that not all peat constituents will be equally affected. Mid-infrared (MIR) vibrational spectroscopy is a fast, cost-efficient technique that provides information on the molecular composition of many different materials, although the interpretation of MIR spectra may be hampered by the compositional complexity of the material analyzed. Chemometric methods, such as the application of principal components analysis (PCA), can help to untangle the MIR spectrum but, to date, most previous studies using this approach only consider a few main vibrations representative of the main peat components (i.e., cellulose, lignin, etc.), losing part of the information contained in the spectra. In this chapter, we explore the application of PCA on

A. Martínez Cortizas (✉) · M. Traoré
CRETUS, EcoPast (GI-1553), Departamento de Edafoloxía e Química Agrícola, Universidade de Santiago de Compostela, Santiago de Compostela, Spain
e-mail: antonio.martinez.cortizas@usc.es

A. Martínez Cortizas · J. K. Sjöström · M. E. Kylander
Bolin Centre for Climate Research, Stockholm University, Stockholm, Sweden

O. López-Costas
CRETUS, EcoPast (GI-1553), Área de Arqueoloxía, Departamento de Historia, Universidade de Santiago de Compostela, Santiago de Compostela, Spain

Archaeological Research Laboratory, Wallenberg Laboratoriet, Stockholm University, 10691 Gothenburg, Sweden

J. K. Sjöström · M. E. Kylander
Department of Geological Sciences, Bolin Centre for Climate Research, Stockholm University, Stockholm, Sweden

A. Núñez-Delgado (ed.), *Frontier Studies in Soil Science*, Frontier Studies in Soil Science, https://doi.org/10.1007/978-3-031-50503-4_11

transposed MIR data (tPCA), using a previously well studied peat sequence from Store Mosse (SM-S_{2008}), one of the largest peatlands in southern Sweden. Here, we exemplify how using the whole spectrum of each sample in the statistical analysis enables a proper characterization of the constituents of the peat. We describe both the scores' spectra of the extracted components and identify the peat constituents they represent. We also describe the records of samples' loadings, which show the variation of the constituents along the peat sequence. To support our interpretations, we compare the tPCA components with those extracted in a previous investigation using direct PCA (dPCA), correlate the tPCA components with elemental (C, N, and C/N ratio) and isotopic ($\delta^{13}C$ and $\delta^{15}N$) properties of the peat, and model these same properties using the principal components. The results show that the major, and more obvious, constituents of the peat are easily identified here, but other relevant signals, more difficult to detect, such as that of microbial biomass are also extracted. Most of the tPCA components were highly correlated with the peat properties and were significant in the statistical models. These findings show that tPCA allows extracting detailed information about peat constituents, which is essential for understanding how peatlands may respond to ongoing and future climate change.

Keywords FTIR-ATR spectroscopy · Mid-infrared · Peatlands · Principal components analysis · Principal components regression · Peat elemental and isotopic composition

1 Introduction

Despite that peatlands only represent about 3% of the total land surface, these water-logged ecosystems act as a major carbon reservoir, that have accumulated the equivalent to the amount in present day atmosphere [5]. Peatlands act as sinks of atmospheric carbon, through the accumulation of partly decomposed plant remains. However, ongoing and projected climate change have raised concerns that boreal peatlands may become sources of carbon to the atmosphere in the future due to increased peat mineralization [27]. In this regard, the characterization of the molecular composition of the peat organic matter is key to understanding how the peatland carbon reservoir may respond to ongoing and projected climate change.

Mid-infrared (MIR) vibrational spectroscopy is a fast, cost-efficient technique that provides information on the molecular composition of many different materials and has been extensively applied to characterize peat organic matter [2, 4, 6, 19, 21, 29]. Recent applications of chemometric approaches to vibrational spectroscopic data have generated greater knowledge of the processes behind peat chemical composition. The use of principal component analysis (PCA) has facilitated extracting details that are not accessible from direct analysis of the bulk peat MIR data. It should be noted that PCA is mostly based on pattern recognition related to the variance within an infrared spectroscopic dataset [7]. To date, PCA has been successfully employed to determine infrared characteristics that are connected to various aspects such as the type of peat, peat decomposition processes, peat forming vegetation, responses to environmental drivers, and so on (see for example [2–4, 19]). Despite these advances, it is obvious that the traditional way of applying PCA, with samples in rows and a selection of infrared bands as columns (i.e., direct matrix, dPCA), tends to reflect the variations of the main peat components at the risk of losing information related to the unused part of the recorded spectra. The recent application of PCA to transposed infrared data matrix (i.e., wavenumbers in rows and samples in columns, tPCA), by using the full spectra—or at least the regions with significant absorbance—as variables, has been shown to be an advantageous way to analyze MIR data. For example, Traoré et al. [28] used this approach to successfully identify holocellulose and Klason lignin in pine wood and provided information about species differences (*Pinus sylvestris vs. Pinus nigra*) and effects related to trees' growing location. In peat studies, we have previously used this approach to get a better understanding of peat mineralogical composition [14, 20], and to determine the timing of dust deposition and its relation to changes in storminess in the northern hemisphere [15].

Nonetheless, tPCA is a seldomly used approach and its application requires reconsideration of the meaning and significance of the PCA output data (i.e., scores, loadings, partial communalities, etc.). At the same time, we are aware that the introduction of such an approach requires justification and demonstration that it is at least as useful as the more traditional dPCA. Thus, this chapter explores these ideas by using over 100 samples from a peat sequence retrieved from Store Mosse bog (Sweden), which has been previously extensively studied [11–13], including a recent investigation by dPCA on the MIR peat properties [19]. Here we aim to: (i) systematize the presentation and description of MIR peat data, (ii) apply tPCA and identify the peat constituents/compounds reflected by the scores' spectra, (iii) describe the depth records of the tPCA loadings of the samples, (iv) compare the results with those of the dPCA previously done, and (v) determine the information value of the tPCA components by correlating them to elemental (C, N, and C/N) and isotopic composition (δ^{13}C and δ^{15}N) of the peat and by developing principal component regression models of these peat properties using the tPCA components. Supported by this information, we argue that tPCA is an important chemometric tool that enables identification of both major and minor peat constituents, which allows advancing knowledge about peat chemistry and the processes governing it.

2 Material and Methods

2.1 Store Mosse Info: Location, Sampling, Peat Preparation

Store Mosse (the "Great Bog") is one of the most extensive bog complexes in southern Sweden, covering nearly 77 km^2 (Fig. 1). A protected national park since 1982, this site's rich scientific history reaches back over one hundred years. The stratigraphy has shown to be widely consistent across the southern portion of the bog complex with the deepest sections reaching nearly 7 m or 10 ka. The area was deglaciated ~13,900 years ago and a large glacial ice lake was left in its wake [18]. With time this system was tipped southwards by isostatic rebound and the lake was drained leaving a flat shallow basin, ideal for peat accumulation [11–13]. Coring was conducted in November 2008, using two bore holes to retrieve eight overlapping peat cores (100 × 7.5 cm), to a total depth of 568 cm [11]. The peat cores were sub-sampled, freeze dried, and analyzed for bulk density on cm resolution (n = 568). The bulk density results were used to align the cores and construct a composite sequence ($SM\text{-}S_{2008}$), for further details the reader is referred to Kylander et al. [11]. Macrofossil work has identified five main developmental stages. This starts with the lake and fen phases at the base followed by a bog phase which is characterized by a dry *Sphagna* stage, an oligotrophic reversal and a wet *Sphagna* stage [25]. The dry *Sphagna* stage, in particular, is of interest as it overlaps with a high peat accumulation event (HPAE) where rates increase five-fold over background values over just a short period of time [11–13, 19].

2.2 Elemental and Isotopic Data

For stable isotope analysis, samples were combusted with a Carlo Erba NC2500 analyzer connected via a split interface to reduce the gas volume to a Thermo Delta V advantage mass spectrometer. From these measurements the reproducibility was calculated to be better than 0.15‰ for $\delta^{13}C$ and $\delta^{15}N$. Carbon and nitrogen values were determined simultaneously when measuring the isotope ratios. The relative error was <1% for both measurements.

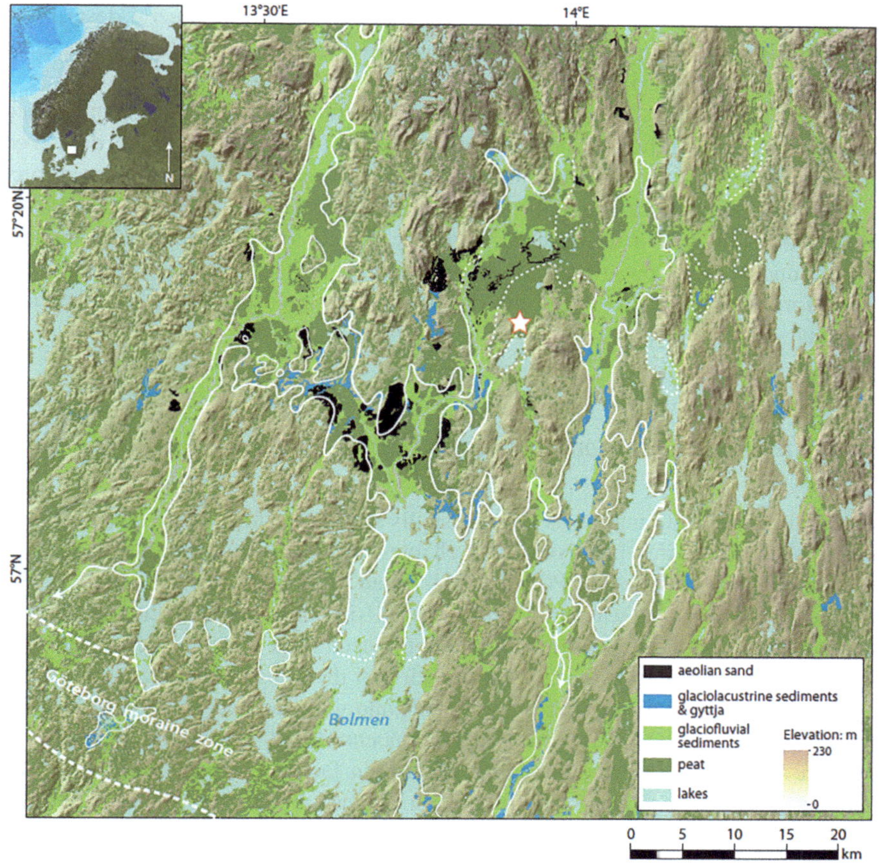

Fig. 1 Store Mosse bog is located in south central Sweden in a peat rich area. When the area was deglaciated it was covered by a large glacial ice lake (extent shown in white, with dashed lines being more uncertain), which provided an optimal habitat for peat accumulation. The core was taken from the southern end of the bog complex of Store Mosse (star) (© Geological Survey of Sweden)

2.3 *FTIR-ATR Analysis*

Finely milled bulk peat samples (105) from the SM-S_{2008} sequence were analyzed by Fourier-transform infrared spectroscopy in attenuated total reflectance mode (FTIR-ATR), using a Gladi-ATR spectrometer (PIKE Technologies, Inc., USA), at the IR-Raman facility of the RIAIDT of the *Universidade de Santiago de Compostela*, Spain. Spectra were acquired in the mid-infrared region ranging from 4000 to 400 cm^{-1} at a resolution of 4 cm^{-1} with an average of 100 scans per sample. During the analytical process, the equipment was thoroughly cleaned, and a new background was collected between samples. To avoid bias in the spectroscopic signal due to scattering, reflection, temperature, concentration, or instrument anomalies, the spectra were baseline corrected using the R package {*baseline*} [17]. For a general description of the results the average, standard deviation, and second derivative spectra were calculated using the R package {*andurinha*} [1].

For a systematic description and interpretation of the results, we also synthesized the MIR spectrum of the peat samples into eight significant regions (as it is described in the result and discussion section below), and these are the regions that are represented in the figures concerning the tPCA results (from 3700 to 2700 cm^{-1}, and from 2300 to 700 cm^{-1}). For the sake of simplicity, we avoided to include regions with background or very low overall absorbance that do not represent relevant absorption information associated with the chemical structures of peat organic chemical contents.

2.4 Numerical Methods: PCA on Whole Spectra

Principal components analysis (PCA) is usually applied to MIR spectral data in direct matrices (dPCA). The objective of this chapter, as indicated in the introduction, is to explore the application of PCA on transposed MIR data matrices (tPCA), with samples in columns and wavenumbers in rows. Using this approach, the whole spectrum of every sample is considered, without the need for a previous identification of the significant wavenumbers and quantification of their absorbance/transmittance.

The tPCA was applied in correlation mode and with varimax rotation. It is worth noting that a non-rotated solution of tPCA concentrates almost all the spectral MIR variance in the first PCA component, which is in fact the average spectrum of all samples. Even when using a varimax rotation, and as we describe below, the first two components tend to account for a large proportion of the variance, so we recommend fixing the number of components to be extracted to 10 or more to be able to further explore the spectra of the peat samples in depth—i.e., going beyond the obvious.

In the tPCA approach, scores are assigned to the wavenumbers and loadings are assigned to the samples. For any component, the variation of score values along the MIR region (4000–400 cm^{-1}) can be represented as a sort of scores' spectrum. This spectrum will contain positive and negative values, indicative of vibrations with absorbances above average and absorbances below average. In this way, vibrations belonging to a peat constituent or group of constituents that covary between samples will end up in the same component spectrum. Meaning that the interpretation of the components is based on the spectrum of scores instead of the covariation between selected peaks.

The loading of a sample in a component is a measure of the weight of that component in the sample MIR spectral variance, and its square value (i.e., partial communality) indicates the proportion of the sample's variance allocated to that component. Variations in loadings and/or partial communalities can be taken as a proxy for changes in the relative abundance of the peat constituent represented by the principal component. In peat sequences, every tPCA component can be depicted as a depth (or time) record of loadings, which enables to get insights into the changes of relative abundance of the peat constituent(s) reflected by each component.

It is important to note that tPCA explores the spectral variability of the samples, decomposing the sample spectrum into components and assigning each sample a loading value that corresponds to the proportion of the spectral variance of the sample of a given constituent or group of constituents as indicated by the scores' spectrum.

3 Results and Discussion

3.1 Elemental and Isotopic Data

Carbon, nitrogen, C/N, $\delta^{13}C$ and $\delta^{15}N$ ($n = 105$) content for the SM-S_{2008} sequence has previously been reported [13], we therefore only briefly describe these results here (Fig. 2). Carbon concentration ranges from 45 to 58% (average 49 ± 6%, all standard deviations here reported as 2σ) while nitrogen ranges from 0.3 to 2% (average: 0.8 ± 0.7). Overall higher values are observed in the fen (C average: 53 ± 5; N average: 1.2 ± 0.7, n = 25) compared to the bog (C average: 47 ± 4%; N average: 0.7 ± 0.5%, n = 81). During the HPAE somewhat lower values are observed between 438–402 cm compared the upper part (397–308 cm). The C/N ratio ranges from 28 to 158 (average: 77 ± 57), displaying a generally increasing trend from the lowermost part, towards the fen bog transition. Following the transition to ombrotrophic conditions, the C/N ratio increases sharply in the HPAE, reaching maxima values at 397 cm (158), followed by a decreasing trend. After HPAE, the C/N ratio is less variable, but lower values are recorded at 274 (36), 215 (47), 156 (48) and 134 (52 cm), while high values occur at 287, (76), 256 (59), 195 (92), 176 (95) and between 127 and 176 cm (average: 90 ± 21, n = 13).

The $\delta^{13}C$ values range from −28.8 to −21.9‰ (average: 26.1 ± 2.5‰). Generally, more depleted, and less variable, $\delta^{13}C$ values are observed in the fen (−27.2 ± 1.3, n = 25) compared to the bog (−25.7 ± 2.4). During the HPAE less depleted values are recorded in the lowermost part of the section, compared to the upper part. Following the HPAE, overall greater variability is observed, with peaks noted at 278 (−21.9‰), 253 (−22.1‰), and 185 cm (23.5‰). More depleted values occur at 185 (−23.8‰) and 61 cm (−23.5‰). The $\delta^{15}N$ values range from −4.1 to 1.3‰ (average: 1.7 ± 2.5) with higher values observed in the fen (average: −0 ± 1.5‰, n = 25) compared to the bog (average: −2.3 ± 1.7‰, n = 81). During the HPAE more depleted values are noted in the lowermost part of the section, followed by a variable, but generally

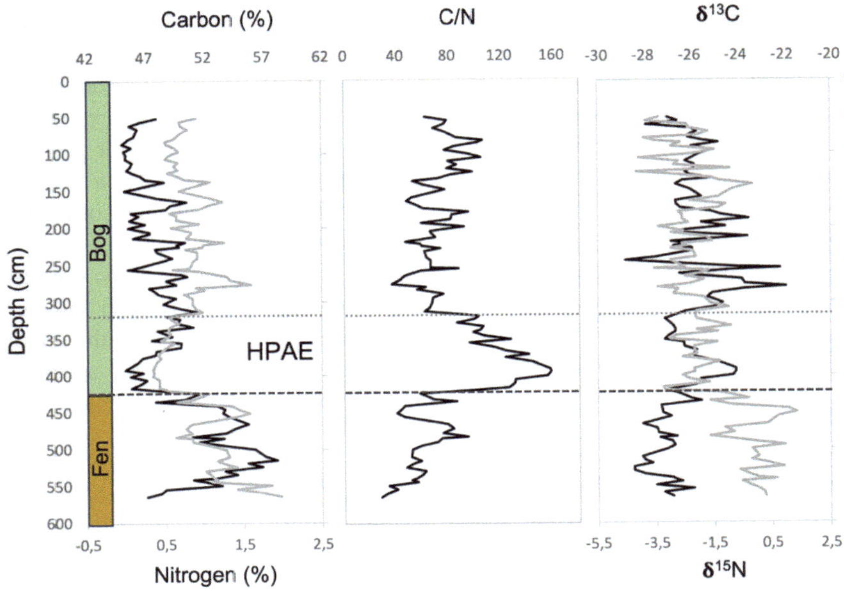

Fig. 2 Depth records of C (black line), N (grey line), C/N, $\delta^{13}C$ (black line) and $\delta^{15}N$ (grey line) in the SM-S_{2008} peat sequence

increasing trend. Between 292 and 256 cm the values stay around the bog average, followed by a period of less depleted values between 171 and 131 (>2.3‰). Between 127 and 51 cm the $\delta^{15}N$ oscillated between high and low, with lower values observed at 114 (−4.2‰), 106 (4.1‰), 91–79 (average: −3.7‰, n = 4) and 66–51 (−3.0‰, n = 4). Less depleted values are recorded at 121 (−0.9‰), 96 (1.5‰) and around 96 cm (1.5‰).

3.2 *MIR Spectra: Average, Standard Deviation and Second Derivative Spectra*

To better describe the MIR variation, the samples' absorbance spectra, together with the standard deviation spectrum and the second derivative spectrum are represented in Fig. 3. We opted for dividing the mid-infrared range into eight distinctive regions representing the main vibrations of the functional groups and their relative variations in absorbance (see Larkin [16], and Pavia et al. [22], for a similar approach to the MIR spectrum), in an attempt to synthesize the spectrum of peat material. Some of the characteristic peaks of each region can be found in Table 1 and those found in the SM-S_{2008} sequence are found in Table 2. The range of wavenumbers, bond vibrations and associated peat constituents of the regions are the following:

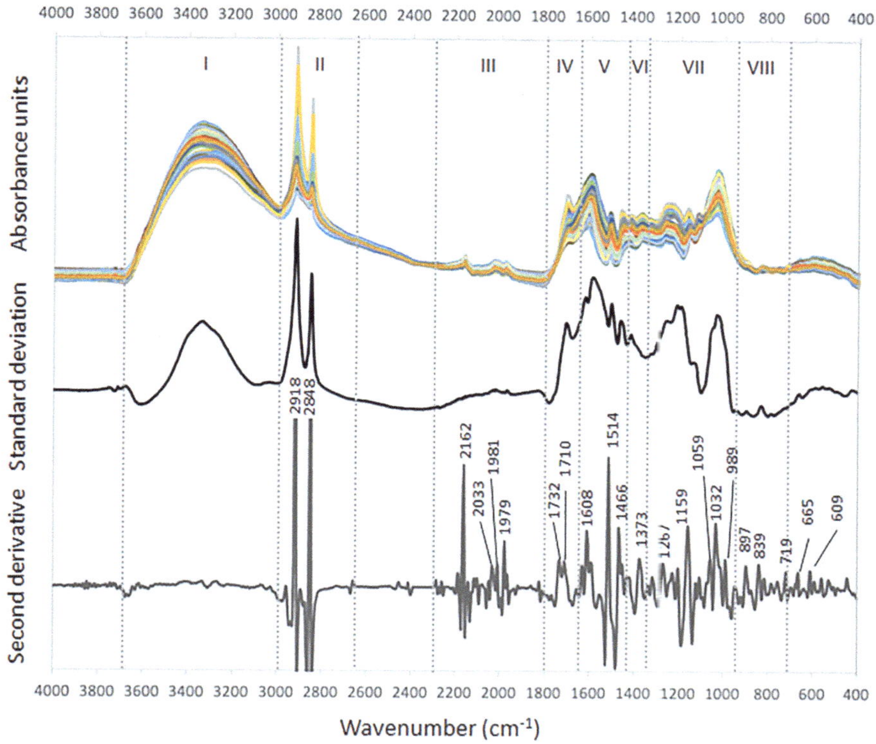

Fig. 3 FTIR spectra, standard deviation spectrum, and second derivative spectrum (reversed values) of the SM-S_{2008} peat sequence. This composed figure provides an easy way to summarize most of the MIR spectral variability

Table 1 Regions of the peat MIR spectrum with characteristic bonds and corresponding vibrations in peat samples (assignments of the characteristic bands are provided in Table 2)

IR region	Type of vibration	Characteristic IR bands
I (3700–3000)	Hydroxyl (O–H) and N–H	3300
II (3000–2650)	C–H stretching	2918, 2848
III (2300–1800)	C≡N, C≡C, and X=C=Y	2162, 2033, 1981, 1979
IV (1800–1650)	Carbonyl (C=O) group	1732, 1710
V (1650–1450)	Aromatic C=C, skeletal vibrations, and Amide I and II	1628, 1514, 1466
VI (1450–1350)	C–H and O–H deformations	1373
VII (1350–950)	C–O stretching	1267, 1159, 1059, 1032, 989
VIII (950–700)	C–H out of plane vibration	897, 839, 719

Table 2 The characteristic IR bands of the studied peat samples and their assignments according to references in the literature (1. Chapman et al. [4]; 2. Heller et al. [8]; 3. Cocozza et al. [6]; 4. Artz et al. [2]; 5. Stuart [26]; 6. Martínez Cortizas et al. [19])

Bands (cm^{-1})	Assignment
3300	O–H stretching, hydrogen bonds in cellulose, N–H bond vibrations (1, 2)
2918	Asymmetric stretching of aliphatic C–H, in fats, wax, lipids (1, 3, 4)
2848	Asymmetric stretching of aliphatic C–H, in fats, wax, lipids (13, 4)
2162	Aliphatic isonitrile –N≡C stretching (5, 6)
2033	Aliphatic isonitrile –N≡C stretching (5, 6)
1981	Overtone and combination bands (5, 6)
1979	Overtone and combination bands (5, 6)
1732	C=O stretching of carbonyl functions, aldehydes, ketones and carboxyl groups (3)
1710	C=O stretch of COOH in free organic acids (4)
1608	Aromatic C=C stretching and/or asymmetric C–O stretch in COO– (3, 4)
1514	Aromatic skeletal vibrations, to conjugated C=N systems and amino functionalities, lignin or phenolic backbone (3, 4)
1466	(amide-II) Aromatic skeletal vibrations, to conjugated C=N systems and amino functionalities, lignin or phenolic backbone (5)
1373	C–H deformations of phenolic and aliphatic groups (3, 4)
1267	C–O stretching of ethers and/or carboxyl groups, indicative of lignin backbone (3, 4)
1159	C–O stretching of polysaccharide structures (6)
1059	C–O stretching of polysaccharide structures (1, 3)
1032	C–O stretching of polysaccharide structures (3)
989	C–O stretching of polysaccharide structures (6)
897	O–H, carbohydrates (5)
839	Aromatic CH out of plane (4)
719	CH_2 wagging (4)

- Region I (3700–3000 cm^{-1}): vibrations corresponding to O–H, from alcohols, phenols, carboxylic acids, and polysaccharides, N–H vibrations from primary and secondary amides, and even C-H from alkynes. In SM-S_{2008} peat samples it is characterized by one broad absorption peak near 3300 cm^{-1} that is mainly related to the hydroxyl (O–H) and N–H bonds vibration (Fig. 3, Table 2). These molecular vibrations in peat material are usually associated with hydrogen bonds in cellulose compounds [4] and the presence of nitrogenated compounds.
- Region II (3000–2700 cm^{-1}): C-H stretching vibrations of aromatics, alkanes, alkenes and aldehydes. In SM-S_{2008} peat it is mainly characterized by two well-defined peaks at around 2919 and 2850 cm^{-1}, which are related to C-H vibrations in aliphatic structures (Fig. 3, Table 2). The aliphatic signature can be associated

with various organic components such as fats, wax, and lipids that derive from the original vegetation [4, 6].

- Region III (2300–1800 cm^{-1}): triple bond vibrations of alkynes and nitriles, but also multiple double bonds in alkenes, ketones, isocyanates and isothiocyanates [22, 26]. S–H stretching of mercaptans also occurs in this region even if it is not commonly found in peat infrared data. In contrast to the other spectral regions, this one shows very low absorbances in SM-S_{2008} peat that can be assigned to the C≡N and C≡C bonds at bands around 2162, 2033, 1981 and 1979 cm^{-1}, but also to the presence of multiple double bonds (X=C=Y). The standard deviation spectrum of the peat samples did not suggest much variation in this region along the peat core (Fig. 3).
- Region IV (1800–1650 cm^{-1}): related to the carbonyl bond (C=O) vibrations of anhydrides, aldehydes, ketones, carboxylic acids, imines, oximes and even alkenes (C=C). In SM-S_{2008} peat, the second derivative spectrum locates two peaks around 1732 and 1710 cm^{-1} (Fig. 3, Table 2). Both can be related to the polysaccharide contents (hemicellulose), but also to carboxylate and carboxylic groups that are formed upon peat oxidation reactions [2, 4].
- Region V (1650–1450 cm^{-1}): vibrations of N–H in primary and secondary amides, C=C in aromatics, C–H bending, and N=O in nitro compounds. In SM-S_{2008} peat, these vibrations mainly correspond to chemical compounds with aromatic structures like lignin and other phenolic compounds [6, 26]. Note that the amide I and II vibrations of nitrogenated compounds also occur in this region (Table 2).
- Region VI (1450–1350 cm^{-1}). Peaks in this region are related to a variety of vibrations that can be due to bond vibrations in different chemical structures and functional groups. For example, in SM-S_{2008} peat the peak near 1373 cm^{-1} can be related to O–H as well as C–H deformations in phenolic and aliphatic groups [2, 6].
- Region VII (1350–950 cm^{-1}). It is mostly characterized by the C-O vibrations. In SM-S_{2008} peat spectra, the peak at 1032 cm^{-1} resulting from the combination of C–O stretching and O–H deformation in polysaccharide compounds [6] tends to be the dominant one. Nevertheless, in this region certain bands (i.e., 1315, 1267 and 1159 cm^{-1}) are also related to bond vibrations in (G and S)-lignin structures [6, 19].
- Region VIII (950–700 cm^{-1}). It is mostly characterized by bands associated with C–H out-of-plane vibrations in lignin, with also O–H stretching in polysaccharides [2, 19]. In SM-S_{2008} peat samples, it is one of the regions showing very low absorbances.

The standard deviation spectrum (Fig. 3), an indication of the relative variability in peat organic matter composition, is similar to that of the peat spectra: it shows high variability (i.e., compositional differences between samples) in regions II and IV–VII. In regions IV–VII maximum variability occurs at 1600–1500 cm^{-1} and minimum at 1200–1100 cm^{-1}. Moderate to high variability is also found in region I (at about 3300 cm^{-1}). Although in general most of the molecular vibrations are located within the so-called fingerprint region (1800–700 cm^{-1}, regions IV–VIII), the standard deviation spectrum demonstrates that other regions are also of importance for a full description of the MIR spectral properties of peat materials.

3.3 PCA on Transposed MIR Data Matrix (tPCA)

3.3.1 Identification of Components

The first two components usually account for almost all the spectral variance when conducting PCA on transposed MIR data matrices of relatively homogeneous materials, such as peat. This is also the case for the SM-S_{2008} sequence, with Cp1 and Cp2 accounting for more than 97% of the MIR spectral variance of the peat samples. In most instances, these two components are the only ones considered as they represent the main, and obvious, constituents of the peat—as we describe below. But focusing only on Cp1 and Cp2 fails to consider other components and changes in peat composition that may also be relevant for understanding peat chemistry ("going beyond the obvious"). The latter is the *leitmotiv* of this chapter and the reason we consider here up to the 11th component, knowing beforehand that these higher order components (Cp3 to Cp11) only represent up to 3% of the total MIR spectral variance. Even a minor component of the MIR spectral variance can be important if it can unveil crucial changes in peat formation and degradation. For example, microbial biomass will only make up a minor part of the peat mass but may have significant effects on peat mineralization. The eleven components scores' spectra are illustrated in Fig. 4, together with the average spectrum of the peat core, as it enables to evaluate the overall absorbance of the spectral regions of the whole MIR spectrum extracted by each component.

A detailed identification of the main regions, characteristic vibrations (with positive and negative scores), and corresponding OM compounds can be found in Table 5. The scores' spectra of the principal components can be summarized as follows:

- Cp1: elevated scores' values in regions I (3320 cm^{-1}) and VII (1072, 1057, 1034, 1011, 989 cm^{-1}) that are mainly assigned to vibrations in polysaccharides structures.
- Cp2: high values in regions II (2918 and 2850 cm^{-1}), IV (1734, 1710 cm^{-1}), V (1628, 1608, 1593, 1577, 1552 cm^{-1}) and part of region VII (1267 and 1223 cm^{-1}). These peaks correspond to bond vibrations in aliphatic (i.e., alkanes, alkenes), carbonyl (i.e., acids, esters, aldehydes, and ketones) and aromatic, compounds.

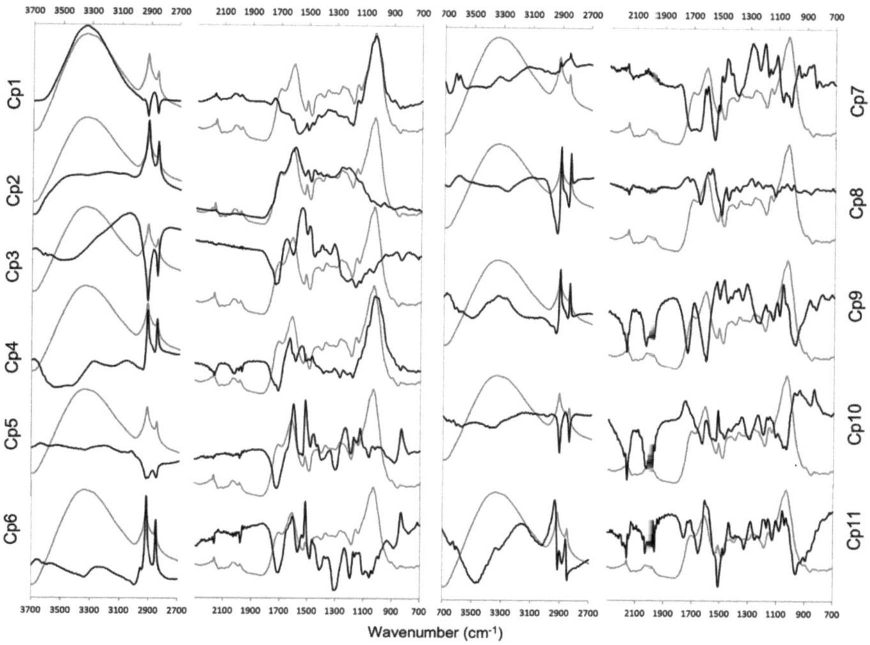

Fig. 4 Scores' spectra of the extracted principal components (black lines). Gray lines correspond to the average spectrum of all peat samples

They mainly represent spectroscopic characteristics of functional groups in lignin and organic acids.

- Cp3: high positive values in regions V (1540 cm^{-1}) and VI (1400, 1315 cm^{-1}), where the absorptions of aromatics and nitrogenated compounds occur. Large negative scores are shown in regions II (2918, 2850 cm^{-1}), IV (1730 cm^{-1}) and VII (1260, 1220, 1159, 1124, 1105, 1034 cm^{-1}) corresponding to aliphatic and carbonyl-rich compounds; possibly associated with organic acids.
- Cp4: high scores in regions II (2918, 2850 cm^{-1}), V (1628, 1552 cm^{-1}) and VII (1107, 1072, 1034, 1011, 989 cm^{-1}), indicating an OM enriched in aliphatic, N-rich compounds (i.e., proteins) and polysaccharides.
- Cp5: large positive peaks in regions V (1593, 1510, 1452 cm^{-1}) and VII (1226, 1169, 1122 cm^{-1}), associated to enrichments in N-compounds, lignin and polysaccharides, and large negative peaks in regions II (2918, 2850 cm^{-1}), IV (1714 cm^{-1}), VII (1296 cm^{-1}) and VIII (874 cm^{-1}), associated to enrichments in aliphatics and carboxylates.
- Cp6: large positive scores in regions I (3643 cm^{-1}), II (2918, 2850 cm^{-1}), V (1610, 1514 cm^{-1}) and VIII (719 cm^{-1}), corresponding to clay minerals, aliphatics and aromatics; and large negative loadings in regions IV (1705 cm^{-1}), V (1480 cm^{-1}), VI (1410 cm^{-1}) and VII (1300, 1194, 1099, 1055, 1030 cm^{-1}), of carboxylated OM, N-compounds, lignin and polysaccharides.

- Cp7: large positive peaks in regions II (2918, 2850 cm^{-1}), V (1460 cm^{-1}), and VII (1280, 1255, 1194, 1109 cm^{-1}), reflecting enrichment in aliphatics, lignin, and polysaccharides, and large negative peaks in regions IV (1734–1655 cm^{-1}), V (1545 cm^{-1}), VI (1514 cm^{-1}), and VII (1072, 1011 cm^{-1}) due mostly to carboxylated compounds.
- Cp8: only shows significant positive peaks in region II (2918, 2850 cm^{-1}), IV (1766, 1750, 1710 cm^{-1}) and V (1620, 1605, 1570 cm^{-1}), due to aliphatics and carboxylated OM, and negative peaks in region IV (1655 cm^{-1}) and V (1525, 1504, 1452 cm^{-1}) due to other aliphatics (possibly alkenes) and lignin.
- Cp9 to Cp11: the spectra of these components have in common the presence of large peaks, both positive and negative, in region IV (2200–1900 cm^{-1}) which correspond to absorbances in compounds with alkyne structures as well as nitro functional groups with multiple double bonds. In Cp9 they anti-covary with aliphatic compounds (vibrations in region II), while in Cp10 and Cp11 they covary with aliphatics (Fig. 4).

The Cp1 spectrum fits rather well with the spectrum of holocellulose extracted from wood, while the Cp2 spectrum overlaps only partially with the spectrum of Klason lignin (Fig. 8), reinforcing the idea that it is a combined signal of the less degradable OM compounds. Thus, components Cp1 and Cp2 account, respectively, for changes in the labile (cellulose and hemicellulose) and recalcitrant (lignin and aliphatic) compounds of the SM-S_{2008} peat OM.

In the higher order components (Cp3–Cp11), large positive and negative values in the samples' scores spectra are an indication of exclusion between compounds, i.e., samples with positive loadings are enriched in compounds with positive scores, and samples with negative loadings are enriched in compounds with negative scores (as shown below). Cp3 accounts for the opposed enrichment in aromatics and N-compounds versus aliphatic, carboxylated OM and polysaccharides. Cp4 reflects opposed enrichments in N-compounds (peaks in amide I–III), and polysaccharides and, interesting enough, its spectrum is quite similar to that of microbial biomass [10, 23]. Cp5 reflects changes between enrichments in N-compounds, lignin, and polysaccharides versus enrichments in aliphatics and carboxylates. The Cp6 spectrum is dominated by the opposed enrichment in aliphatics and aromatics versus carboxylates, N-compounds and polysaccharides; it is also one of the few components showing vibrations compatible with inorganics (clay minerals; see Table 5). The Cp7 MIR signal is dominated by lignin vibrations versus carboxylates. The spectrum of this component fits well with that of the Klason lignin for regions V, VI and VII (except for the polysaccharides maximum in Klason lignin at ~1000 cm^{-1}), which suggests it represents a secondary variation in peat lignin content (Fig. 8). Cp8 reflects opposed enrichments in different aliphatics, possibly between alkanes (positive values) and alkenes (negative values). As already mentioned, components C9 to Cp11 are dominated by the signal related to alkynes (and compounds with multiple double bonds) reflecting their changes in peat.

These data indicate that the spectral signal of the SM-S_{2008} peat samples is dominated by vibrations related to organic components. The presence of inorganics is only suggested by peaks in region I (at 3700–3600 cm^{-1}) in components Cp6 to Cp9 and in region VII (doublet at 798–777 cm^{-1}), indicating the possible presence of clay minerals and quartz [11–13, 15, 20]. The fact that they are captured in very high order components points to very low amounts of inorganics, in line with the low ash content of the peat [15, 20].

Another recurrent feature is the large score values (negative and positive) of the aliphatic vibrations (2918, 2850 cm^{-1}, related to alkanes and alkenes) in all components (Fig. 4). This suggests that the distribution of aliphatics' is complex, possibly resulting from interaction between peat decomposition and changes in OM sources (i.e., changes in peat forming vegetation) [9]. The novelty of the tPCA approach here is that carboxylates (1700–1600 cm^{-1}) also show a complex distribution, something that went undetected in the previous dPCA study [19], and is related to the relatively large number of peat compounds (anhydrides, esters, aldehydes, ketones, carboxylic acids) with carbonyl groups. In the previous dPCA analysis, a few specific bands were selected to represent these compounds while in the tPCA the whole region (IV) is taken into consideration. The new tPCA analysis also enabled the identification of other important MIR signals, as those of microbial biomass (Cp4) and specific lignin variations (Cp7), by comparison with reference spectra. The tPCA provides, to some extent, an easier way to identify compounds that are present in lower amounts in the peat, since a whole spectrum lends more support to interpretation than single vibrations or even fractions of the variance of single vibrations, as it occurs in high order components in dPCA. Another advantage of the tPCA scores' spectra is that they can be compared to reference spectra, obtained from literature or MIR reference databases [20], or validated with alternative techniques. For example, in a core from South Africa [24] found a MIR signal in the ninth tPCA component compatible with charcoal that was corroborated by pyrolysis GC–MS analyses. While it is true that compounds reflected by high order components may represent small proportions of the peat mass, this does not mean they are not important. As in a cooking recipe, there are main ingredients but also small amounts of secret ingredients which are key to the flavor.

3.3.2 Depth Records of the Components

As indicated in the methods section, tPCA produces loadings for the samples that provide an estimation of the weight (i.e., proportion of variance) of each component (i.e., identified peat constituents) on the spectral MIR signal of the peat OM. The depth distribution of the loadings is a good way to evaluate the variability of the constituents throughout the peat sequence (Fig. 5).

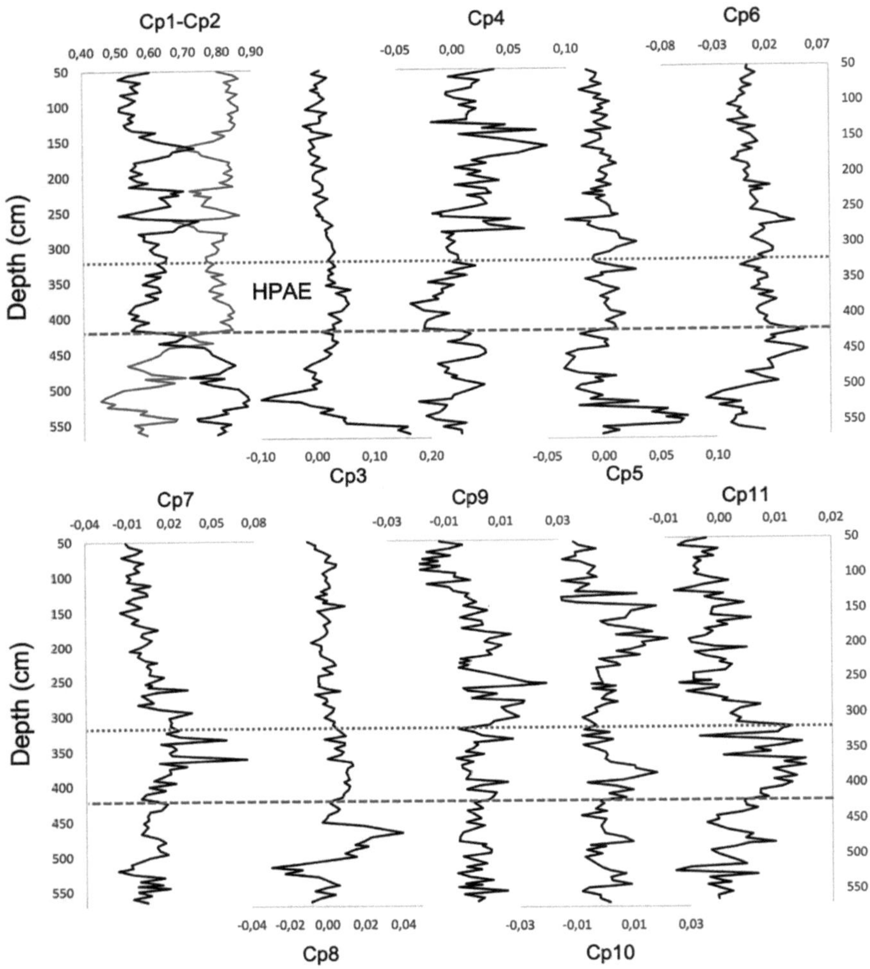

Fig. 5 Depth records of the loadings of the tPCA components. Cp1–Cp11 depict relative changes in peat OM constituents throughout SM-S_{2008} sequence

Cp1 and Cp2 loadings' records are anti-correlated, with Cp2 displaying high values in the fen peat, abruptly decreasing during the fen-bog transition, followed by low and variable values. Most of the variation of the Cp3 record occurs in the fen peat (Fig. 5), with large positive values at the base of the core that rapidly decrease to sequence minima in the middle of the fen section, followed by recovering values. In the bog peat Cp3 shows much less variability, with a decreasing trend from the HPAE section towards the surface. Cp5 and Cp6 also show larger variations in the fen peat, with positive (Cp5) and negative (Cp6) values in the lower half of the section. In contrast, Cp4, Cp9, and Cp10 show much higher variability in the bog peat, after the HPAE, with large positive excursions at 300–250 cm and 175–100 cm in Cp4,

300–200 cm in Cp9, and 250–100 cm in Cp10. Cp7 and Cp11 show the largest values in the HPAE section.

In a previous FTIR-ATR study of the SM-S_{2008} sequence, in which we applied the dPCA approach [19], we identified the first five principal components as representing long-term peat decomposition (dCp1), abundance of compounds with triple bonds and multiple double bonds (dCp2), lignin and organic acids versus N-compounds (dCp3), mineral components (dCp4), and secondary variation of aliphatic compounds (dCp5). Cp1 and Cp2 of the tPCA are highly correlated to dCp1 (r Pearson correlation coefficient, −0.96 and 0.95 respectively); Cp3, is highly correlated (r −0.88) to dCp3, the negative sign of the correlation due to vibrations of N-compounds showing a negative loading in dPCA; Cp4, is moderately correlated (r 0.66) to dCp4, Cp5 and Cp6 are highly correlated to dCp5 (r 0.76 and −0.77, respectively), and Cp9 is moderately correlated (r −0.61) to dCp2 (Table 3).

These results show that both statistical approaches will identify the same main peat constituents but tPCA, as applied here, will also allow identification of minor occurrences. Thus, tPCA effectively decomposes the signal into separate spectra representing OM compounds involved in peat compositional change. But there are also some differences such as the order of the components. In dPCA alkynes are extracted as the second component while in the tPCA they appear in high order components (Cp9–Cp11). This can be interpreted as dPCA giving more importance to alkynes in terms of peat composition. In most investigations using dPCA only the first two components are considered as they account for the largest proportion of total variance. This is because the number of highly covarying variables drives the dPCA extraction. In dPCA, peaks have the same weight/importance (the variance of each one is set to 1) and extraction of components is driven by covariation between MIR vibrations. To some extent, the weight of a component is conditioned by the

Table 3 Correlation (Pearson correlation coefficients) between the depth records of extracted components in tPCA (tCp1–tCp11) and in dPCA (dCp1–dCp5; data obtained from Martínez Cortizas et al. [19])

	dCp1	dCp2	dCp3	dCp4	dCp5
tCp1	−0.96	0.15	−0.15	−0.03	0.08
tCp2	0.95	−0.17	0.12	0.02	−0.13
tCp3	0.16	0.25	−0.86	−0.03	0.03
tCp4	−0.12	−0.16	0.25	0.66	−0.30
tCp5	0.13	−0.18	−0.21	−0.29	0.76
tCp6	−0.07	−0.10	−0.31	−0.17	−0.77
tCp7	0.10	0.19	−0.31	−0.33	−0.18
tCp8	0.00	0.06	−0.19	−0.41	−0.42
tCp9	−0.05	−0.61	−0.23	−0.14	0.08
tCp10	−0.03	−0.48	−0.18	−0.02	0.00
tCp11	0.03	−0.09	−0.35	−0.46	−0.17

pre-selection of significant vibrations, and including several vibrations related to a specific peat compound may "artificially" produce a higher weight for that compound. In contrast, in tPCA the extraction is driven by covariation between samples (i.e., the whole spectrum is considered). Thus, tPCA better reflects the relative abundance of the OM compounds in peat. In addition, the spectrum-based interpretation of the tPCA has enabled the identification of a component reflecting microbial biomass and another component reflecting secondary variations in lignin, which are more difficult to assign in dPCA as they may appear as residual variance of specific vibrations and are interpretation-dependent.

3.4 Correlation and Regression Between MIR Data, Peat OM Elemental Composition, and Peat Isotopic Composition

The agreement between the results obtained with tPCA and dPCA lends support to the use of whole spectra in transposed matrices as an efficient way to explore changes in peat OM composition. An additional line of evidence can be provided by also determining if the extracted components are relevant for understanding other peat properties, i.e., if they are correlated or are of use to model their changes with depth/age; a practice we highly recommend. We approached this by correlating and regressing (by principal components linear regression, PCR) elemental composition (total C and N, C/N) and isotopic composition (δ^{13}C and δ^{15}N) of the peat using the loadings of the tPCA extracted components (Tables 4 and 6).

Table 4 Correlation (Pearson correlation coefficients) between the depth records of extracted components in tPCA and depth records of the elemental (C, N, C/N), and isotopic composition (δ^{13}C and δ^{15}N) of SM-S_{2008} peat

	C	N	C/N	δ^{13}C	δ^{15}N
Cp1	−0.90	−0.71	0.53	0.59	−0.76
Cp2	0.91	0.72	−0.55	−0.60	0.77
Cp3	−0.23	0.19	0.02	0.12	0.15
Cp4	−0.03	0.35	−0.49	−0.06	−0.04
Cp5	0.00	0.03	0.02	0.09	0.09
Cp6	0.05	0.05	0.04	0.11	0.06
Cp7	0.13	−0.24	0.32	−0.02	0.00
Cp8	0.04	−0.33	0.39	−0.01	0.04
Cp9	0.06	0.10	−0.08	0.35	0.04
Cp10	−0.04	−0.09	0.14	0.15	0.04
Cp11	0.06	−0.35	0.53	−0.06	−0.01

Cp1 and Cp2, interpreted previously as reflecting opposed changes in labile OM (i.e., polysaccharides) and recalcitrant OM (i.e., aliphatics, lignin) due to peat decomposition, are highly correlated (positive or negative) to C, N, and $\delta^{15}N$, and moderately correlated to C/N and $\delta^{13}C$ (Table 4). This indicates that the main MIR signals (>97% of the MIR spectral variation in the samples), reflected by Cp1 and Cp2 anti-covariation, are truly connected to C and N variability and isotopic changes in peat. Enrichment in lignin and aliphatics and depletion of polysaccharides result in increasing C, and N concentrations together with an enrichment in ^{15}N in the peat. Different peat compounds (e.g., polysaccharides and lignin) have different susceptibility to microbial decomposition and typically contain different amounts of C and N (e.g., Hodgkins et al. [9]. Peat decomposition also results in C/N decreases and ^{13}C enrichment, however, since the correlations are only moderate (Table 4), other processes than peat decomposition should be involved in changes in C/N and ^{13}C. A moderate positive correlation was also found between Cp11, i.e., enrichment in alkynes and multiple double bond compounds, and C/N (r 0.53); as well as weak correlations between Cp4 (r 0.35 and −0.49), and N and C/N; Cp7 (r 0.32) and C/N; Cp8 with N (r −0.33) and C/N (r 0.39); Cp9 and $\delta^{13}C$ (0.35); and Cp11 and N (r − 0.35). Examples of the stratigraphical correlation between the records of the highly correlated tPCA loadings and the peat properties are shown in Fig. 6.

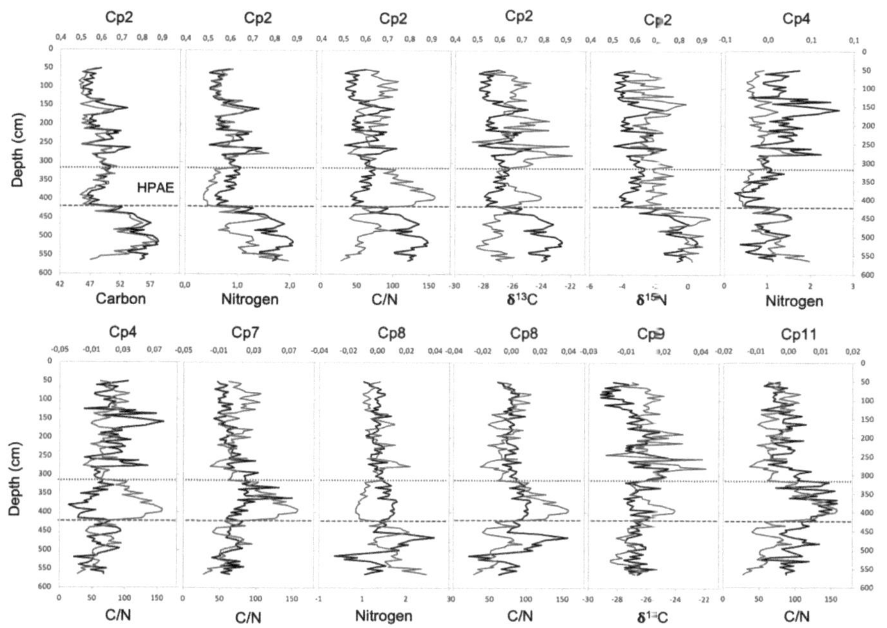

Fig. 6 Stratigraphical correlation between the depth records of the peat properties (elemental and isotopic composition; grey lines) and the depth records of principal components (Cp2, Cp4, Cp7–Cp9, and Cp11; black lines) which are highly correlated to peat properties

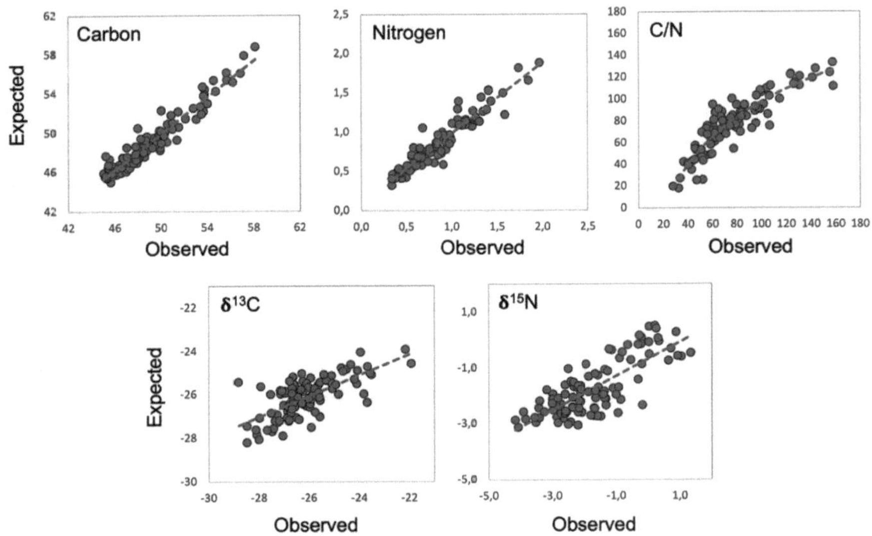

Fig. 7 Fitting between observed and expected values of the PCR models obtained using the loadings of the extracted principal components (tPCA)

The PCR statistics showed that the regression models based on the tPCA loadings values accounted for 48 (for $\delta^{13}C$) to 92% (total C) of the variance of the variables selected to represent the elemental and isotopic composition of the peat (correlation coefficients between 0.69 and 0.96; Table 6). All components were significant in at least one model, except for Cp1, which is redundant with Cp2, and Cp10. Figure 7 shows the fitting between observed and expected values.

Total carbon concentration was associated to four components (Cp2–Cp4, and Cp7; R 0.96), of which Cp2 had the largest positive weight on C concentration, as indicated by the elevated standardized beta coefficient (b-coef 0.911; Table 6), which is three times larger than that of any other predictor. Cp7 had a moderate to low positive weight (b-coef 0.18) and Cp3 and Cp4 showed negative moderate to low weights (b-coef −0.316 and −0.09). Thus, the accumulation of recalcitrant OM compounds due to peat decomposition (Cp2) is the main mechanism controlling carbon content, followed by secondary enrichment in lignin (Cp7).

The regression model for total N included 9 out of the 11 principal components (R 0.90; Table 6), suggesting that more complex mechanisms are involved in nitrogen accumulation compared to C content. Again, Cp2 had the largest positive weight (b-coef 0.72). Cp3 and Cp4 showed moderate positive weights (b-coef 0.34, 0.30), and Cp6 had a low to moderate weight (b-coef 0.18). Negative, moderate to low weights were found for Cp7, Cp8, and Cp11 (b-coef −0.22, −0.22, −0.20). The weights of Cp5 and Cp9 were very low (b-coef 0.01, 0.08). Thus, nitrogen content in

SM-S_{2008} peat is also driven by peat decomposition, with moderate influence from the accumulation of some nitrogenated compounds and microbial biomass.

The regression model of C/N (R 0.88) also resulted in a large number of associated components (Table 6), possibly due to the complexity introduced by N distribution in the SM-S_{2008} peat sequence. The largest negative weight was that of Cp2 (b-coef −0.55), followed by Cp4 (b-coef −0.38), Cp3 (b-coef −0.21) and Cp9 (b-coef −0.14). Cp11 (b-coef 0.37), Cp8 (b-coef 0.12), and Cp7 (b-coef 0.14), had positive weights. Thus, larger values of components reflecting an increase in N-compounds (Cp2, peat decomposition; Cp4, microbial biomass, and Cp3, nitrogenated compounds) result in a decrease in C/N, while larger values of components related to carbon-rich compounds (Cp7–Cp9) are related to higher C/N ratios. It is interesting to note that the fitting between observed and expected values for C/N suggests a non-linear relationship (Fig. 7).

As for the peat isotopic composition, the regression models were quite simple, involving only two variables each, but also of lower fitting ability (Table 6). Nitrogen isotopic composition (R 0.78) was found to depend on the abundance of recalcitrant OM (Cp2 b-coef 0.766), and marginally on nitrogenated compounds (Cp3 b-coef 0.14), the enrichment in ^{15}N increasing with the weight of these two components. While C isotopic composition (R 0.69) was also controlled by peat decomposition (Cp2 b-coef −0.60) and by the anti-covariation between alkynes and aliphatics (Cp9 0.36). These results indicate that ^{13}C is depleted in peat as the accumulation of aliphatics increases. Nevertheless, for both δ^{13}C and δ^{15}N, other processes should be involved as the regression models failed to account for significant proportions of the variance (52 and 39% respectively).

4 Conclusions

Peat is a complex organic material formed under waterlogged conditions at centennial to millennial scales, resulting from the interaction of multiple factors such as succession, vegetation, hydrology, temperature, nutrients deposition, microbial communities, etc. In this study, we showcase how principal components analysis on the transposed MIR spectral matrix (tPCA) allows a good characterization of the molecular composition of a peat sequence from a boreal peatland. Our results indicate that this approach is effective in extracting detailed molecular information from the MIR spectra of bulk peat samples and helps to untangle some of the complex processes driving peat OM change. In fact, single absorption bands can be associated with more than one chemical compound in peat, but the use of the whole MIR spectrum improves data interpretation, as tPCA decomposes the raw spectra into single spectra corresponding to specific compounds. The first two tPCA components (Cp1 and Cp2), which reflect the labile and recalcitrant constituents of the peat, were associated with peat formation and decomposition, whereas the higher order principal components, from Cp3 to Cp11, were associated to OM transformation products, microbial biomass, and peat mineral content. Most important, for each principal component, the tPCA results facilitate the visualization of vertical trends along the peat core. Cp1 and Cp2 accounted for whole-sequence trends, but the higher order components reflected more localized changes in the fen, the high peat accumulation event, and the bog sections.

Furthermore, the tPCA factors were shown to be significantly related to peat elemental composition (C, N, C/N), as well as isotopic composition ($\delta^{13}C$ and $\delta^{15}N$). This fact supports the conclusion that the extracted tPCA components can be used to get detailed understanding of peat material, beyond its´ obvious organic characteristics. We also compared the extracted components to those of isolated compounds (cellulose and Klason lignin), finding large similarities and corroborating the interpretation of tPCA components. In future investigations, we will progress in this line of research by physically and chemically extracting peat components (plant remains and organic compounds), obtaining their MIR spectra, and using these spectra in the tPCA to improve the interpretation of the components, i.e., a kind of "supervised" tPCA extraction.

Acknowledgements This research was developed with funding by projects "Grupos de Referencia Competitiva (ED431C 2021/32, Xunta de Galicia)", Vetenskapsrådet (the Swedish Research Council grant number 2019-03434) and Formas (Research Council for Sustainable Development grant number 2020-01536). MT is funded by Plan Galego I2C Modalidade A /ED481B-2022-017) and OLC is funded by Ramón y Cajal Grant 2020 (RYC2020-030531-I). Authors would like to thank the use of RIAIDT-USC analytical facilities.

Supporting Material

See Tables 5, 6 and Fig. 8.

Table 5 Principal components (+, positive loading, −, negative loading), regions, vibrations, and corresponding compounds included in the principal components extracted with tPCA. Bold wavenumbers and compounds are those of vibrations showing large (positive >1, and negative < −1) scores values

PCA	Region	Vibrations	Compounds
Cp1+	I	**3320**	**p+s (primary and secondary) amides, polysaccharides, carboxylic acids**
	VII	1157, 1107, **1072, 1057, 1034, 1011, 989**	**Polysaccharides**
Cp2+	II	**2918, 2850**	**Alkanes, aldehydes**
	IV	1734, 1710	Aldehydes, ketones, carboxylic acids
	V	**1628, 1608, 1593, 1577, 1552**, 1508, 1466, 1452	**p+s amides, aromatics**, aliphatics
	VI	1417, 1373	Lignin
	VII	1331, 1315, **1267, 1223**, 1124	**Lignin (G+S)**, amines, polyscacharides
Cp3+	II	3040	Alkenes
	V	1655, **1540**, 1485	Alkenes, p+s amides, **aromatics**
	VI	1400, 1315	Lignin
Cp3−	II	**2918, 2850**	**Alkanes, aldehydes**
	IV	**1730**	**Aldehydes, ketones**
	V	1608, 1452	Aromatics
	VII	**1260, 1220, 1159, 1124, 1105, 1034**	**Lignin (G), polysacharides**
	VII	835, 719	Lignin, alkanes
Cp4+	II	**2918, 2850**	**Alkanes, aldehydes**
	IV	1655	Alkenes
	V	**1628**, 1608, **1552**, 1491, 1466	**p+s amides**, aromatics, aliphatics
	VII	1124, **1107, 1072, 1034, 1011, 989**	**Polysaccharides**
	VIII	945, 922, 895	Polysaccharides, aromatics
Cp4−	I	**3520-3420**	**Polysaccharides, p+s amides**
	III	2162, 2027, 1967	Alkynes, nitriles, ketones
	IV	**1766, 1734, 1709**	**Anhydrides, esters, ketones, carboxylic acids**
	VI	1388	N-compounds (nitro)
	VII	1331, 1267, 1223, 1180	Lignin (G), polysaccharides
Cp5+	IV	1655	Alkenes
	V	**1593, 1510, 1452**	**Amides, lignin**
	VI	1420	Alcohols, phenols
	VII	1340, **1226, 1169, 1122**	**Lignin, polysaccharides**

(continued)

Table 5 (continued)

PCA	Region	Vibrations	Compounds
	VIII	**831**	**Lignin**
Cp5−	II	**2918, 2850**	**Alkanes, aldehydes**
	IV	1714	Carboxylic acids, ketones
	VI	1392	N-compounds (nitro)
	VII	**1296**, 1190, 1060	**Carboxylic acids, amines, polysaccharides**
	VIII	945, **874**	Polysaccharides, **aliphatics**
Cp6+	I	**3643**	**Clays**
	II	2956, **2918, 2850**	**Alkanes, aldehydes**
	IV	1655	Alkenes
	V	1628, **1610**, 1593, 1535, **1514**	**p+s amides (N-H), aromatics**
	VIII	835, **719**	Lignin, **alkanes**
Cp6−	IV	**1705**	**Ketones, carboxylic acids**
	V	1570, **1480**	**p+s amides**
	VI	**1410**	**Lignin**
	VII	**1300, 1194, 1099, 1055, 1030**, 985	**Lignin, polysaccharides**
	VIII	895, 874	Polysaccharides, aliphatics
Cp7+	I	3624, 3602, 3332, 3128	Clays, alcohols, phenols, amines, amides, carboxylic acids
	II	**2918, 2850**	**Alkanes, aldehydes**
	V	1500, **1460**	**Lignin, aliphatics**
	VI	1423	Alcohols, phenols
	VII	**1280, 1255, 1194**, 1161, 1130, **1109**	**Lignin, Polysaccharides**
	VIII	960, 926, 885, 858	Polysaccharides, aromatics
Cp7−	I	3660, 3550–3425	Clays, p + s amides
	IV	**1734-1655**	**Aldehydes, amides, alkenes**
	V	**1545**	**Lignin**
	VI	**1514**	**Lignin**
	VII	1373, **1072, 1011**	**Lignin, polysaccharides**
	VIII	835, 787, 754	Aromatics
Cp8+	I	**3600, 3100, 3040**	**Clays, alcohols, phenols, p+s amides, carboxylic acids**

(continued)

Table 5 (continued)

PCA	Region	Vibrations	Compounds
	II	**2918, 2850**	**Alkanes, aldehydes**
	IV	**1766, 1750, 1710**	**Anhydrides, esters, carboxylic acids**
	V	**1620, 1605, 1570**	**p+s amides**
	VII	1288, 1250, 1190, 1080, 1040, 989	lignin, polysaccharides
Cp8−	I	3280	p+s amides, polysaccharides, carboxylic acids
	II	**2960, 2937, 2862**	**Alkanes, aldehydes**
	III	2162, 2033–1967	Alkynes, nitriles, ketones
	IV	**1655**	**Alkenes**
	V	**1525, 1504, 1452**	**Lignin**
	VI1	417, 1373	Lignin
	VII	**1120**, 1093, 1011	Polysaccharides
	VIII	945, **874**	Polysaccharides, **aliphatics**
Cp9+	I	**3683**, 3402, 3332, 3280	**Clays**, p+s amides, polysacchaarides, carboxylic acids
	II	**2916, 2848**	**Alkanes, aldehydes**
	III	**1900-1800**	**Alkynes, nitriles, ketones**
	IV	1687	Amides
	V	**1540, 1525, 1493, 1473**	**Lignin**
	VI	**1431, 1400, 1373**	**Lignin**
	VII	**1315**, 1182, 1099, **1057**, 1034	**Lignin, polysaccharides**
	VIII	874, 839, **791, 777, 714**	**Aromatics, minerals (quartz)**
Cp9−	I	3500	Alcohols, phenols
	II	**2974, 2940**	**Alkanes, aldehydes**
	III	**2183, 2162, 2148, 2027, 2009, 1967**	**Alkynes, nitriles, ketones**
	IV	**1734**	**Esters, aldehydes**
	V	**1610, 1599**	**p+s amides**
	VII	1240, 1223, 1128, **1001, 984, 966**	Lignin, **Polysaccharides**
	VIII	930	Polysaccharides
Cp10+	IV	**1743, 1720**, 1700	**Esters, ketones, carboxylic acids**
	V	1593, **1514**	p+s amides, **lignin**
	VI	1380	N-compounds (nitro)

(continued)

Table 5 (continued)

PCA	Region	Vibrations	Compounds
	VII	1344, 1230, 1172, 1149, 1130, 1011, **978**	Lignin, **polysaccharides**
	VIII	**937, 910, 895, 835**	**Polysaccharides, aromatics**
	I	3672, 3645, 3610, 3570, 3370	Clays, alcohols, phenols, p+s amides, carboxylic acids
Cp10−	II	**2918, 2850**	**Alkanes, aldehydes**
	III	**2175, 2156, 2141, 2038, 2029, 2017, 1975, 1961**	**Alkynes, nitriles, ketones**
	V	1628, 1552, 1530, **1466**	p+s amides, lignin, **aliphatics**
	VI	1417	Lignin
	VII	1315, **1286**, 1201, 1188, 1107, 1072, **1057, 1034**	**Lignin, polysaccharides**
Cp11+	I	3330, **3200-3100**	**p+s amides, polysaccharides, carboxylic acids**
	II	**2960, 2933**	**Alkanes, aldehydes**
	III	**1944-1800**	**Alkynes, nitriles, ketones**
	IV	1734, 1710	Aldehydes, ketones, carboxylic acids
	V	**1610, 1577**	**p+s ammides**
	VI	1435	Lignin
	VII	1280, 1186, 1157, 1107, **1057, 1034**, 1011	Lignin, **polysaccharides**
	VIII	814, 783, 771, **719**	Aromatics, **aliphatics**
Cp11−	I	**3500-3400**	**Alcohols, phenols, polysaccharides**
	II	**2918**, 2885, **2850**	**Alkanes, aldehydes**
	III	**2161, 2027, 1967**	**Alkynes, nitriles, ketones**
	IV	1760, **1655**	Anhydrides, **alkenes**
	V	**1537, 1510, 1469**	**Lignin**
	VII	1331, 1253, 1223, 1124, **989, 964**	Lignin, **polysaccharides**
	VIII	**930, 895, 854**, 835	**Polysaccahrides, aromatics**

Table 6 Statistics of the principal components' regression models obtained using the loadings of the extracted components in tPCA and the peat variables related to elemental and isotopic composition

												U-std coeff		Std. coeff		
	R	R square	Adj R square	Std. error		Sum of squares	df	Mean square	F	Sig.		B	Std error	Beta	t	Sig.
Carbon	0.963	0.928	0.925	0.88	Regression	982.210	4	245.552	317.902	<0001	Constant	31.55	0.53		58.85	<0.001
					Residual	76.469	99	0.772			Cp2	27.58	0.81	0.911	33.70	<0.001
					Total	1058.679	103				Cp3	−30.55	2.80	−0.316	−10.91	<0.001
											Cp7	40.03	6.61	0.175	6.05	<0.001
											Cp4	−13.13	3.97	−0.094	−3.30	<0.001
Nitrogen‘	0.948	0.898	0.888	0.11	Regression	10.9333	9	1.215	91.981	<0001	Constant	−0.69	0.07		−9.92	<0.001
					Residual	1.241	94	0.013			Cp2	2.31	0.10	0.715	21.66	<0.001
					Total	12.175	103				Cp11	−12.00	2.43	−0.199	−4.93	<0.001
											Cp3	3.54	0.40	0.342	8.85	<0.001
											Cp4	4.46	0.56	0.299	7.85	<0.001
											Cp8	−8.67	1.62	−0.222	−5.32	<0.001
											Cp7	−5.37	0.97	−0.219	−5.53	<0.001
											Cp9	3.56	1.64	0.084	2.16	0.033
											Cp6	3.71	1.00	0.179	3.67	<0.001
											Cp5	1.73	0.87	0.099	2.01	0.047
C/N	0.884	0.781	0.765	13.8	Regression	66022.221	7	9431.746	48.924	<0001	Constant	174.73	8.42		20.62	<0.001
					Residual	18507.332	96	192.75			Cp2	−149.10	12.90	−0.551	−11.53	<0.001
					Total	84529.553	103				Cp11	1871.57	293.52	0.373	6.37	<0.001
											Cp4	−467.63	64.32	−0.376	−7.27	<0.001

(continued)

Table 6 (continued)

												U-std coeff		Std. coeff		
	R	R square	Adj R square	Std. error		Sum of squares	df	Mean square	F	Sig.		B	Std error	Beta	t	Sig.
											Cp3	−182.95	44.40	−0.212	−4.12	<0.001
											Cp8	387.52	182.37	0.119	2.12	0.036
											Cp9	−485.23	173.88	−0.137	−2.79	0.006
											Cp7	281.61	113.29	0.137	2.48	0.015
$\delta^{13}C$	0.694	0.482	0.472	0.93	Regression	81.855	2	40.927	47.039	<0001	Constant	−21.47	0.56		−37.90	<0.001
					Residual	87.877	101	0.870			Cp2	−7.24	0.88	−0.598	−8.35	<0.001
					Total	169.731	103				Cp9	56.77	11.39	0.357	4.98	<0.001
$\delta^{15}N$	0.781	0.610	0.602	0.79	Regression	96.228	2	49.114	78.859	<0001	Constant	−7.64	0.40		−15.91	<0.001
					Residual	62.904	101	0.623			Cp2	9.04	0.74	0.766	12.31	<0.001
					Total	161.132	103				Cp3	5.37	2.38	0.142	2.28	0.024

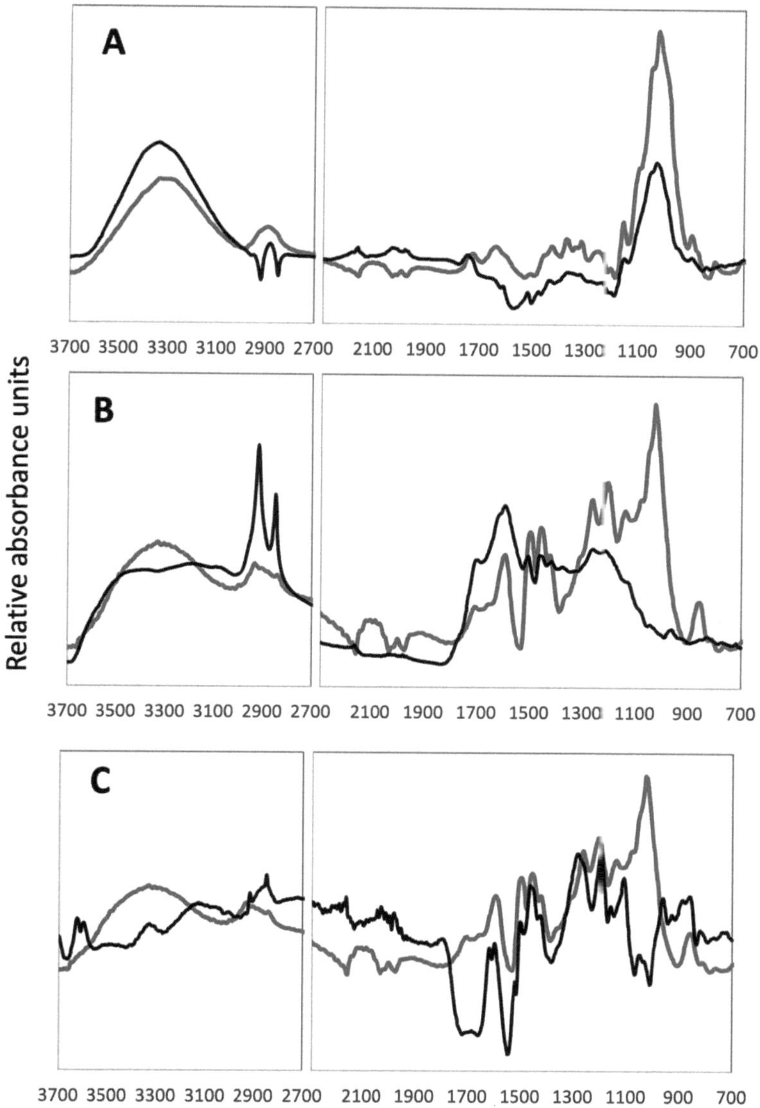

Fig. 8 Spectra of A) Cp1 (black lines) and holocellulose (gray line), B) Cp2 (black line) and Klason lignin (gray line), and C) Cp7 (black line) and Klason lignin (gray line). Spectra of holocellulose and Klason lignin obtained from [28]

References

1. Álvarez Fernández N, Martínez Cortizas A (2020) Andurinha: make spectroscopic data processing easier; R package version 0.0.2.; R Foundation for Statistical Computing, Vienna, Austria
2. Artz RR, Chapman SJ, Robertson AJ, Potts JM, Laggoun-Défarge F, Gogo S, Comont L, Disnar J-R, Francez AJ (2008) FTIR spectroscopy can be used as a screening tool for organic matter quality in regenerating cutover peatlands. Soil Biol Biochem 40:515–527
3. Biester H, Knorr KH, Schellekens J, Basler A, Hermanns YM (2014) Comparison of different methods to determine the degree of peat decomposition in peat bogs. Biogeosciences 11:2691–2707
4. Chapman SJ, Campbell CD, Fraser AR, Puri G (2001) FTIR spectroscopy of peat in and bordering Scots pine woodland: relationship with chemical and biological properties. Soil Biol Biochem 33:1193–1200
5. Clymo RS (1987) The ecology of peatlands. Sci Prog Oxf 71:593–614
6. Cocozza C, D'Orazio V, Miano TM, Shotyk W (2003) Characterization of solid and aqueous phases of a peat bog profile using molecular fluorescence spectroscopy, ESR and FT-IR, and comparison with physical properties. Org Geochem 34:49–60
7. Eriksson L, Byrne T, Johansson E, Trygg J, Vikström C (2013) Multi-and megavariate data analysis basic principles and applications, vol 1. Umetrics Academy
8. Heller C, Ellerbrock RH, Roßkopf N, Klingenfuß C, Zeitz J (2015) Soil organic matter characterization of temperate peatland soil with FTIR-spectroscopy: effects of mire type and drainage intensity. Eur J Soil Sci 66:847–858
9. Hodgkins SB, Richardson CJ, Dommain R, Wang H, Glaser PH, Verbeke B, Winkler BR, Cobb AR, Rich VI, Missilmani M, Flanagan N, Ho M, Hoyt AM, Harvery CF, Vining SR, Hough MA, Moore TR, Richard PJ, De la Cruz FB, Toufaily J, Hamdan R, Cooper WT, Chanton JP (2018) Tropical peatland carbon storage linked to global latitudinal trends in peat recalcitrance. Nat Commun 9:3640
10. Kamnev AA, Dyatlova YA, Kenzhegulov OA, Vladimirova AA, Mamchenkova PV, Tugarova AV (2021) Fourier transform infrared (FTIR) spectroscopic analyses of microbiological samples and biogenic selenium nanoparticles of microbial origin: sample preparation effects. Molecules 26:1146
11. Kylander ME, Bindler R, Cortizas AM, Gallagher K, Mörth CM, Rauch S (2013) A novel geochemical approach to paleorecords of dust deposition and effective humidity: 8500 years of peat accumulation at Store Mosse (the "Great Bog"), Sweden. Quatern Sci Rev 69:69–82
12. Kylander ME, Martinez-Cortizas A, Bindler R, Greenwood SL, Mörth CM, Rauch S (2016) Potentials and problems of building detailed dust records using peat archives: an example from Store Mosse (the "Great Bog"), Sweden. Geochim Cosmochim Acta 190:156–174
13. Kylander ME, Martínez-Cortizas A, Bindler R, Kaal J, Sjöström JK, Hansson SV, Silva-Sánchez N, Greenwood SL, Gallagher K, Rydberg J, Mörth CM, Rauch S (2018) Mineral dust as a driver of carbon accumulation in northern latitudes. Sci Rep 8:6876
14. Kylander ME, Holm M, Fitchett J, Grab S, Martinez Cortizas A, Norström E, Bindler R (2021) Late glacial (17,060–13,400 cal yr BP) sedimentary andpaleoenvironmental evolution of the Sekhokong Range (Drakensberg), southern Africa. PLoS One 16:e0246821
15. Kylander ME, Martínez-Cortizas A, Sjöström JK, Gåling J, Gyllencreutz R, Bindler R, Gallagher K (2023) Storm chasing: tracking Holocene storminess in southern Sweden using mineral proxies from inland and coastal peat bogs. Quatern Sci Rev 299:107854
16. Larkin PJ (2011) IR and Raman Spectroscopy, principles and spectral interpretation. Elsevier, Amsterdam, 228 pp
17. Liland KH, Mevik BH, Canteri R (2022) Baseline correction of spectra. R Package version 1.3-4
18. Lundqvist J, Wohlfarth B (2001) Timing and east-west correlation of Swedish ice marginal lines during the late Weichselian. Quatern Sci Rev 20:1127–1148

19. Martínez Cortizas A, Sjöström JK, Ryberg EE, Kylander ME, Kaal J, López-Costas O, Álvarez-Fernández N, Bindler R (2021) 9000 years of changes in peat organic matter composition in Store Mosse (Sweden) traced using FTIR-ATR. Boreas 50:1161–1178
20. Martínez Cortizas A, López-Merino L, Silva-Sánchez N, Sjöström JK, Kylander ME (2021) Investigating the mineral composition of peat by combining FTIR-ATR and multivariate analysis. Minerals 11:1084
21. Niemeyer J, Chen Y, Bollag JM (1992) Characterization of humic acids, composts, and peat by diffuse reflectance Fourier-transform infrared spectroscopy. Soil Sci Soc Am J 56:135–140
22. Pavia DL, Lampman GM, Kriz GS, Vyvyan JA (2014) Introduction to spectroscopy. Cengage learning
23. Pistorius AM, DeGrip WJ, Egorova-Zachernyuk TA (2009) Monitoring of biomass composition from microbiological sources by means of FT-IR spectroscopy. Biotechnol Bioeng 103:123–129
24. Sjöström JK, Martínez Cortizas A, Nylund A, Hardman A, Kaal J, Smittenberg RH, Risberg J, Schillereff D, Norström E (2023) Complex evolution of Holocene hydrocliamte, fire and vegetation revealed by molecular, minerogenic and biogenic proxies, Marais Geluk wetland, eastern Free State, South Africa. Quat Sci Rev 314:108216
25. Ryberg EE, Valiränta M, Martínez Cortizas A, Erhlen J, Sjöström JK, Kylander ME (2022) Postglacial peatland vegetation succession in Store Mosse bog, south-central Sweden: an exploration of factors driving species change. Boreas. https://doi.org/10.1111/bor.12580
26. Stuart BH (2004) Infrared spectroscopy: fundamentals and applications. Wiley
27. Tarnocai C, Stolbovoy V (2006) Northern peatlands: their characteristics, development and snesitivity to climate change. In: Martini IP, Martínez Cortizas A, Chesworth W (eds) Peratlands: evolution and records of environmental and climate changes, developments in earth surface processes, vol 9, pp 17–51
28. Traoré M, Kaal J, Martínez Cortizas A (2018) Differntiation between pine woods according to species and growing location using FTIR-ATR. Wood Sci Technol 52:487–504
29. Zaccone C, Miano TM, Shotyk W (2007) Qualitative comparison between raw peat and related humic acids in an ombrotrophic bog profile. Org Geoch 38:151–160

Application of Biochar in Agricultural Soil Green Remediation and Sustainable Development

Limei Chen, Chaoran Sun, Yaoyu Zhou, Songlin Sun, and Yuchen Zhuo

Abstract Sustainable agriculture, characterized by agriculture profitably while derogating destroy to the environment, is not easy to practice. For this purpose, biochar has attracted enormous attention because of its abilities to change soil properties, improve soil fertility, remove pollutants, and reduce greenhouse gas emissions, thereby enhancing agricultural sustainable development. This paper aims to discuss the influence of biochar on agricultural soil fertility and crop productivity. The agricultural productivity can be increased by biochar play various beneficial roles as a fertilizer and soil amendment. Furthermore, biochar in the remediation of soil contamination were discussed. At the same time, application of biochar in agriculture may have an important result on reducing global warming through the decrease of greenhouse gas (GHG) emissions and the sequestering of atmospheric carbon (C) into soil. Biochar is considered as one of the optimal management strategies to address agriculture challenges. However, the effects of a combination of factors should be considered carefully prior to application.

Keywords Biochar · Sustainable agriculture · Crop productivity · Soil remediation

1 Introduction

Soils provide the necessary spaces for plant growth, and they also form the material basis for plant nutrient sources; thus, they play crucial roles in the nutrient cycle. The accumulation of soil nutrients determines the status of soil fertility. Similarly, as fundamental nutrients are required for crop growth, the soil nutrient supply capacity

L. Chen · C. Sun (✉) · S. Sun
College of Mechanical and Electrical Engineering, Hunan Agricultural University, Changsha 410128, China
e-mail: scr@hunau.edu.cn

Y. Zhou (✉) · Y. Zhuo
College of Resources and Environment, Hunan Agricultural University, Changsha 410128, China
e-mail: zhouyy@hunau.edu.cn

A. Núñez-Delgado (ed.), *Frontier Studies in Soil Science*, Frontier Studies in Soil Science, https://doi.org/10.1007/978-3-031-50503-4_12

has differing levels of impact on the physicochemical and biological nature of the soil nutrient content, soil moisture, enzyme activity, and microbial community structure. Nevertheless, soil degradation is a major problem that severely threatens soil health and food security worldwide. Additionally, many human activities have not only damaged soil environmental quality but also contaminated soils with organic pollutants, inorganic pollutants, and heavy metals [5, 12]. Both degraded and polluted soils have been reported to inhibit and prevent crop growth [121]. China has one of the largest agricultural economies in the world, and edatopes have changed due excessive fertilizer application, which influences crop growth. Accordingly, a noteworthy research trend involves the adoption management measures that retain the health of cropland soils, ameliorate the growth environment of crops, and promote sustainable agricultural development. For this purpose, biochar has been proven to be more suitable than other organic soil amendments due to its unique characteristics that confer many advantages [81].

As a soil amendment, biochar has received worldwide attention for promoting green and sustainable agricultural development through improving soil structure, decreasing soil nutrient leaching, regulating soil pH, immobilizing heavy metals, and enhancing agricultural productivity while sequestering C and thereby alleviating global climate changes [83]. Biochar is a stabilized organic C compound that is produced at a temperature of 300–1000 °C from biomass under low oxygen or anoxia conditions [13, 21]. The special properties of biochar, such as its porous structure and high adsorption capacity, have important effects on its ability to improve the sustainable development of agricultural soils [104–105, 107]. However, the amount of biochar added hinges on the raw materials used to produce it and on production factors [62], which result in diversities in their properties and amendment values [52]. Biochar has been increasingly used for soil amendment because it can stabilize organic matter and improve soil properties for crop growth. The application of biochar to a soil may decrease CH_4 and CO_2 emissions, sequester increasing amounts of organic C [66, 67], change the properties of the soil, improve soil productivity, and promote crop yields.

Biochar has recently attracted increasing amounts of attention because of its significant agricultural benefits. Biochar is a C-based solid that is generally derived from C-rich wastes, particularly agricultural residues. Its elemental composition comprises C, nitrogen (N), hydrogen (H), potassium (K), and magnesium, all of which are primary nutrients for crop growth. Biochar can improve soil quality and structure and change soil physicochemical and biological properties, thereby altering agricultural soil environments. Therefore, the main objectives of this chapter are to (i) provide the characteristics of biochar, (ii) review the impact of biochar application as a soil ameliorant in agriculture, (iii) introduce biochar application for the remediation of contaminated soils, and (iv) propose possible issues and challenges for future research.

2 Biochar Characteristics

Numerous classes of biomass are widely utilized in the preparation of biochar, including crop biomass, animal and human wastes, sewage sludge, forestry biomass, and kitchen wastes [36]. The main thermochemical technologies for biochar preparation involve slow and fast pyrolysis, torrefaction, hydrothermal carbonization (HTC), and gasification. The reaction conditions and related product contents of several pyrolysis procedures are generalized in Table 1 [81]. The different production technologies influence the physical and chemical characteristics of biochar, e.g., output, ash, pore structure, stamp, pH, and cation exchange capacity (CEC).

Biochar is a C-rich matter derived from the pyrolysis of various organic waste feedstocks. Biochar composed of stabilized or fixed C, unstable C and other volatile compounds, moisture, and ash [81]. The physicochemical characteristics of biochar (e.g., particle density, surface area, porosity, pH, total C, CEC, etc.) usually rely on the applied pyrolysis conditions and on the characteristics biochar feedstock and thus are broad [27, 77, 131]. In addition, biochar contains major elements (such as N, phosphorus (P) and K) and trace elements.

2.1 Biochar Microstructure

The microstructure of biochar depends on the original feedstock, and the surface of biochar is usually porous. The microporous structure of biochar plays a significant role in the enhancement of its specific surface area and pore structure. Macropores can improve the permeability and water content of soils and provide space for the growth and reproduction of microorganisms, while the adsorption capacity of biochar is increased by small pores. Biochar can also introduce hazardous and harmful matter into soils via its pores. In addition, the mobility of water in soils will be affected by the

Table 1 The reaction conditions and product distribution of several patterns of pyrolysis [81, 115, 130]

Process	Temperature (°C)	Residence time	Yield (%)		
			Biochar	Bio-oil	Syngas
Slow pyrolysis	300–700	Hours–days	35	30	35
Fast pyrolysis	500–1000	<2 s	12	75	13
Gasification	750–900	10–20 s	10	5	85
Hydrothermal carbonization (HTC)	180–300	1–16 h	50–80	5–20	2–5
Torrefaction	~290	10–60 min	80	0	20
Microwave pyrolysis	550–700	5–20 min	34	/	/
Flash carbonization	300–600	~30 min	37–50	/	/

pore structure of the biochar applied, thus affecting the retention of soil nutrients [98]. However, some microporous structures on the surface of biochar may decompose when the temperature exceeds 700 °C, and the biochar C structure changes when the temperature exceed 800 °C [21].

2.2 Elemental Content

Biochar normally comprises C, hydrogen (H), oxygen (O), N, sulfur (S), P, K, calcium (Ca), magnesium (Mg), sodium (Na), silicon (Si), and other elements. Among the biochar, the high C content in biochar (Table 2) and the C content of wood biochar and bamboo biochar are 56% higher than those of other biomasses [44, 98, 123]. The C in the biochar is primarily present as stable aromatic C in irregular aromatic chain stacks. In addition to C, the H and O contents are relatively high in biochar. The mineral elements are mainly retained in the ash. N mainly exists on the surface of biochar, with a C–N heterocyclic composition, and the available N content in biochar is less than 3% [55]. The total P content is low in biochar. The available P of different biochar varies greatly and is negatively related to the pyrolysis temperature. This phenomenon may be affected by the high pH value of biochar and by phosphates containing Ca and Mg [20]. The contents of K, Ca, Mg and Na in different biochar are different, among which their contents in livestock manure biochar are the highest, followed by those in wood biochar. Low-valence metal ions (e.g., K and Na) are more easily found in biochar than high-valence metal ions (e.g., Al, Ca, and Mg). In general, the elemental content of biochar is affected by materials, pyrolysis conditions, pH value and other factors [89].

2.3 Physical and Chemical Properties

Biochar physical properties are characterized by its yield and surface characteristics, while its central chemical properties involve pH, total C and total N, total P, CEC, nutrients, and microelement contaminants [4]. Biochar is mostly alkaline, which is chiefly associated with the feedstock material (Table 2); inorganic substances are present (e.g., carbonates and phosphates) with the ash formed during carbonization [98]. Besides this, the carbonization temperature also affects the pH value of biochar, which increases with temperature (Table 2). Moreover, increasing the residence time can improve the pH value, while the heating rate appears to have little effect on the pH value [10].

The porosity of biochar changes with the volatilization of gas during pyrolysis, so the total surface area of biochar also changes. The specific surface area of biochar is usually in the range of 1.5–500 $m^2\ g^{-1}$ [57, 58]. Its specific surface area increases with increasing temperature, within limits [109]. The surface area of biochar can

Table 2 Characteristics of biochar produced from different feedstocks

Feedstock	Pyro. tem. (°C)	pH	Elemental analysis (%)				References
			C	N	P	K	
Maize straw	400	9.6	68.5	1.34	0.15	1.34	[63]
Peanut shell	550 ± 50	10.1	67.4	1.3	/	/	[113]
Corn straw	500	10	58	2.3	/	/	[114]
Wheat straw	450	9.92	45.7	0.89	/	/	[122]
Maize cobs	360	/	65.7	0.91	0.08	1.6	[39]
Rice straw	400–500	9.02	63.27	1.28	0.73	1.46	[23]
Wheat straw	350–550	9.9	46.72	0.59	1.44	1.15	[1]
Maize straw	300–600	9.84–11.37	48.9–62.9	1.21–1.25	0.06	3.34–6.61	[91]
Rice straw	400–450	9.6	88.82	0.25	0.08	1.17	[125, 129]
Cotton straw	800	8.6	68.7	0.325	0.12	1.6	[87]
Tobacco straw	550	11.2	22.74	1.31	/	/	[110]
Canola straw	400–500	9.49	49	2.3	0.55	/	[132]
Corn straw	450	7.94	71.5	1.54	0.78	1.68	[64]
Peanut shell	500	9.16	64.7	1.522	/	/	[68]
Wheat straw	400	9.2	48.65	0.75	0.17	4.92	[112]
Corn straw	400–450	8.56	50.6	1.4	0.46	1.78	[29]
Maize straw	500	9.2	66	1.27	0.887	/	[104]
Rice/ canola straw	350/550	9.94–11.11	42.01–60.32	/	0.16–0.31	2.53–4.06	[117]
Wheat straw	550	9.94	68.26	1.39	0.14	1.56	[66, 67]

Notes "/" not measured

reach 800 m^2 g^{-1} at approximately 700 °C [75]. Additionally, an increase in residence time can lead to a further increase in surface area, but compared with an increase in temperature, even a very long residence time (several hours) has less of an influence on the increase in surface area [108]. If the pyrolysis temperature is low, the tar and other products generated during the decomposition of biomass will remain in the internal pores of the biochar, thus reducing the specific surface area. When the temperature is increased, the volatile substances in the biomass are decomposed to form an increased number of denser pore structures, thus yielding a larger specific surface area. For some substances, a surface area of hundreds of m^2 g^{-1} can be achieved under suboptimal production conditions. For sewage sludge, the surface area remains below 100 m^2 g^{-1}, regardless of the pyrolysis conditions [109]. However, when the pyrolysis temperature reaches its peak value, the specific surface area of biochar also reaches its peak value. Due to the destruction of the biomass cell structure and the enlargement of micropores, when the peak value is exceeded, the specific surface area of biochar decreases with increasing temperature. The pore structure in biochar can be divided into macropores (with a pore diameter of 1000–0.05 μm), mesopores (0.05–0.002 μm), and micropores (0.05–0.0001 μm) [18]. The porosity of biochar generally increases with increasing temperature. The microporous structure of biochar contains large pores, and the microporous area constitutes 80% or more of the surface area. In contrast, the number of micropores in noncarbonized crop straw is smaller.

In general, as the temperature is increased, the degree of aromatization and hydrophobicity of biochar increase, while the number of O- and N-containing functional groups and the water-holding capacity of biochar decrease. The hydrophobicity of biochar increases with the increase of temperature because the polar functional groups in the biomass decrease [8, 28]. Studies have illustrated that straw biochar carbonization at 300 °C yield a water-holding capacity of 13×10^{-4} ml m^{-2}; however, it is reduced to 4.1×10^{-4} ml m^{-2} when the carbonization temperature is increased to 700 °C [21]. Hydrophobicity is influence by surface functional groups, while the porosity and pore connectivity of biochar determine its water holding capacity. The hydrophobic surfaces within the pores of biochar can prevent water from entering the porous C structure [33]. Therefore, studies have confirmed that biochar produced at high temperatures contains more water in its porous structure [33, 124]. Biochar with a porous structure may also be produced under low temperature conditions, but the pore sizes are small, the level of interconnection is low, and the byproduct tar produced during the preparation process blocks the pores, so it is difficult to obtain porous structured biochar under low temperature conditions [55].

The CEC of biochar affects the CEC of the soil to which it is applied, thus changing the physical and chemical properties of the soil (Table 3). Biochar CEC is related to biomass type and pyrolysis temperature [93]. When biochar is added to a soil, the soil CEC will increase with the oxidation of functional groups on the biochar surface. To an extent, biochar CEC decreases with increasing temperature [49], the O-containing functional groups are destroyed, the passive charges on the surfaces of biochar decreases, and the O/C ratio decreases [94]. In addition, the increase in temperature also leads to an increase in alkali metals (such as K, Ca, and Mg) in biochar, which may cause an increase in CEC [49].

3 Biochar Application for Agricultural Soil Improvement

The inherent architecture and physicochemical natures of biochar can affect the soil structure, porosity, CEC, and nutrient cycle and influence the soil microecological environment directly or indirectly [125, 129]. Biochar, as a carbon-rich biomass, increases soil organic C via its C content. Moreover, the soil water holding capacity significantly improves with high porosity biochar [31]. As a soil conditioner, biochar is more suitable for making improvements to acidic soils by increasing the nutrient content of the soil [86]. Therefore, biochar can change the limiting factors in soils, reduce the absorption of harmful pollutants by crops, and promote high-quality growth in crops [51]. Biochar can be used as a carrier of slow-release fertilizer because of its ability to delay the release of nutrients while improving the retention capacity and nutrient utilization efficiency of soil nutrients. Therefore, many researchers have studied biochar-based fertilizers that are suitable for different crops.

Soil fertility mirrors the generative ability of soils for plant growth, which can be enhanced by the addition of biochar. The effect of biochar on soil can be analyzed from the aspects of nutrient cycling, crop productivity, soil pH, CEC, N, P, microbial community, water conservation and carbon sequestration (Table 3). The application of biochar as a soil amendment helps enhance the characteristics of nutrient-depleted and degraded soils. Previous studies have shown that the application of biochar to low-fertility soil is conducive to a reduction in soil bulk density, increase in soil total porosity, improvements to water holding capacity, neutralization of acidic soils, and improvements to soil nutrients, particularly in poor and degraded soils.

3.1 Physicochemical Properties of Soils

Biochar modification can reduce soil bulk density and particle density [121]. The high porosity of biochar (70–90%) leads to an increase in soil porosity, which is also leads to a decrease in soil bulk density, an increase in soil aggregation and interactions with mineral soil particles [17]. Improvements to the water, heat and gas content of soils can be achieved through decreases in soil bulk density and boosts in soil porosity. Studies have shown an increase in the resistance of clay agglomerates due to enhanced internal cohesion by combined mineral C particles, and biochar macropores and mesopores in clay can effectively improve soil water retention [65]. In addition, biochar addition enhances soil structure and soil aggregation [17]. Biochar can reduce soil tensile strength and increase the soil water holding capacity. Studies have shown that biochar-added soils are more brittle and less compact than soils without biochar. Most studies show that soil organic C accumulation is central to the improvement

Table 3 Changes in soil properties with biochar application

Biochar source	Soil type	Treatment	Impact on soil properties	References
Maize straw	Loamy sand	0, 3.2, 16, 32 t ha^{-1} BC + 7.8 t ha^{-1} corn straw	Increase the SOC concentration by 31–298%	[63]
Wheat straw	Silty clay loam	0.028, 0.084, 0.28 kg pot^{-1}	The SOC was increased by 13.2–155.1% on average. Addition of 0.028 and 0.084 kg pot^{-1} BC increased MBC content but decreased DOC content	[46]
Rice straw	Sand loam	CK, urea, $(NH_4)_2SO_4$, 2% BC, 2% BC + urea, 2% BC + $(NH_4)_2SO_4$	The application of only N fertilizer and N fertilizer + BC significantly increased the abundance of ammonia-oxidizing bacteria. The effect of combined application was optimal	[14]
Rice straw	Silt loam	0/2.25 t ha^{-1} BC + 120/180 kg ha^{-1} N fertilizer + 37.5/67.5 P fertilizer	BC application increased in the soil pH but decreased in CEC. The interactive effect of N fertilizer, P fertilizer and BC was significant for soil TN	[88]
Maize straw	Sand loam	0, 10, 20, 40, 60 t ha^{-1}BC	Addition of BC reduced the bulk density and increased the porosity of soil. Increase in soil pH, AN, AP, AK, and SOC	[57, 58]
Rice straw	Silt loam	8 t ha^{-1} BC + 8 t ha^{-1} steel slag	The SOC concentrations, carbon pool index was 28.7% and 42.2, 22.4 and 40.1% higher in the early and late crops	[101, 106]
Wheat straw	Silt loam	0, 5, 20, 40 t ha^{-1} BC + N fertilizer (0, 60, 90, 120 kg ha^{-1})	High dose of BC (20 and 40 t ha^{-1}) combined with N fertilizer can significantly increase soil pH and AP in the first year; however, the effectiveness on these metrics weakened over time	[48]
Maize straw	/	5 g kg^{-1}BC + 0.5 mg kg^{-1} oxalic acid	Increase in soil urease, polyphenol oxidase and dehydrogenase enzymes. Soil microbial biomass and the abundances of genera were significantly enhanced	[59]

(continued)

Table 3 (continued)

Biochar source	Soil type	Treatment	Impact on soil properties	References
Wheat straw	Sand loam	2% BC + NPK fertilizer	BC amendment significantly increased the concentrations of NH_4^+-N, TN, AP, and AK in soil	[69]
Rice/wheat straw	Sand loam	Inorganic fertilizer, straw, BC	Enhance in SOC, TN, and aggregation in 0–20 and 20–40 cm soil. Straw return increased fungal community richness and diversity, while BC decreased	[11]
Peanut shell	/	4.5 t ha^{-1} BC + 9 t ha^{-1} chicken manure	Increasing in soil pH by 0.02 units. Increasing SOC, AP, AK, and MBC by 21.76, 174, 13.92, and 198.19%, respectively. Decreasing soil AN	[68]
Maize straw	Clay loam	CK, 6 t ha^{-1} straw, 2.4 t ha^{-1} manure, 0.6 t ha^{-1} BC	Original straw return only decreased soil bulk density, while manure and BC return increased soil pH, SOC, TN, and AK significantly	[82]
Cottonseed husk and rice husk	/	50 t ha^{-1}BC + mineral fertilizer or + 0.9 t ha^{-1} organic fertilizer	Soil MBC was significantly increased with the application of BC plus organic fertilizer, but not affect the amount of MBN	[97]
Rice/canola straw	Ultisol	21, 15.5, 17, 14.5 g pot^{-1}	BC increased the total porosity, total pore volume, average pore diameter and the pores in soils. Soil pore structure changes induced by BC affect microbial diversity and community structure	[117]

Notes Control (CK); soil organic carbon (SOC); microbial biomass carbon (MBC); dissolved organic carbon (DOC); cation exchange capacity (CEC); total nitrogen (TN); available phosphorus (AP); available potassium (AK); available nitrogen (AN); microbial biomass nitrogen (MBN). The same below

of soil quality and that low-fertility soils benefit more from the addition of biochar than fertile or high-yielding soils [17].

Biochar application has been indicated to significantly increase the chemical properties of poor soils. Excessive applications of chemical fertilizers can lead to soil acidification, while the application of alkaline biochar can improve acidic soils. These results are attributed to the alkalinity of biochar, the influence of functional groups and the silicon (Si) effect [121]. The enhancement in the pH value of acidic soils is closely related to the alkalinity of biochar. Mineral elements (such as Ca, K, Mg, Na, and Si) in the biomass form carbonates or oxides during pyrolysis, which counteract the H^+ and monomeric Al species in acidic soils by decreasing exchangeable acidity and improving pH [25]. Additionally, the biochar functional groups –COO^- and –O^- also benefit greatly from biochar alkalinity when biomass is pyrolyzed at 300–500 °C [123]. The enhancement in CEC upon the addition of biochar contributes to an enhanced H buffering capacity. The cations released by biochar are the main factors that cause an increase in pH, e.g., K, Ca, Mg and Na. In addition, the reaction of soluble silicon with the protons in biochar also leads to an increase in pH. Compared with other materials used to neutralize soil pH, biochar can exist in soils for long periods of time, thereby keeping the soil pH at desired levels. At present, there is little research on the application of biochar to alkaline soils. However, biochar produces acids that lower the pH of soils. Carbonization temperature is the main factor that affects the production of acidic substances in biochar. Similarly, the application of acidic biochar to alkaline soils, such as salt-affected soils, improves soil pH [45].

The use of biochar can promote the maintenance of soil EC by 2–85% [76]. Additionally, biochar that has added ash that contains soluble salts can lead to mild EC enhancements [100]. Most of the research indicates that EC percentages change with biochar addition. The raw substrates, carbonization requirements, production methods, and addition ratios are the controlling factors that influence the EC of soils.

Soil CEC can be modified substantially by biochar application, and this depends on feedstock type and the pyrolysis temperature. For instance, the addition of biochar that is generated at low temperatures results in a comparatively high CEC its latent role in soil quality enhancement [43, 93]. However, biochar produced from poplar and Douglas fir bark at high temperature is not suitable for low CEC soils. In contrast, many studies have shown that the impact of biochar on soil CEC is essentially positive [93]. The oxidation of aromatic C and the generation of carboxyl groups in biochar may be the basis for biochar application to improve soil CEC [76]. In addition, over time, the aging of biochar in soils may lead to an increase in its surface area and the generation of higher CEC [92]. Accordingly, the addition of biochar has been proposed for low-fertility soils to boost crop productivity and nutrient preservation.

Biochar has the capacity to markedly reduce soil erosion [60]. Soil erosion decreased by 50 and 64% in highly weathered soils when biochar was applied at 33.75 and 67.5 t ha^{-1}, respectively [47]. Likewise, biochar is able to decrease additive run-off, run-off rates, and nutrient losses in sloping highlands [60]. Biochar promotes the formation of soil microaggregates, which is a possible mechanism for reducing soil loss [84]. Studies have shown that biochar reduces the occurrence of soil erosion due to an increase in soil water-stable aggregates. Moreover, biochar addition results in an approximately 55% reduction in the run-off time at a rainfall rate of 50 mm h^{-1} [84]. These findings suggest that biochar's beneficial effects on soil runoff and erosion can be ascribed to biochar's water holding capacity and other soil-related physical properties that prevent encrustation and high levels of infiltration. These beneficial effects also demonstrate that the addition of biochar in sloping uplands may be a valid way to reduce crop damage due to soil erosion, specifically during the rainy season.

3.2 Biological Properties of Soils

Biochar improves soil microbial activity and provides suitable habitats [72, 121]. The high surface area and hydrophobicity of biochar mainly contribute to its enhanced microbial activity and diversity [133, 134]. The addition of biochar to soil positively affects arbuscular mycorrhizal fungi. Because the pore structure of biochar provides a habitat for fungi and other filamentous microorganisms to infiltrate, the yeast- and sugar-rich biochar stimulates the growth of soil fungi and gram-negative bacteria. Compared with gram-negative bacteria, the growth of gram-positive bacteria benefits from the background C content of biochar; biochar lacks carbohydrates, amino acids, and small proteins that are essential for the growth of gram-negative bacteria [9]. However, other research has shown that alkaline biochar may promote the growth of both gram-positive and gram-negative bacteria. At the same time, the pH value of biochar decreases after aging, which also promotes the survival of fungi within the pores of biochar.

Bacteria and fungi can use their extracellular enzymes to decompose biomass (including biochar derived C) into smaller molecules, and cells can absorb small molecules more easily to meet the energy needs of their metabolic activities [24]. Microbial biomass C is an effective indicator for measuring changes in soil organic carbon decomposition. Therefore, any process that changes the soil C content will affect the microbial community structure.

3.3 Nutrient Cycling

Nutrient cycling in soil systems consists of many complex components that depend on a variety of external factors (such as rainfall patterns, climatic leaching, and type of fertilization) and internal factors (such as vegetation species, plant root characteristics, and soil structure). The addition of biochar could lead to significant changes in soil ecosystems, improving degraded agricultural soils to achieve higher crop yields. The effect of biochar on soil nutrient cycling has been studied under different environmental conditions, and it has been found that the factors that affect nutrient cycling in biochar-improved soils include soil type, raw material properties and biochar pyrolysis conditions [9].

The nutrient content of biochar varies greatly due to factors such as the biomass raw materials used and production technology. When biochar is added to a soil, the nutrient content of biochar is responsible for regulating the interactions between biochar and plant roots, affecting root growth and the overall performance of plants by directly serving as the source of nutrients and changing the availability of soil nutrients [80]. The soil nutrient content is determined by the amendment quantity, elemental components, and biodegradability of biochar [43]. Biochar amendment influences soil pH, nutrient content, and microbial architecture, thus changing soil C-cycling, N-cycling, and P-cycling [35]. For soil C-cycling, most studies have given attention to the alterations in total C, organic C, and inorganic C in the soil. Although most studies show that the C in biochar is relatively stable, biochar-C can promote long-term C-cycling by regulating soil C mineralization, CO_2 release and soil organic C accumulation. C is a meaningful element in soil ecosystems. C storage has become extremely prominent topic due to climate change. Biochar applications facilitate the sequestration of atmospheric C dioxide in soils. Corncob-derived biochar reduced soil respiration in a seven-year study. These results illustrate that biochar has the potential to sequester C in temperate farmland soils [37]. Release of NH_3 and denitrification are immediately affected by the adsorption of gaseous N composites and the ionic characteristics of N in biochar. The addition of biochar and compost markedly decreased the leaching of nutrients (such as NO_3^--N and P) from a soil while boosting maize crop growth [2]. Moreover, the addition of rice straw-derived biochar at 13.5 and 67.5 t ha^{-1} improved nitrification and decreased N_2O emissions [37]. In addition, the application of willow biochar improved the contents of NH_4^+ and NO_3^- in corn-cultivation soils. The application of willow biochar slightly increased soil NH_4^+ and NO_3^-, while mixed biochar and compost showed a significant increase in inorganic nitrogen content [3]. The combined application of biochar and compost was suitable for achieving improvements to soil NH_4^+ and NO_3^-. However, another study showed that the addition of wheat straw biochar increased the emissions of NH_3 (32.4–68.2%) and N_2O (9.4–35.2%) [38]. The study showed that the significant increase in soil pH caused by biochar resulted in NH_3 loss. Compared with NH_3, the emissions of N_2O were lower due to the low availability of N in the soil due to the physical adsorption or microbial fixation induced by biochar. The addition of biochar promotes soil nutrient storage, improves soil structure, decreases GHG emissions, alters the geochemical cycle of C, N, and P, and boosts the effectiveness of C, N, and P in soils, thus increasing crop yield.

4 Impact on Crop Productivity

Biochar ameliorant can significantly increase soil fertility and crop productivity. The application quantity and soil type to which biochar is applied have strong impacts on crop yield, especially when biochar is applied to low fertility or degraded soils [27] (Table 4). A meta-analysis showed that biochar-based fertilizers applied at very low application rates (average 0.9 t ha^{-1}) promoted crop productivity by 10 and 186%, respectively, relative to fertilized and unfertilized controls. This means that when biochar is used as a soil conditioner, the enhancement in crop productivity is considerable (i.e., 15–30 t ha^{-1} increases crop productivity by 10%). This increase in crop yield recommends that the use of biomass-based char-based fertilizers increases fertilizer utilization efficiency to a greater extent than the use of chemical fertilizers alone. A study reported that biochar-based fertilizers had the potential to enhance crop productivity by 15% when the addition of conventional fertilizers did not significantly alter the soil. Furthermore, biochar-based fertilizers produced at pyrolysis temperatures >400 °C increased crop productivity by 12%, while charring temperatures ≤400 °C showed no significant change. Biochar-based fertilizers with a C content >30% increased crop yield by 17%, while fertilizers with a C content of ≤30% had no effect [71]. Another study showed that the application of biochar remarkably improved crop yield compared with a control by 34 and 30% in culture and field trial studies, respectively, but not in laboratory studies. The increase in crop yield in indoor and field studies was similar. With respect to the type of feedstock, the study found that biochar made from lignocellulosic waste (35%) and herbaceous (53%) materials significantly increased crop yield, while wood biochar (17%) had little effect on yield. Besides, biochar application significantly increased crop yield in a fine soil (40%) and coarse textured soil (57%) but not in a medium textured soil (9%). However, the authors of the present study, came to the opposite conclusion: compared with a control, biochar prepared at low pyrolysis temperatures increased crop yield by 43%, not at high pyrolysis temperatures. Besides this, most studies show that crop yields increase significantly at low (30%) and moderate (40%) application rates, while at high application rates, there is a yield reduction effect [90].

5 The Interactions of Biochar with Soil Contaminants

Biochar can be a suitable choice for improving the soil fertility of croplands. More and more evidence suggest that biochar ameliorant improve problematic soils and promote crop growth, although some negative effects have been reported. As a soil amendment, biochar can reduce the toxicity of soil contaminants to soil microorganisms [54]. Willow biochar can increase the number of microorganisms present, promote reproduction of Folsomia candida in soils polluted by heavy metals and organic pollutants, and reduce the toxicity of leachate in Vibrio Fischer. The main reason for the decrease in the toxicity of soil pollutants to microorganisms and the increase in microbial biomass may be that soil pollutants (including heavy metals, such as Al, Cd, Co, Cr, Mn and Ni, and organic pollutants, such as PAHs) can be fixed in the pores of biochar, thus reducing their bioavailability [85, 137]. The organic-complexed fractions of heavy metals (Cd, Cu, Pb and Zn) can be increased by 68% with an application rate of 5% rice straw biochar [70]. When heavy metals have adverse effects on N-fixing bacteria, they can further promote N storage during plant growth [85]. The enhancement in microbial activity by biochar addition can be attributed both to decrease in the toxicity of pollutants to soil microorganisms and to the fate of soil pollutants (containing their immobilization and degradation) because of interactions between biochar and soil pollutions.

Table 4 Summaries of responses of crops to different sources of biochar applications

Biochar source	Crop	Treatment	Response	References
Maize straw	Wheat	0, 3.2, 16, 32 t ha^{-1} BC + 7.8 t ha^{-1} corn straw	Biochar increased grain mass per plant of the wheat by 27.7%	[63]
Wheat straw	Maize	24 t ha^{-1} BC + NPK fertilizer	BC amendment increased the maize biomass and the N use efficiency in the red soil	[135]
Maize cobs	Wheat–maize	4.5, 9 Mg ha^{-1} $year^{-1}$ BC; 15 Mg ha^{-1} $year^{-1}$ straw	No significant differences in the grain yields of wheat, maize and the annually total productivity (wheat + maize) were observed between treatments	[39]
Wheat straw	Maize	0, 20, 40 t ha^{-1} BC	Increase maize yield and partial nutrient productivity	[126]
Corn straw	Tobacco	1, 5, 10, 20 Mg ha^{-1} BC	The application of BC decreased the total sugar and reducing sugar contents and increased the leaf K content, thus greatly improving the leaf quality	[127]
Rice straw	Lettuce	0, 10, 20, 30, 40 Mg ha^{-1} + 1010 kg N fertilizer	A significant 17–29% increase in vegetable yield was induced by BC, which due to improvements in SOC and the CEC	[56]
Corn straw	Soybean	0, 2.5, 5, 10% BC	Single application of BC positively and significantly improved soybean productivity and quality attributes	[64]

(continued)

Table 4 (continued)

Biochar source	Crop	Treatment	Response	References
Rice straw	Rice	0.12 t ha^{-1} N fertilizer + 30% urea + 2.8 t ha^{-1} BC	Increase rice grain yield, effective panicles, total biomass accumulation, the greenness of the leaves, and recovery efficiency of applied N	[119]
Corn straw	Corn	0, 15, 30, 45 t ha^{-1} BC	Application of 30 t ha^{-1} BC significantly increased corn yields	[118]
Wheat straw	Rice–wheat	0.25 t ha^{-1} N fertilizer + 20/40 t ha^{-1} BC	Increase crop (rice and wheat) root (by 3–19%), straw (by 10–19%) and grain (by 10–16%) biomasses, as well as grain NUE (by 20–53%) and PUE (by 38–230%)	[128]
Wheat straw	Wheat–maize	8 or 16 t ha^{-1} BC + fertilizer	Increase in plant height, aboveground biomass, grains per spike, and 1000-grain weight for both crops. The four-year average crop yields increased by 65.4–81.7%	[42]
Enteromorpha prolifera/rich corn straw	Cherry tomato	1% BC	The cherry tomato seedlings with BC addition were much stronger and taller, and exhibited more leaves and a taller and stronger shoot	[96]

(continued)

Table 4 (continued)

Biochar source	Crop	Treatment	Response	References
Maize straw	Beans and maize	0, 2.1, 4.2, 8.3 g kg^{-1}	Application of BC increased all plant growth parameters (height, stem diameter, leaf number, and fresh and dry weight) for both beans and maize	[50]
Empty fruit bunch	Cotton-sugar beet	0, 10, 50, 100 t ha^{-1} in 2018; 0, 10, 25, 50, 100 t ha^{-1} in 2019; and 0, 10, 25, 30 t ha^{-1} in 2020	The yield of cotton and sugar beet first increased and then decreased with increasing amount of BC	[105]
Corn straw	Maize	0, 10, 20, 40 t ha^{-1}	Application of BC increased maize yield in different degrees. Application of 20 t ha^{-1} BC tended to delay maize leaf senescence	[116]

Notes Control (CK); nitrogen use efficiency (NUE); phosphorus use efficiency (PUE)

5.1 Heavy Metal Contamination in Soils

The removal of heavy metals from soils by biochar chiefly includes two pathways: (1) porous-structured biochar adsorbs heavy metals in its pores, thereby reducing the residual amount in the soil and (2) certain components of biochar interact with the heavy metals. The ions undergo ion exchange or redox reactions to reduce toxicity by the formation of heavy metal precipitates or by their conversion to low valence heavy metal ions [120]. Biochar removal mechanisms for different types of heavy metals are different. Possible mechanisms involve a combination of interactions, such as electrostatic attraction, ion exchange, physical adsorption, surface complexation, and precipitation. As a remediation agent, biochar has a large surface area and abundant micropores, and abundant functional groups are considered to be excellent adsorbents for the effective removal of various heavy metals from polluted soils [7, 73, 78]. Of course, the fixation of heavy metals in biochar may be determined by the electrostatic interactions and adsorption deposition behavior between biochar functional groups and heavy metals. Therefore, compared with the specific surface area, the adsorption capacity of heavy metals depends more on surface functional groups. In addition, ion exchange, metal ligand complexation, cation bonding and surface coprecipitation are conducive to the preservation of metals in biochar [26]. Taking

Pb^{2+} as an example, its adsorption to biochar includes the exchange of heavy metals with Ca^{2+} and Mg^{2+} in biochar and the interaction of heavy metals with different functional groups on the surface of biochar. The adsorption of Cd on biochar was also discovered to be principally controlled by ion exchange [61]. Generally, biochar obtained by carbonization at low and medium temperatures has strong adsorption potential for metal cations [111]. For other inorganic contaminants, such as F^-, ClO^{4-}, PO_4^{3-}, Cr(III), and Cr(VI), their adsorption capacity can be enhanced by biochar pyrolyzed at higher temperatures. In addition, alkaline biochar may reduce the leaching and bioavailability of heavy metals in soils through some physical and chemical reactions [101, 106]. The study also found that the addition of biochar increased the soil pH value, which led to heavy metal precipitation.

5.2 *Soils Polluted Organic Compounds*

In recent years, biochar has been widely utilized to remove various organic pollutants from soils. These organic pollutants include agricultural chemicals (such as pesticides, herbicides, and fungicides), antibiotics/drugs (such as penicillin, oxacillin, cephalexin, and cefadroxil), industrial chemicals (phenol, acetone, toluene, etc.), volatile organic compounds (such as chlorobenzene, methanol, and styrene) [74]. Biochar is considered a suitable material for removing various pollutants from soils due to its large specific surface area and high affinity for the active groups in pollutants. The adsorption mechanisms for organic pollutants by biochar can be summarized as the accumulation and concentration of organic pollutants on biochar. The removal mechanisms for organic pollutants mainly include chemisorption (electrophilic interactions) and physical adsorption, such as hydrophobicity, electrostatic attraction by π-π electron donor-acceptors, pore diffusion, interparticle diffusion, carboxylic acids, hydrogen bonds on diols, and alcohol functional groups [6, 74, 99].

The use of pesticides, fungicides and herbicides in agriculture can contaminate soils. The stress and toxicity of pollutants on plant roots can significantly affect plant growth. The results of several studies suggest that biochar diminishes the bioavailability of contaminants in the soil [15, 34]. This usually hinges on the sorption level of the pollutants and the characteristics of the soil and biochar utilized for remediation. The porosity of biochar strongly affects the retention time of contaminants [136]. The utilization of biochar in agricultural soils increases the adsorption of organic pesticides, which is beneficial on the one hand because it reduces the hazards associated with pesticide residues, but on the other hand, it also reduces the efficacy of herbicides, so the application rates must be evaluated in different agricultural soils [22, 86]. It is also worth noting that unstable components in biochar may reduce the adsorption of pesticides. For example, one study reported that labile organic carbon components in biochar block biochar pores and thus reduce atrazine adsorption [86]. Another study also verified that biochar could decrease the efficacy of herbicides by sorption mechanisms [22]. The carbonization temperature of biochar also affects adsorption efficiency. Generally, the adsorption efficiency of biochar increases with increasing pyrolysis temperature. This can be explained by the electrostatic attraction between ionic organic pollutants and the charged surfaces of biochar [5, 32]. Therefore, the properties of biochar should be carefully considered to improve their adsorption capacity before use. In particular, biochar with low adsorption properties may be more effective when preemergence herbicides are agronomically desired, whereas high adsorption properties are required to reduce the bioavailability of polluting organic compounds.

5.3 *Potential Risks Associated with Biochar*

Biochar could have positive impact on one crop and may have a negative impact on another crop. Biochar could contain other organic compounds, including biooil, toxic elements, pesticides, PAHs, volatile organic compounds (VOCs), and dioxins. These contaminants could either be naturally present in the materials used for biochar production or could be generated during pyrolysis [40]. Pollutant availability and content depend mainly on the feedstock and pyrolysis conditions of biochar [30, 79]. Interestingly, the contents of potentially toxic elements, such as Cr, Ni, Cu and Zn, increases with increasing pyrolysis temperature [19]. The application of biochar containing potentially toxic compounds is an important issue related to soil contamination and grain security.

Although biochar has substantial advantages for improving soils, the long-term effect of these pollutants in biochar on soils is still unclear. Biochar can reduce the bioavailability of pollutants and thus be used for the remediation of contaminated soils, but biochar may have a negative influence on soil functions due to the presence of toxic compounds [16, 53, 102]. As discussed previously, although the application of biochar can immobilize pollutants, the immobilization effect of biochar is generally effective for 2–3 years and decreases over time [102]. In addition, biochar produced from crop straw may contain heavy metals, which could be released when applied to soils. Over time, biochar undergoes physical, chemical, and biological degradation, thus altering the microbial community in soils. However, due to long-term interactions between these compounds in biochar and the soil, the major mechanisms of microbial transformation still need to be elucidated. Therefore, to make better use of biochar and decrease its possible risks, proper measures must be taken before biochar can be used in agricultural soils.

6 Climate Change Mitigation

The process of converting C into a stable form to prevent C emissions is called C storage. During the Earth's C cycle, plants use photosynthesis to absorb C dioxide and store it in their own cells. When plants die and are decomposed, some unstable C eventually returns to the atmosphere in the form of C dioxide. If plants are converted into biochar, they can be stored in the soil in a more stable form. Due to biochar's large microporous structure, it can immobilize C for long periods of time. Returning biochar to the field reduces greenhouse gas emissions such as CH_4, N_2O and CO_2. Biochar-improved soils reduce CO_2 emissions due to C sequestration. In one study, the use of biochar in paddy fields increased the overall C input while reducing N_2O emissions [95]. The addition of biochar to paddy fields increased CO_2 sequestration by 47–55% compared to control. Under realistic conditions, biochar application has been shown to improve ecosystem quality by 18%, mitigate climate change by 15%, and improve resource utilization by 13% [41].

In general, estimates of emitted CO_2, N_2O, and CH_4 rely on the physical and chemical characteristics of biochar, the soil type, soil microorganisms, and water and fertilizer regimes [81]. The strong adsorption capacity of biochar is the main reason for the reduction in greenhouse gas emissions. For example, increased adsorption of NO_3^- via biochar reduces N_2O emissions and affects N conversion and soil N-cycling patterns, thereby reducing N_2O emissions. In addition, studies have shown that biochar addition to soils can inhibit the mineralization of native soil organic C, thereby reducing CO_2 emissions. However, this phenomenon usually occurs over the short term, that is, within the first few months after biochar addition. From the perspective of climate change mitigation, the stability of biochar helps extend the timeline for its positive agronomic effects. The long-term stability of biochar in soils is a prominent factor that reduces greenhouse gas emissions.

7 Strategies for Biochar Application to Soils

It is very important to determine the best application strategy for biochar to eliminate the potentially adverse effects of biochar and reduce the cost of biochar. Biomass raw materials and pyrolysis conditions determine the properties of biochar, and these properties determine the potential of biochar to improve soil fertility. Appropriate applications of biochar can improve soil fertility and quality and boost crop yields. Before biochar is applied, we should determine the elemental composition and surface properties of the target biochar to select appropriate raw materials. Then, the pyrolysis conditions must be optimized to obtain selective high-quality biochar. Figure 1 illustrates the modification process for biochar to obtain different target products.

Generally, the higher the pyrolysis temperature is, the fewer the nutrients in the biochar, the greater the pH value and the greater the surface area. As a soil conditioner, biochar has strong adsorption properties and C fixation potential [27]. Biochar produced by pyrolysis at high temperatures may be more suitable for improving acidic soils and increasing biomass. However, increases in the pyrolysis temperature can lead to losses in easily decomposed components and to the disappearance of acidic functional groups, which is not conducive to the fixation of some organic pollutants.

Therefore, biochar can be mixed or applied with other substances, such as straw when it is returned to fields after composting, and applied with fertilizers, beneficial

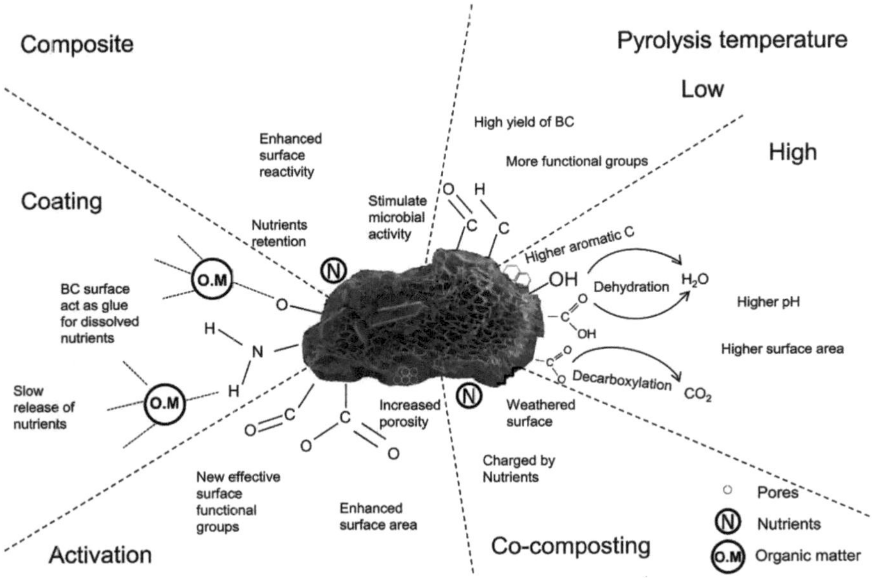

Fig. 1 Improvement in biochar surface properties by different modification methods [27]

bacteria, or other organic materials, including clay minerals [103]. The application of biochar to soils as a mixed additive containing other beneficial substances has recently been recognized as an effective method for enhancing soil nutrient preservation, improving the fertility of low-fertility soils, remediating polluted soils, and improving the growth of crops in poor soils. In addition, combining biochar with nanotechnology can produce hybrid nanomaterial biochar composites, which are environmentally friendly and have the latent capacity to promote soil fertility and remediate various pollutants.

8 Conclusions and Future Perspectives

Healthy and fertile soils are the basis for stabilizing crop productivity. However, due to various phenomena induced by human activities such as soil acidification, salinization, nutrient loss, heavy metals contaminations, and organic pollutant pollution caused by the excessive application of chemical fertilizers and pesticides, large portions of farmland show poor potential for food productivity. Biochar has been proven to be an effective method to resolve soil-related issues. However, the current challenges involve the selection of appropriate raw materials and preparation procedures to achieve specific uses for biochar. Biomass can be transformed or industrialized to improve targeting, and its application costs should be considered to meet the needs of specific soils.

At present, there are many studies on the application of biochar for soil improvement, but few soil management plans include the selection of suitable customized biochar at optimal dosages to satisfy management objectives and soil characteristics. Future research should consider soil type and crop growth characteristics for the selection of biochar, should better elucidate the properties of biochar, and should seek new methods to remedy the current shortcomings to fully utilize the properties of specific biochar in different soils. At the same time, long-term field experiments should be carried out to explore the combined application of biochar and other materials (such as compost, zeolite, controlled release fertilizer, organic fertilizer, or beneficial bacteria), to comprehensively consider the various factors that affect crop growth, to analyze the adverse effects of biochar aging, and to maximize the potential of biochar as a soil conditioner to improve soil properties and increase crop productivity more effectively.

Acknowledgements This study was supported by the Guangdong Province Key Field R&D Program Project (2020B0202010007).

References

1. Agbna GHD, She DL, Liu ZP, Elshaikh NA, Guangcheng S, Timm LC (2017) Effects of deficit irrigation and biochar addition on the growth, yield, and quality of tomato. Sci Hortic 222:90–101. https://doi.org/10.1016/j.scienta.2017.05.004
2. Agegnehu G, Bass AM, Nelson PN, Muirhead B, Wright G, Bird MI (2015) Biochar and biochar-compost as soil amendments: effects on peanut yield, soil properties and greenhouse gas emissions in tropical North Queensland, Australia. Agr Ecosyst Environ 213:72–85. https://doi.org/10.1016/j.agee.2015.07.027
3. Agegnehu G, Bass AM, Nelson PN, Bird MI (2016) Benefits of biochar, compost and biochar-compost for soil quality, maize yield and greenhouse gas emissions in a tropical agricultural soil. Sci Total Environ 543:295–306. https://doi.org/10.1016/j.scitotenv.2015.11.054
4. Agegnehu G, Srivastava AK, Bird MI (2017) The role of biochar and biochar-compost in improving soil quality and crop performance: a review. Appl Soil Ecol 119:156–170. https://doi.org/10.1016/j.apsoil.2017.06.008
5. Ahmad M, Rajapaksha AU, Lim JE, Zhang M, Bolan N, Mohan D, Vithanage M, Lee SS, Ok YS (2014) Biochar as a sorbent for contaminant management in soil and water: a review. Chemosphere 99:19–33. https://doi.org/10.1016/j.chemosphere.2013.10.071
6. Ahmad M, Usman ARA, Rafique MI, Al-Wabel MI (2019) Engineered biochar composites with zeolite, silica, and nano-zerovalent iron for the efficient scavenging of chlortetracycline from aqueous solutions. Environ Sci Pollut Res 26:15136–15152. https://doi.org/10.1007/s11356-019-04850-7
7. Ahmad Z, Gao B, Mosa A, Yu H, Yin X, Bashir A, Ghoveisi H, Wang S (2018) Removal of Cu(II), Cd(II) and Pb(II) ions from aqueous solutions by biochars derived from potassium-rich biomass. J Clean Prod 180:437–449. https://doi.org/10.1016/j.jclepro.2018.01.133
8. Al-Wabel MI, Al-Omran A, El-Naggar AH, Nadeem M, Usman AR (2013) Pyrolysis temperature induced changes in characteristics and chemical composition of biochar produced from conocarpus wastes. Biores Technol 131:374–379. https://doi.org/10.1016/j.biortech.2012.12.165

9. Al-Wabel MI, Hussain Q, Usman ARA, Ahmad M, Abduljabbar A, Sallam AS, Ok YS (2018) Impact of biochar properties on soil conditions and agricultural sustainability: a review. Land Degrad Dev 29:2124–2161. https://doi.org/10.1002/ldr.2829
10. Angin D (2013) Effect of pyrolysis temperature and heating rate on biochar obtained from pyrolysis of safflower seed press cake. Biores Technol 128:593–597. https://doi.org/10.1016/j.biortech.2012.10.150
11. Bai N, Zhang H, Li S, Zheng X, Zhang J, Zhang H, Zhou S, Sun H, Lv W (2019) Long-term effects of straw and straw-derived biochar on soil aggregation and fungal community in a rice-wheat rotation system. PeerJ 6:e6171. https://doi.org/10.7717/peerj.6171
12. Bartoli M, Giorcelli M, Jagdale P, Rovere M, Tagliaferro A (2020) A review of non-soil biochar applications. Materials (Basel) 13:261. https://doi.org/10.3390/ma13020261
13. Basak BB, Sarkar B, Saha A, Sarkar A, Mandal S, Biswas JK, Wang H, Bolan NS (2022) Revamping highly weathered soils in the tropics with biochar application: what we know and what is needed. Sci Total Environ 822:153461. https://doi.org/10.1016/j.scitotenv.2022.153461
14. Bi QF, Chen QH, Yang XR, Li H, Zheng BX, Zhou WW, Liu XX, Dai PB, Li KJ, Lin XY (2017) Effects of combined application of nitrogen fertilizer and biochar on the nitrification and ammonia oxidizers in an intensive vegetable soil. AMB Expr 7:198. https://doi.org/10.1186/s13568-017-0498-7
15. Bian R, Li L, Bao D, Zheng J, Zhang X, Zheng J, Liu X, Cheng K, Pan G (2016) Cd immobilization in a contaminated rice paddy by inorganic stabilizers of calcium hydroxide and silicon slag and by organic stabilizer of biochar. Environ Sci Pollut Res 23:10028–10036. https://doi.org/10.1007/s11356-016-6214-3
16. Bielska L, Skulcova L, Neuwirthova N, Cornelissen G, Hale SE (2018) Sorption, bioavailability and ecotoxic effects of hydrophobic organic compounds in biochar amended soils. Sci Total Environ 624:78–86. https://doi.org/10.1016/j.scitotenv.2017.12.098
17. Blanco-Canqui H (2017) Biochar and soil physical properties. Soil Sci Soc Am J 81:687–711. https://doi.org/10.2136/sssaj2017.01.0017
18. Brewer CE, Chuang VJ, Masiello CA, Gonnermann H, Gao X, Dugan B, Driver LE, Panzacchi P, Zygourakis K, Davies CA (2014) New approaches to measuring biochar density and porosity. Biomass Bioenerg 66:176–185. https://doi.org/10.1016/j.biombioe.2014.03.059
19. Buss W, Graham MC, Shepherd JG, Masek O (2016) Risks and benefits of marginal biomass-derived biochars for plant growth. Sci Total Environ 569–570:496–506. https://doi.org/10.1016/j.scitotenv.2016.06.129
20. Cao X, Harris W (2010) Properties of dairy-manure-derived biochar pertinent to its potential use in remediation. Biores Technol 101:5222–5228. https://doi.org/10.1016/j.biortech.2010.02.052
21. Chen W, Meng J, Han X, Lan Y, Zhang W (2019) Past, present, and future of biochar. Biochar 1:75–87. https://doi.org/10.1007/s42773-019-00008-3
22. Clay SA, Krack KK, Bruggeman SA, Papiernik S, Schumacher TE (2016) Maize, switchgrass, and ponderosa pine biochar added to soil increased herbicide sorption and decreased herbicide efficacy. J Environ Sci Health B 51:497–507. https://doi.org/10.1080/03601234.2016.1170540
23. Cui Y, Meng J, Wang Q, Zhang W, Cheng X, Chen W (2017) Effects of straw and biochar addition on soil nitrogen, carbon, and super rice yield in cold waterlogged paddy soils of North China. J Integr Agric 16:1064–1074. https://doi.org/10.1016/s2095-3119(16)61578-2
24. Dahlawi S, Naeem A, Rengel Z, Naidu R (2018) Biochar application for the remediation of salt-affected soils: challenges and opportunities. Sci Total Environ 625:320–335. https://doi.org/10.1016/j.scitotenv.2017.12.257
25. Dai Z, Zhang X, Tang C, Muhammad N, Wu J, Brookes PC, Xu J (2017) Potential role of biochars in decreasing soil acidification—a critical review. Sci Total Environ 581–582:601–611. https://doi.org/10.1016/j.scitotenv.2016.12.169
26. Ding Z, Wan Y, Hu X, Wang S, Zimmerman AR, Gao B (2016) Sorption of lead and methylene blue onto hickory biochars from different pyrolysis temperatures: importance of

physicochemical properties. J Ind Eng Chem 37:261–267. https://doi.org/10.1016/j.jiec.2016.03.035
27. El-Naggar A, Lee SS, Rinklebe J, Farooq M, Song H, Sarmah AK, Zimmerman AR, Ahmad M, Shaheen SM, Ok YS (2019) Biochar application to low fertility soils: a review of current status, and future prospects. Geoderma 337:536–554. https://doi.org/10.1016/j.geoderma.2018.09.034
28. Fang Q, Chen B, Lin Y, Guan Y (2014) Aromatic and hydrophobic surfaces of wood-derived biochar enhance perchlorate adsorption via hydrogen bonding to oxygen-containing organic groups. Environ Sci Technol 48:279–288. https://doi.org/10.1021/es403711y
29. Feng W, Yang F, Cen R, Liu J, Qu Z, Miao Q, Chen H (2021) Effects of straw biochar application on soil temperature, available nitrogen and growth of corn. J Environ Manage 277:111331. https://doi.org/10.1016/j.jenvman.2020.111331
30. Figueredo NAD, Costa LMD, Melo LCA, Siebeneichlerd EA, Tronto J (2017) Characterization of biochars from different sources and evaluation of release of nutrients and contaminants. Revista CiÊncia AgronÔmica 48. https://doi.org/10.5935/1806-6690.20170046
31. Fischer BMC, Manzoni S, Morillas L, Garcia M, Johnson MS, Lyon SW (2019) Improving agricultural water use efficiency with biochar—a synthesis of biochar effects on water storage and fluxes across scales. Sci Total Environ 657:853–862. https://doi.org/10.1016/j.scitotenv.2018.11.312
32. Gonzaga MIS, Mackowiak CL, Comerford NB, Moline EFDV, Shirley JP, Guimaraes DV (2017) Pyrolysis methods impact biosolids-derived biochar composition, maize growth and nutrition. Soil Till Res 165:59–65. https://doi.org/10.1016/j.still.2016.07.009
33. Gray M, Johnson MG, Dragila MI, Kleber M (2014) Water uptake in biochars: the roles of porosity and hydrophobicity. Biomass Bioenerg 61:196–205. https://doi.org/10.1016/j.biombioe.2013.12.010
34. Gu J, Zhou W, Jiang B, Wang L, Ma Y, Guo H, Schulin R, Ji R, Evangelou MW (2016) Effects of biochar on the transformation and earthworm bioaccumulation of organic pollutants in soil. Chemosphere 145:431–437. https://doi.org/10.1016/j.chemosphere.2015.11.106
35. Gul S, Whalen JK (2016) Biochemical cycling of nitrogen and phosphorus in biochar-amended soils. Soil Biol Biochem 103:1–15. https://doi.org/10.1016/j.soilbio.2016.08.001
36. Haider FU, Coulter JA, Cai L, Hussain S, Cheema SA, Wu J, Zhang R (2022) An overview on biochar production, its implications, and mechanisms of biochar-induced amelioration of soil and plant characteristics. Pedosphere 32:107–130. https://doi.org/10.1016/S1002-0160(20)60094-7
37. He L, Liu Y, Zhao J, Bi Y, Zhao X, Wang S, Xing G (2015) Comparison of straw-biochar-mediated changes in nitrification and ammonia oxidizers in agricultural oxisols and cambosols. Biol Fertil Soils 52:137–149. https://doi.org/10.1007/s00374-015-1059-3
38. He T, Liu D, Yuan J, Luo J, Lindsey S, Bolan N, Ding W (2018) Effects of application of inhibitors and biochar to fertilizer on gaseous nitrogen emissions from an intensively managed wheat field. Sci Total Environ 628–629:121–130. https://doi.org/10.1016/j.scitotenv.2018.02.048
39. He X, Du Z, Wang Y, Lu N, Zhang Q (2016) Sensitivity of soil respiration to soil temperature decreased under deep biochar amended soils in temperate croplands. Appl Soil Ecol 108:204–210. https://doi.org/10.1016/j.apsoil.2016.08.018
40. Hilber I, Bastos AC, Loureiro S, Soja G, Marsz A, Cornelissen G, Bucheli TD (2017) The different faces of biochar: contamination risk versus remediation tool. J Environ Eng Landsc Manag 25:86–104. https://doi.org/10.3846/16486897.2016.1254089
41. Homagain K, Shahi C, Luckai N, Sharma M (2015) Life cycle environmental impact assessment of biochar-based bioenergy production and utilization in Northwestern Ontario, Canada. J For Res 26:799–809. https://doi.org/10.1007/s11676-015-0132-y
42. Hu Y, Sun B, Wu S, Feng H, Gao M, Zhang B, Liu Y (2021) After-effects of straw and straw-derived biochar application on crop growth, yield, and soil properties in wheat (*Triticum aestivum* L.)–maize (*Zea mays* L.) rotations: a four-year field experiment. Sci Total Environ 780:146560. https://doi.org/10.1016/j.scitotenv.2021.146560

43. Igalavithana AD, Ok YS, Usman ARA, Al-Wabel MI, Oleszczuk P, Lee SS (2015) The effects of biochar amendment on soil fertility. In: Agricultural and environmental applications of biochar: advances and barriers, pp 123–144. https://doi.org/10.2136/sssaspecpub63.2014.0040
44. Ippolito JA, Cui L, Kammann C, Wrage-Mönnig N, Estavillo JM, Fuertes-Mendizabal T, Cayuela ML, Sigua G, Novak J, Spokas K, Borchard N (2020) Feedstock choice, pyrolysis temperature and type influence biochar characteristics: a comprehensive meta-data analysis review. Biochar 2:421–438. https://doi.org/10.1007/s42773-020-00067-x
45. Ippolito JA, Ducey TF, Cantrell KB, Novak JM, Lentz RD (2016) Designer, acidic biochar influences calcareous soil characteristics. Chemosphere 142:184–191. https://doi.org/10.1016/j.chemosphere.2015.05.092
46. Ji Q, Zhao S-X, Li Z-H, Ma Y-Y, Wang X-D (2016) Effects of biochar-straw on soil aggregation, organic carbon distribution, and wheat growth. Agron J 108:2129–2136. https://doi.org/10.2134/agronj2016.02.0121
47. Jien S-H, Wang C-S (2013) Effects of biochar on soil properties and erosion potential in a highly weathered soil. CATENA 110:225–233. https://doi.org/10.1016/j.catena.2013.06.021
48. Jin Z, Chen C, Chen X, Jiang F, Hopkins I, Zhang X, Han Z, Billy G, Benavides J (2019) Soil acidity, available phosphorus content, and optimal biochar and nitrogen fertilizer application rates: a five-year field trial in upland red soil, China. Field Crop Res 232:77–87. https://doi.org/10.1016/j.fcr.2018.12.013
49. Kalinke C, Oliveira PR, Oliveira GA, Mangrich AS, Marcolino-Junior LH, Bergamini MF (2017) Activated biochar: preparation, characterization and electroanalytical application in an alternative strategy of nickel determination. Anal Chim Acta 983:103–111. https://doi.org/10.1016/j.aca.2017.06.025
50. Kang MW, Yibeltal M, Kim YH, Oh SJ, Lee JC, Kwon EE, Lee SS (2022) Enhancement of soil physical properties and soil water retention with biochar-based soil amendments. Sci Total Environ 836:155746. https://doi.org/10.1016/j.scitotenv.2022.155746
51. Kavitha B, Reddy PVL, Kim B, Lee SS, Pandey SK, Kim KH (2018) Benefits and limitations of biochar amendment in agricultural soils: a review. J Environ Manag 227:146–154. https://doi.org/10.1016/j.jenvman.2018.08.082
52. Khan MA, Mahmood-ur-Rahman, Ramzani PMA, Zubair M, Rasool B, Khan MK, Ahmed A, Khan SA, Turan V, Iqbal M (2020) Associative effects of lignin-derived biochar and arbuscular mycorrhizal fungi applied to soil polluted from Pb-acid batteries effluents on barley grain safety. Sci Total Environ 710. https://doi.org/10.1016/j.scitotenv.2019.136294
53. Khan N, Chowdhary P, Gnansounou E, Chaturvedi P (2021) Biochar and environmental sustainability: emerging trends and techno-economic perspectives. Biores Technol 332:125102. https://doi.org/10.1016/j.biortech.2021.125102
54. Koltowski M, Charmas B, Skubiszewska-Zieba J, Oleszczuk P (2017) Effect of biochar activation by different methods on toxicity of soil contaminated by industrial activity. Ecotoxicol Environ Saf 136:119–125. https://doi.org/10.1016/j.ecoenv.2016.10.033
55. Lehmann J, Gaunt J, Rondon M (2006) Bio-char sequestration in terrestrial ecosystems—a review. Mitig Adapt Strat Glob Change 11:403–427. https://doi.org/10.1007/s11027-005-9006-5
56. Li B, Huang W, Elsgaard L, Yang B, Li Z, Yang H, Lu Y (2020) Optimal biochar amendment rate reduced the yield-scaled N_2O emissions from Ultisols in an intensive vegetable field in South China. Sci Total Environ 723:138161. https://doi.org/10.1016/j.scitotenv.2020.138161
57. Li C, Xiong Y, Qu Z, Xu X, Huang Q, Huang G (2018) Impact of biochar addition on soil properties and water-fertilizer productivity of tomato in semi-arid region of Inner Mongolia, China. Geoderma 331:100–108. https://doi.org/10.1016/j.geoderama.2018.06.014
58. Li J, Cao L, Yuan Y, Wang R, Wen Y, Man J (2018) Comparative study for microcystin-LR sorption onto biochars produced from various plant- and animal-wastes at different pyrolysis temperatures: Influencing mechanisms of biochar properties. Biores Technol 247:794–803. https://doi.org/10.1016/j.biortech.2017.09.120

59. Li X, Song Y, Wang F, Bian Y, Jiang X (2019) Combined effects of maize straw biochar and oxalic acid on the dissipation of polycyclic aromatic hydrocarbons and microbial community structures in soil: a mechanistic study. J Hazard Mater 364:325–331. https://doi.org/10.1016/j.jhazmat.2018.10.041
60. Li Z, Gu C, Zhang R, Ibrahim M, Zhang G, Wang L, Zhang R, Chen F, Liu Y (2017) The benefic effect induced by biochar on soil erosion and nutrient loss of slopping land under natural rainfall conditions in central China. Agric Water Manag 185 145–150. https://doi.org/10.1016/j.agwat.2017.02.018
61. Lian F, Xing B (2017) Black carbon (biochar) in water/soil environments: molecular structure, sorption, stability, and potential risk. Environ Sci Technol 51:13517–13532. https://doi.org/10.1021/acs.est.7b02528
62. Lin G, Yang H, Hu J, Luo Y, Shao J, Wang X, Chen H (2016) Effects of the physicochemical properties of biochar and soil on moisture sorption. J Renew Sustain Energy 8:225. https://doi.org/10.1063/1.4967706
63. Lin XW, Xie ZB, Zheng JY, Liu Q, Bei QC, Zhu JG (2015) Effects of biochar application on greenhouse gas emissions, carbon sequestration and crop growth in coastal saline soil. Eur J Soil Sci 66:329–338. https://doi.org/10.1111/ejss.12225
64. Liu D, Feng Z, Zhu H, Yu L, Yang K, Yu S, Zhang Y, Guo W (2020) Effects of corn straw biochar application on soybean growth and alkaline soil properties. https://doi.org/10.15376/biores.15.1.1463-1481
65. Liu L, Shen G, Sun M, Cao X, Shang G, Chen P (2014) Effect of biochar on nitrous oxide emission and its potential mechanisms. J Air Waste Manag Assoc 64:894–902. https://doi.org/10.1080/10962247.2014.899937
66. Liu M, Zhu J, Yang X, Fu Q, Hu H, Huang Q (2022) Biochar produced from the straw of common crops simultaneously stabilizes soil organic matter and heavy metals. Sci Total Environ 828:154494. https://doi.org/10.1016/j.scitotenv.2022.154494
67. Liu X, Manevski K, Liu F, Andersen MN (2022) Biomass accumulation and water use efficiency of faba bean-ryegrass intercropping system on sandy soil amended with biochar under reduced irrigation regimes. Agric Water Manag 273:107905. https://doi.org/10.1016/j.agwat.2022.107905
68. Liu X, Zhang D, Li H, Qi X, Gao Y, Zhang Y, Han Y, Jiang Y, Li H (2020) Soil nematode community and crop productivity in response to 5-year biochar and manure addition to yellow cinnamon soil. BMC Ecol 20:39. https://doi.org/10.1186/s12898-020-00304-8
69. Liu Y, Zhu J, Gao W, Guo Z, Xue C, Pang J, Shu L (2019) Effects of biochar amendment on bacterial and fungal communities in the reclaimed soil from a mining subsidence area. Environ Sci Pollut Res Int 26:34368–34376. https://doi.org/10.1007/s11356-019-06567-z
70. Lu K, Yang X, Gielen G, Bolan N, Ok YS, Niazi NK, Xu S, Yuan G, Chen X, Zhang X, Liu D, Song Z, Liu X, Wang H (2017) Effect of bamboo and rice straw biochars on the mobility and redistribution of heavy metals (Cd, Cu, Pb and Zn) in contaminated soil. J Environ Manag 186:285–292. https://doi.org/10.1016/j.jenvman.2016.05.068
71. Melo LCA, Lehmann J, Carneiro JSDS, Camps-Arbestain M (2022) Biochar-based fertilizer effects on crop productivity: a meta-analysis. Plant Soil 472:45–58. https://doi.org/10.1007/s11104-021-05276-2
72. Mia S, van Groenigen JW, van de Voorde TFJ, Oram NJ, Bezemer TM, Mommer L, Jeffery S (2014) Biochar application rate affects biological nitrogen fixation in red clover conditional on potassium availability. Agric Ecosyst Environ 191:83–91. https://doi.org/10.1016/j.agee.2014.03.011
73. Mosa A, El-Banna MF, Gao B (2016) Biochar filters reduced the toxic effects of nickel on tomato (Lycopersicon esculentum L.) grown in nutrient film technique hydroponic system. Chemosphere 149:254–262. https://doi.org/10.1016/j.chemosphere.2016.01.104
74. Oni BA, Oziegbe O, Olawole OO (2019) Significance of biochar application to the environment and economy. Ann Agric Sci 64:222–236. https://doi.org/10.1016/j.aoas.2019.12.006

75. Peter Q, Kathrin W, Herstellung (2016) Eigenschaften und Verwendung von Biomassekarbonisaten. Springer Vieweg. https://doi.org/10.1007/978-3-658-03689-8
76. Palansooriya KN, Ok YS, Awad YM, Lee SS, Sung JK, Koutsospyros A, Moon DH (2019) Impacts of biochar application on upland agriculture: a review. J Environ Manage 234:52–64. https://doi.org/10.1016/j.jenvman.2018.12.085
77. Pandey D, Daverey A, Arunachalam K (2020) Biochar: production, properties and emerging role as a support for enzyme immobilization. J Clean Prod 255:120267. https://doi.org/10.1016/j.jclepro.2020.120267
78. Patra JM, Panda SS, Dhal NK (2017) Biochar as a low-cost adsorbent for heavy metal removal: a review. Int J Res Biosci 6:1–7
79. Pituello C, Francioso O, Simonetti G, Pisi A, Torreggiani A, Berti A, Morari F (2014) Characterization of chemical–physical, structural and morphological properties of biochars from biowastes produced at different temperatures. J Soils Sediments 15:792–804. https://doi.org/10.1007/s11368-014-0964-7
80. Prendergast-Miller MT, Duvall M, Sohi SP (2014) Biochar–root interactions are mediated by biochar nutrient content and impacts on soil nutrient availability. Eur J Soil Sci 65:173–185. https://doi.org/10.1111/ejss.12079
81. Qambrani NA, Rahman MM, Won S, Shim S, Ra C (2017) Biochar properties and eco-friendly applications for climate change mitigation, waste management, and wastewater treatment: a review. Renew Sustain Energy Rev 79:255–273. https://doi.org/10.1016/j.rser.2017.05.057
82. Qiao Y, Miao S, Zhong X, Zhao H, Pan S (2020) The greatest potential benefit of biochar return on bacterial community structure among three maize-straw products after eight-year field experiment in Mollisols. Appl Soil Ecol 147:103432. https://doi.org/10.1016/j.apsoil.2019.103432
83. Rombola AG, Torri C, Vassura I, Venturini E, Reggiani R, Fabbri D (2022) Effect of biochar amendment on organic matter and dissolved organic matter composition of agricultural soils from a two-year field experiment. Sci Total Environ 812:151422. https://doi.org/10.1016/j.scitotenv.2021.151422
84. Sadeghi SH, Hazbavi Z, Harchegani MK (2016) Controllability of runoff and soil loss from small plots treated by vinasse-produced biochar. Sci Total Environ 541:483–490. https://doi.org/10.1016/j.scitotenv.2015.09.068
85. Seneviratne M, Weerasundara L, Ok YS, Rinklebe J, Vithanage M (2017) Phytotoxicity attenuation in Vigna radiata under heavy metal stress at the presence of biochar and N fixing bacteria. J Environ Manage 186:293–300
86. Shaaban M, Van Zwieten L, Bashir S, Younas A, Nunez-Delgado A, Chhajro MA, Kubar KA, Ali U, Rana MS, Mehmood MA, Hu R (2018) A concise review of biochar application to agricultural soils to improve soil conditions and fight pollution. J Environ Manage 228:429–440. https://doi.org/10.1016/j.jenvman.2016.07.024
87. Shi Y, Liu X, Zhang Q, Gao P, Ren J (2019) Biochar and organic fertilizer changed the ammonia-oxidizing bacteria and archaea community structure of saline–alkali soil in the North China Plain. J Soils Sediments 20:12–23. https://doi.org/10.1007/s11368-019-02364-w
88. Si L, Xie Y, Ma Q, Wu L (2018) The short-term effects of rice straw biochar, nitrogen and phosphorus fertilizer on rice yield and soil properties in a cold waterlogged paddy field. Sustainability 10:537. https://doi.org/10.3390/su10020537
89. Silber A, Levkovitch I, Graber AER (2010) PH-dependent mineral release and surface properties of cornstraw biochar: agronomic implications. Environ Sci Technol 44:9318. https://doi.org/10.1021/es101283d
90. Singh H, Northup BK, Rice CW, Prasad PVV (2022) Biochar applications influence soil physical and chemical properties, microbial diversity, and crop productivity: a meta-analysis. Biochar 4:8. https://doi.org/10.1007/s42773-022-00138-1
91. Song D, Tang J, Xi X, Zhang S, Liang G, Zhou W, Wang X (2018) Responses of soil nutrients and microbial activities to additions of maize straw biochar and chemical fertilization in a calcareous soil. Eur J Soil Biol 84:1–10. https://doi.org/10.1016/j.ejsobi.2017.11.003

92. Suddick EC, Six J (2013) An estimation of annual nitrous oxide emissions and soil quality following the amendment of high temperature walnut shell biochar and compost to a small scale vegetable crop rotation. Sci Total Environ 465:298–307. https://doi.org/10.1016/j.scitotenv.2013.01.094
93. Suliman W, Harsh JB, Abu-Lail NI, Fortuna A-M, Dallmeyer I, Garcia-Perez M (2016) Influence of feedstock source and pyrolysis temperature on biochar bulk and surface properties. Biomass Bioenerg 84:37–48. https://doi.org/10.1016/j.biombioe.2015.11.010
94. Suliman W, Harsh JB, Fortuna AM, Garcia-Perez M, Abu-Lail NI (2017) Quantitative effects of biochar oxidation and pyrolysis temperature on the transport of pathogenic and nonpathogenic Escherichia coli in biochar-amended sand columns. Environ Sci Technol 51:5071–5081. http://pubs.acs.org. Accessed 31 March 2017
95. Sun X, Han X, Ping F, Zhang L, Zhang K, Chen M, Wu W (2018) Effect of rice-straw biochar on nitrous oxide emissions from paddy soils under elevated CO2 and temperature. Sci Total Environ 628–629:1009–1016. https://doi.org/10.1016/j.scitotenv.2018.02.046
96. Suo F, You X, Yin S, Wu H, Zhang C, Yu X, Sun R, Li Y (2021) Preparation and characterization of biochar derived from co-pyrolysis of Enteromorpha prolifera and corn straw and its potential as a soil amendment. Sci Total Environ 798:149167. https://doi.org/10.1016/j.scitotenv.2021.149167
97. Tan G, Wang H, Xu N, Junaid M, Liu H, Zhai L (2021) Effects of biochar application with fertilizer on soil microbial biomass and greenhouse gas emissions in a peanut cropping system. Environ Technol 42:9–19. https://doi.org/10.1080/09593330.2019.1620344
98. Tan Z, Lin CSK, Ji X, Rainey TJ (2017) Returning biochar to fields: a review. Appl Soil Ecol 116:1–11. https://doi.org/10.1016/j.apsoil.2017.03.017
99. Tan Z, Zhang X, Wang L, Gao B, Luo J, Fang R, Zou W, Meng N (2019) Sorption of tetracycline on H2O2-modified biochar derived from rape stalk. Environ Pollut Bioavailab 31:198–207. https://doi.org/10.1080/26395940.2019.1607779
100. Usman ARA, Al-Wabel MI, Ok YS, Al-Harbi A, Wahb-Allah M, El-Naggar AH, Ahmad M, Al-Faraj A, Al-Omran A (2016) Conocarpus biochar induces changes in soil nutrient availability and tomato growth under saline irrigation. Pedosphere 26:27–38. https://doi.org/10.1016/S1002-0160(15)60019-4
101. Wang H, Gao B, Fang J, Ok YS, Xue Y, Yang K, Cao X (2018) Engineered biochar derived from eggshell-treated biomass for removal of aqueous lead. Ecol Eng 121:124–129. https://doi.org/10.1016/j.ecoleng.2017.06.029
102. Wang J, Shi L, Zhai L, Zhang H, Wang S, Zou J, Shen Z, Lian C, Chen Y (2021) Analysis of the long-term effectiveness of biochar immobilization remediation on heavy metal contaminated soil and the potential environmental factors weakening the remediation effect: a review. Ecotoxicol Environ Saf 207:111261. https://doi.org/10.1016/j.ecoenv.2020.111261
103. Wang L, Ok YS, Tsang DCW, Alessi DS, Rinklebe J, Mašek C, Bolan NS, Hou D (2021) Biochar composites: emerging trends, field successes and sustainability implications. Soil Use Manag 38:14–38. https://doi.org/10.1111/sum.12731
104. Wang Q, Wang X, Zhang D, Zhou X, Mak-Mensah E, Zhao X, Zhao W, Ghanney P, Haider FU, Liu Q, Li G, Li X, Li Y, Majeed Y (2022) Selection of suitable type and application rate of biochar for alfalfa (*Medicago sativa* L.) productivity in ridge-furrow rainwater-harvesting in semiarid regions of China. Field Crops Res 277:108428. https://doi.org/10.1016/j.fcr.2021.108428
105. Wang Q, Yuan J, Yang X, Han X, Lan Y, Cao D, Sun Q, Cui X, Meng J, Chen W (2022) Responses of soil respiration and C sequestration efficiency to biochar amendment in maize field of Northeast China. Soil Till Res 223:105442. https://doi.org/10.1016/j.still.2022.105442
106. Wang W, Lai D, Abid A, Neogi S, Xu X, Wang C (2018) Effects of steel slag and biochar incorporation on active soil organic carbon pools in a subtropical paddy field. Agronomy 8:135. https://doi.org/10.3390/agronomy8080135
107. Wang XF, Li Y, Wang HR, Wang YZ, Biswas A, Chau HW, Liang JP, Zhang FC, Bai YG, Wu SF, Chen JY, Liu HG, Yang G, Pulatov A (2022) Targeted biochar application alters physical, chemical, hydrological and thermal properties of salt-affected soils under cotton-sugarbeet intercropping. CATENA 216:106414. https://doi.org/10.1016/j.catena.2022.106414

108. Wang Y, Hu Y, Zhao X, Wang S, Xing G (2013) Comparisons of biochar properties from wood material and crop residues at different temperatures and residence times. Energy Fuels 27:5890–5899. http://pubs.acs.org. Accessed 17 Sep 2013
109. Weber K, Quicker P (2018) Properties of biochar. Fuel 217:240–261. https://doi.org/10.1016/j.fuel.2017.12.054
110. Wu Y, Sun Y, Guo X, Huang K, Zhang H, Zhang S (2019) Impact of wheat straw and tobacco straw biochar on nitrification of nitrogenous fertilizer and emission of N2O and NH3. J Northwest Agric For Univ 47:0060–0067.
111. Xiao X, Chen B, Chen Z, Zhu L, Schnoor JL (2018) Insight into multiple and multi-level structures of biochars and their potential environmental applications: a critical review. Environ Sci Technol 52:5027–5047. https://doi.org/10.1021/acs.est.7b06487. http://pubs.acs.org. Accessed 10 Apr 2018
112. Xie Y, Dong C, Chen Z, Liu Y, Zhang Y, Gou P, Zhao X, Ma D, Kang G, Wang C, Zhu Y, Guo T (2021) Successive biochar amendment affected crop yield by regulating soil nitrogen functional microbes in wheat-maize rotation farmland. Environ Res 194:110671. https://doi.org/10.1016/j.envres.2020.110671
113. Xu CY, Hosseini-Bai S, Hao Y, Rachaputi RC, Wang H, Xu Z, Wallace H (2015) Effect of biochar amendment on yield and photosynthesis of peanut on two types of soils. Environ Sci Pollut Res Int 22:6112–6125. https://doi.org/10.1007/s11356-014-3820-9
114. Xu N, Tan G, Wang H, Gai X (2016) Effect of biochar additions to soil on nitrogen leaching, microbial biomass and bacterial community structure. Eur J Soil Biol 74:1–8. https://doi.org/10.1016/j.ejsobi.2016.02.004
115. Yaashikaa PR, Kumar PS, Varjani S, Saravanan A (2020) A critical review on the biochar production techniques, characterization, stability and applications for circular bioeconomy. Biotechnol Rep 28:e00570. https://doi.org/10.1016/j.btre.2020.e00570
116. Yan S, Zhang S, Yan P, Aurangzeib M (2022) Effect of biochar application method and amount on the soil quality and maize yield in Mollisols of Northeast China. Biochar 4:56. https://doi.org/10.1007/s42773-022-00180-z
117. Yang C, Liu J, Ying H, Lu S (2022) Soil pore structure changes induced by biochar affect microbial diversity and community structure in an Ultisol. Soil Till Res 224:105505. https://doi.org/10.1016/j.still.2022.105505
118. Yang W, Feng G, Miles D, Gao L, Jia Y, Li C, Qu Z (2020) Impact of biochar on greenhouse gas emissions and soil carbon sequestration in corn grown under drip irrigation with mulching. Sci Total Environ 729:138752. https://doi.org/10.1016/j.scitotenv.2020.138752
119. Yang X, Liu H, Mao X, Deng J, Haefele SM (2020) Non-flooding rice yield response to straw biochar and controlled-release fertilizer. Agron J 112:4799–4809. https://doi.org/10.1002/agj2.20430
120. Yang X, Zhang S, Ju M, Liu L (2019) Preparation and modification of biochar materials and their application in soil remediation. Appl Sci 9:1365. https://doi.org/10.3390/app9071365
121. Yu H, Zou W, Chen J, Chen H, Yu Z, Huang J, Tang H, Wei X, Gao B (2019) Biochar amendment improves crop production in problem soils: a review. J Environ Manage 232:8–21. https://doi.org/10.1016/j.jenvman.2018.10.117
122. Yu L, Lu X, He Y, Brookes PC, Liao H, Xu J (2016) Combined biochar and nitrogen fertilizer reduces soil acidity and promotes nutrient use efficiency by soybean crop. J Soils Sediments 17:599–610. https://doi.org/10.1007/s11368-016-1447-9
123. Yuan JH, Xu RK, Zhang H (2011) The forms of alkalis in the biochar produced from crop residues at different temperatures. Biores Technol 102:3488–3497. https://doi.org/10.1016/j.biortech.2010.11.018
124. Zhang J, You C (2013) Water holding capacity and absorption properties of wood chars. Energy Fuels 27:2643–2648. https://doi.org/10.1021/ef4000769
125. Zhang C, Liu L, Zhao M, Rong H, Xu Y (2018) The environmental characteristics and applications of biochar. Environ Sci Pollut Res 25:21525–21534. https://doi.org/10.1007/s11356-018-2521-1

126. Zhang D, Pan G, Wu G, Kibue GW, Li L, Zhang X, Zheng J, Zheng J, Cheng K, Joseph S, Liu X (2016) Biochar helps enhance maize productivity and reduce greenhouse gas emissions under balanced fertilization in a rainfed low fertility inceptisol. Chemosphere 142:106–113. https://doi.org/10.1016/j.chemosphere.2015.04.088
127. Zhang J, Huang Y, Lin J, Chen X, Li C, Zhang J (2019) Biochar applied to consolidated land increased the quality of an acid surface soil and tobacco crop in Southern China. J Soils Sediments 20:3091–3102. https://doi.org/10.1007/s11368-019-02531-z
128. Zhang Q, Song Y, Wu Z, Yan X, Gunina A, Kuzyakov Y, Xiong Z (2020) Effects of six-year biochar amendment on soil aggregation, crop growth, and nitrogen and phosphorus use efficiencies in a rice-wheat rotation. J Clean Prod 242:118435. https://doi.org/10.1016/j.jclepro.2019.118435
129. Zhang Y, Liu Y, Zhang G, Guo X, Sun Z, Li T (2018) The effects of rice straw and biochar applications on the microbial community in a soil with a history of continuous tomato planting history. Agronomy 8. https://doi.org/10.3390/agronomy8050065
130. Zhang Z, Zhu Z, Shen B, Liu L (2019) Insights into biochar and hydrochar production and applications: a review. Energy 171:581–589. https://doi.org/10.1016/j.energy.2019.01.035
131. Zhao B, O'Connor D, Zhang J, Peng T, Shen Z, Tsang DCW, Hou D (2018) Effect of pyrolysis temperature, heating rate, and residence time on rapeseed stem derived biochar. J Clean Prod 174:977–987. https://doi.org/10.1016/j.jclepro.2017.11.013
132. Zhao WR, Li JY, Deng KY, Shi RY, Jiang J, Hong ZN, Qian W, He X, Xu RK (2020) Effects of crop straw biochars on aluminum species in soil solution as related with the growth and yield of canola (Brassica napus L.) in an acidic Ultisol under field condition. Environ Sci Pollut Res 27:30178–30189. https://doi.org/10.1007/s11356-020-09330-x
133. Zheng H, Wang X, Chen L, Wang Z, Xia Y, Zhang Y, Wang H, Luo X, Xing B (2018) Enhanced growth of halophyte plants in biochar-amended coastal soil: roles of nutrient availability and rhizosphere microbial modulation. Plant Cell Environ 41:517–532. https://doi.org/10.1111/pce.12944
134. Zhou H, Zhang D, Wang P, Liu X, Cheng K, Li L, Zheng J, Zhang X, Zheng J, Crowley D, van Zwieten L, Pan G (2017) Changes in microbial biomass and the metabolic quotient with biochar addition to agricultural soils: a meta-analysis. Agric Ecosyst Environ 239:80–89. https://doi.org/10.1016/j.agee.2017.01.006
135. Zhu Q, Peng X, Huang T (2015) Contrasted effects of biochar on maize growth and N use efficiency depending on soil conditions. Int Agrophys 29:257–266. https://doi.org/10.1515/intag-2015-0023
136. Zhu X, Chen B, Zhu L, Xing B (2017) Effects and mechanisms of biochar-microbe interactions in soil improvement and pollution remediation: a review. Environ Pollut 227:98–115. https://doi.org/10.1016/j.envpol.2017.04.032
137. Zielinska A, Oleszczuk P (2016) Bioavailability and bioaccessibility of polycyclic aromatic hydrocarbons (PAHs) in historically contaminated soils after lab incubation with sewage sludge-derived biochars. Chemosphere 163:480–489. https://doi.org/10.1016/j.chemosphere.2016.08.072

Frontier Studies in Soil Science: A Conclusions Chapter

Avelino Núñez-Delgado

Abstract This chapter shows an overall view of the contents of the book, also including some comments about the future of soil and environmental sciences. The editor and the authors of the different chapters of the book have a positive view on the effects derived from the research to be performed within the broad field of soil and environmental sciences in the coming years, which would help to solve crucial issues affecting our planet.

Keywords Environmental research · Research trends · Soil research · Soil and environmental pollution and treatment

1 Contents in This Book, and Future Trends

The book includes chapters by corresponding authors from a variety of countries (Spain, Italy, China, Lithuania, …), and from various other countries considering the different coauthors of each chapter (Fig. 1).

Specifically, after a Preface and an Introduction chapter, the contents deal with a variety of soil and environmental sciences-related themes: "*Antibiotics as emerging pollutants of soil ecosystems*", "*Rare Earths in Soils*", "*Impact of agricultural wastes on environment and possible management strategies*", "*Organic carbon management and the relations with climate change*", "*Agricultural soil management impacts on soil microbial communities*", "*Antibiotics, other emerging pollutants, and pathogenic microorganisms in raw and treated sewage sludge reaching soils*", "*Remediation of PTEs contaminated soils by washing treatment with biodegradable reagents*", "*Natural barriers to antibiotic entry into the trophic chain: some examples of the role of soils and by-products*", "*Salt-Affected Soils*", "*Beyond the obvious: exploring peat vibrational spectroscopy (FTIR-ATR) data using principal component analysis on transposed data matrix (tPCA), Store Mosse bog (Sweden)*", "*Application of*

A. Núñez-Delgado (✉)
Department of Soil Science and Agricultural Chemistry, Engineering Polytechnic School, University of Santiago de Compostela, 27002 Lugo, Spain
e-mail: avelino.nunez@usc.es

A. Núñez-Delgado (ed.), *Frontier Studies in Soil Science*, Frontier Studies in Soil Science, https://doi.org/10.1007/978-3-031-50503-4_13

Fig. 1 Some soil profiles from Navarra (Spain), and mechanical machinery used to excavate and expose them, as example of starting point for field studies in soil science

biochar in agricultural soil green remediation and sustainable development", and finally this *Conclusions* chapter.

The above-mentioned scientific works constitute just some of the many aspects needing research within this field of knowledge. In fact, there are plenty of potential new and/or complementary frontier studies in soil (and environmental) science. We have previously commented on this (for instance, [3]), and have published recent papers on potential hot new lines of research dealing with soil and environmental sciences (as examples, [1, 4–7]).

Personally, looking back to my thoughts when I was starting my PhD thesis, and, also remembering the lyrics of a popular song, I could say that *I still haven't found what I'm looking for*. In this regard, I would like to see in the coming future new experiments and apparatus to go ahead in soil and environmental research. With a rather "naïf" view, one of my expectations would be having access to devices allowing to achieve some results directly in the field, providing details (at the scale of around one cubic meter) on the complete porous distribution of the soil (for near 1 m^3, as said above), with immediate calculation of the potential flows (also including convective and diffusive) and retentions, as well as giving information on the reactive surfaces corresponding to all these porous, from the biggest to the smallest. Maybe potent computers linked to portable X-ray devices (even of the kind of Computer

Axial Tomography), and/or Nuclear Magnetic Resonance devices, as well as others, would allow any kind of determination in that soil environment, as needed, which together could provide images, quantifications, calculations, and modeling, working in the field with undisturbed soils, at a tridimensional level. So, in situ, for undisturbed soils, and at a tridimensional level, we could have immediate estimations of flows, retentions, interactions, which would affect to nutrients and pollutants. These physical, physicochemical, and chemical results, for rather static interfaces, but also for dynamic-flow situations (during vertical, and also lateral flow events), could provide an integrated knowledge that would sound "symphonic" to me. In addition, all that eventually added to also in situ determinations as regards biological activities (macro and microbiological), enzymatic activities, carbon storage dynamics in a limited time scale, summed to complement the above-mentioned direct results from tridimensional undisturbed soils with rapid calculations and modeling, for changing/dynamic situations, would be great (I would call it the *greatest hits*).

In fact, *this was just a dream* (in the REM phase of sleeping, as well as in some pop song). But, with a justified aim?

Many thousands of years back, primate and hominid were on Earth, evolving, reaching a point when they made artistic/magical representations, some of them remaining in caves. Current researchers indicate that these precursors or precedents started to explore/investigate their environment for survival but also with an incipient use of knowledge derived from repeated trial/error assays (a kind of starting empirical scientific procedure). Closer to now, in historic times, part of the world was rather theocentric, while, more recently, it was the time of higher prevalence of a more anthropocentric world (mainly in western geographic areas/countries). I believe that currently, and for the future, it should be the time for the Earth, *the terra-centric era.*

However, since some decades ago, some people could see our world as "science-centric". In this regard, we could ask: Science is the tool or is the objective?

I think that science could be somehow "pure", and achieving the best science is a desirable aim, but it is not enough. Specifically, in the current situation of the planet Earth, we will deeply try to work in scientific domains that could help to make more sustainable our world.

A sustainable environment could be seen as one of the first and fundamental needs as starting point to achieve a kind of social justice for humans, and environmental equity for as many living beings as possible.

I mean that I agree in the fact that new research is needed, frontier research is needed, but just science is not enough for reaching a sustainable planet. Scientific results giving scientific and technically appropriate procedures should be implemented, and this actual implementation, to be real and have effect, should be carried out by politicians (or anyone having the real power to act effectively) and the whole society [2]. Science is not enough to do that. Science and scientist *are just another brick in the wall.*

Due to that, I would suggest the implementation of the so-called *Armorican Eco-ethical Principles*:

1. You should not investigate, nor promote, anything that would give a negative outcome for the whole Earth.
2. You should not investigate, nor promote, anything that would give a negative outcome for the whole living beings in the Earth.
3. You should not investigate, nor promote, anything that would give a negative outcome for the whole human beings.
4. You should not investigate, nor promote, anything that would give a negative outcome for any of the other living species as a whole, on Earth.

These Principles should be respected in a hierarchical order, being the most important the first one, followed by the second, then the third one, and finally the fourth.

Each of the points would need further development, providing specific details/rules within each of them.

Any case, examples of the use could be the following. First example: A microbial species (considered in point 4) representing evident hazards for humans and/or other animal or vegetal species could be eco-ethically drastically faced (although preferably not till extinction) due to its deleterious effects on aspects of higher rank (at least, rank 3). Another theoretical example could be a situation where humans (rank 3) could suppose a real hazard for rank 2 or rank 1 aspects; in this case, it could not be realistic thinking that researchers would investigate/promote to face the whole human species, but if further rules/details were developed within each of the points, it could result in considering eco-ethically allowable to fight those limited human individuals that would suppose a real danger for the whole living beings (rank 2) or the whole planet (rank 1).

Acknowledgements The scientific editor of this book would like to deeply thank all the Authors and Reviewers who have contributed and collaborated to the volume, as well as to the Springer-Nature staff involved in it.

References

1. Ahmed W, Mehmood S, Núñez-Delgado A, Qaswar M, Ali S, Ying H, Liu Z, Mahmood M, Chen DY (2021) Fabrication, characterization and U(VI) sorption properties of a novel biochar derived from *Tribulus terrestris* via two different approaches. Sci Total Environ 780:146617. https://doi.org/10.1016/j.scitotenv.2021.146617
2. Núñez Delgado A (2023) Our environment: everything is natural on earth, but … editorial piece on current and future soil and environmental research. Processes 11(1):6. https://doi.org/10.3390/pr11010006
3. Núñez-Delgado A, Otero-Pérez XL, Álvarez-Rodríguez E (2023) Editorial: current research on soil science and related aspects of environmental sciences in Galicia. Span J Soil Sci 13.https://doi.org/10.3389/sjss.2023.11485
4. Race M, Ferraro A, Galdiero E, Guida M, Núñez-Delgado A, Pirozzi F, Siciliano A, Fabbricino M (2020) Current emerging SARS-CoV-2 pandemic: potential direct/indirect negative impacts of virus persistence and related therapeutic drugs on the aquatic compartments. Environ Res 188:109808. https://doi.org/10.1016/j.envres.2020.109808

5. Rodríguez-López L, Santás-Miguel V, Cela-Dablanca R, Núñez-Delgado A, Álvarez-Rodríguez E, Rodríguez-Seijo A, Arias-Estévez M (2023) Clarithromycin as soil and environmental pollutant: adsorption-desorption processes and influence of pH. Environ Res 233:116520. https://doi.org/10.1016/j.envres.2023.116520
6. Santás-Miguel V, Laura R-G, Avelino N-D, Esperanza Á-R, Montserrat D-R, Manuel A-E, Fernández-Calviño D (2023) Soil bacterial community tolerance to three tetracycline antibiotics induced by Ni and Zn. Span J Soil Sci 13. https://doi.org/10.3389/sjss.2023.10799
7. Wang ZW, Yang G, Chen J, Zhou Y, Delgado AN, Cui H-L, Duan G-L, Rosen BP, Zhu YG (2024) Fundamentals and application in phytoremediation of an efficient arsenate reducing bacterium Pseudomonas putida ARS1. J Environ Sci 137:237–244. https://doi.org/10.1016/j.jes.2023.02.020

MIX
Papier aus verantwortungsvollen Quellen
Paper from responsible sources
FSC® C105338

If you have any concerns about our products,
you can contact us on
ProductSafety@springernature.com

In case Publisher is established outside the EU,
the EU authorized representative is:
Springer Nature Customer Service Center GmbH
Europaplatz 3, 69115 Heidelberg, Germany

Printed by Libri Plureos GmbH
in Hamburg, Germany